A Practical Guide to Structure Determination in Organic Chemistry

A Practical Guide to Structure Determination in Organic Chemistry

TIMOTHY D. W. CLARIDGE

ANDREW N. BOA

JAMES S. O. MCCULLAGH

OXFORD
UNIVERSITY PRESS

OXFORD
UNIVERSITY PRESS

Great Clarendon Street, Oxford, OX2 6DP,
United Kingdom

Oxford University Press is a department of the University of Oxford.
It furthers the University's objective of excellence in research, scholarship,
and education by publishing worldwide. Oxford is a registered trade mark of
Oxford University Press in the UK and in certain other countries

Published in the United States of America by Oxford University Press
198 Madison Avenue, New York, NY 10016, United States of America

British Library Cataloguing in Publication Data
Data available

Library of Congress Control Number: 2025937743

ISBN 9780198712565

Printed in the UK by
Bell & Bain Ltd., Glasgow

The manufacturer's authorised representative in the EU for product safety is
Oxford University Press España S.A. of El Parque Empresarial San Fernando de Henares,
Avenida de Castilla, 2 – 28830 Madrid (www.oup.es/en or product.safety@oup.com).
OUP España S.A. also acts as importer into Spain of products made by the manufacturer.

Links to third party websites are provided by Oxford in good faith and
for information only. Oxford disclaims any responsibility for the materials
contained in any third party website referenced in this work.

MIX
Paper | Supporting
responsible forestry
FSC
www.fsc.org   FSC® C007785

Contents

About the authors

Tim Claridge is Associate Director, NMR Analytics, at Recursion, an AI-driven Tech-Bio company. Prior to this he was a Professor and Director of NMR Spectroscopy for Organic Chemistry and Chemical Biology in the Department of Chemistry at the University of Oxford, and now retains a Visiting Professor status in the department. During his academic career, he has contributed to over 250 research publications, where his research interests focused on the development and application of solution-state NMR methods in the chemical sciences. He has also written a number of textbooks covering spectroscopy, including *High-Resolution NMR Techniques in Organic Chemistry* and the Oxford Chemistry Primer *Introduction to Organic Spectroscopy* (with Laurence M. Harwood).

Andrew Boa is a Senior Lecturer in Organic Chemistry at the University of Hull. He holds a DPhil in Chemistry from the University of York and has 29 years of experience in teaching organic and bio-organic chemistry. His research interests cover all aspects of organic and bio-organic chemistry and extends to applications for the biosourced polymer sporopollenin.

James McCullagh is Professor of Biological Chemistry and Director of the Mass Spectrometry Research Facility in the Department of Chemistry at the University of Oxford. He has 20 years of experience in the development of analytical chemistry techniques, in particular using mass spectrometry, with applications in chemical, biological, and medical research. He has contributed to over 150 journal publications, book chapters, and articles, and co-written a textbook on *Mass Spectrometry* (with Neil J. Oldham) in the Oxford Chemistry Primers series.

Acknowledgments

The production of this book has been a long-running endeavour and we would like to thank all those who have contributed. In particular, Tim is grateful for the understanding and encouragement given by his wife Rachael and daughter Emma throughout the process, and James would like to thank his wife Yuriria and sons George and Giles for their support, understanding, and encouragement. We also thank our editors, Martha Bailes and Carolin Cichy, for their continued support and invaluable guidance throughout the development of this textbook.

CHAPTER 1

Introduction

1.1 Background

The determination of molecular structure using spectroscopic data is in many ways like solving a jigsaw puzzle. With jigsaws the information comes in little pieces which need to be fitted together in a particular way to reveal the final picture. Molecular structure determination can be thought about in a similar way. Analytical information needs to be organized and then connected to build a coherent 'molecular picture', as illustrated in Figure 1.1. Strategies for solving complex jigsaw puzzles can also be extended to structural elucidation. Some people attempt to solve the puzzle by trying to connect the pieces in a random 'test and reject' fashion. However, this can lead to mistakes, for example when forcing two pieces together that look like they could be a match, but do not actually fit together. An effective approach is often to find the corners and the edge pieces first, without thinking about the overall picture at this stage. This gives you an idea of the overall size and scope of the puzzle and provides a framework in which pieces of similar colour may be associated and finally brought together to form the final picture.

> To solve jigsaw puzzles efficiently, it is important that a systematic method is adopted that is suitable for tackling *any* jigsaw puzzle. This is also true for structure determination using spectroscopic data.

Spectroscopic 'puzzles' present information that doesn't come in ready sorted fragments, which means that errors may occur and effective strategies are required to avoid them. An important skill in structure elucidation is to be able to sort out available information into small blocks first. Analysing these gives you some idea of the size of the molecule (and therefore its complexity), and only then is it sensible to try to fit the small blocks together to build a picture of the overall structure. There are often several possible ways to fit the data together and, like in jigsaws, it is not always immediately obvious which

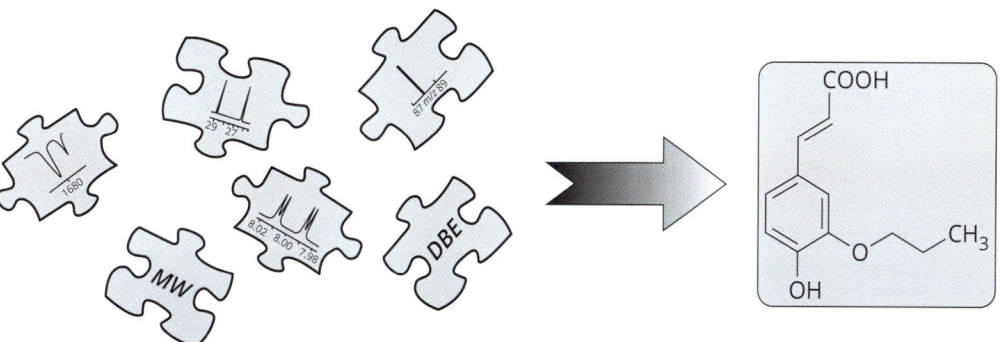

FIGURE 1.1 The jigsaw analogy: the process of deriving a molecular structure from spectroscopic data.

pieces are connected. A systematic method for tackling the problem is needed to become skilled at structure determination and avoid the most common pitfalls. Becoming adept at solving the puzzle of chemical structure by using the information that spectroscopic and spectrometric techniques can provide requires both an effective approach and significant practice. This book introduces you to effective approaches to data interpretation and walks you through many examples that show how these approaches can be applied in practice. It also includes a set of problems that you can work on solving yourself to help you become confident at structure determination.

1.2 About this book

This book provides a general introduction to structure elucidation using mass spectrometry (MS), infrared (IR) spectroscopy, and nuclear magnetic resonance (NMR) spectroscopy, and is designed to be used alongside an introductory lecture course or for revision purposes. It is assumed that you will already have a basic understanding of the physical concepts for each technique, and the book does not attempt to explain the theory behind their function in any great detail. Instead, each technique is introduced with a brief section on 'how it works' which is intended as a prompt or reminder of the physical basis of the method. The focus of this text is very much on how one uses spectroscopic data for problem solving, and you will likely gain most from this book when working through it systematically.

> This book aims to guide you through a systematic 'common sense' approach to organic structure elucidation, exemplified through worked examples and further developed through problems for you to solve yourself.

In the early chapters of the book, we introduce the key analytical techniques in terms of how they can be used to generate relevant information, how this information can be extracted efficiently from a given spectrum, and how this information can then be sorted into various pieces of the structural puzzle. Subsequently, working through the examples in later chapters with guided solutions will enable you to apply these methods and develop your own approach to solving structure determination problems. The examples will also help you develop your understanding of common pitfalls. You can then put these new skills to the test by tackling a second set of problems for which no model answers are provided.

In general, assigning a structure is easier when there is more spectroscopic data available. This book focuses primarily on the use of ^{1}H and ^{13}C NMR spectroscopy, alongside infrared spectroscopy and low-resolution mass spectrometry. We recognize that additional information, for example the molecular formula from high-resolution mass spectrometry (HRMS), can be very useful in finding a solution. But in many cases, candidate solutions can be determined using the techniques covered in this text in combination. Even if there are no data to give direct access to a molecular formula, this can often be quite easily deduced from the other spectral information. For example, use of double bond equivalents derived from molecular formulae can give access to information about the degree of unsaturation or the presence of rings in the structure. It should be noted that there are times when IR and MS data can provide invaluable information to help resolve a potential assignment, and in real-world settings it may not be possible to obtain interpretable NMR spectra due to a lack of detection sensitivity (especially for carbon) or the presence of chemical heterogeneity.

In some examples presented in this book, it will turn out to be impossible to determine a single definitive structure. Most often, this is because alternative structural isomers can be considered consistent with the available spectroscopic data. In later chapters, we introduce carbon spectrum editing and two-dimensional (2D) NMR to illustrate how these more advanced methods can help resolve the structure definitively in such cases and can be used to complete the assignment of all resonances. These later chapters also introduce additional MS approaches to structural elucidation and confirmation using soft ionization techniques combined with tandem mass spectrometry. We hope that these chapters will encourage you to progress your understanding of how these more sophisticated approaches, which are fully described in more advanced texts, can be used for structure elucidation.

Learning to analyse spectra and determine structures can be likened to learning to ride a bicycle. Both involve practice, practice, and more practice that will eventually lead to confidence and speed in completing the task with fewer mistakes along the way.

We hope that this book will set you on the right track and give you confidence to develop your own approaches and preferences for tackling structure determination in organic chemistry.

CHAPTER 2

Analysis of ^{1}H NMR spectra

As a student studying organic spectroscopy, it is likely that, when given a molecular structure, you would feel comfortable identifying the different hydrogen types (environments) present and even predicting their peak multiplicities arising from neighbouring protons. However, when working in the reverse direction from a spectrum to a structure, there is often a tendency to dive into the analysis of coupling patterns first and ignore some of the more straightforward and useful information that can be extracted easily from the spectrum. In this chapter, we guide you through the process of analysing ^{1}H NMR spectra without becoming entrapped by the wealth of data NMR can provide.

This chapter provides a summary of the key information that we can derive from a ^{1}H NMR spectrum and provides a foundation from which to address the examples presented in the book, and subsequently in real-life problems. This process involves the systematic analysis of the spectroscopic data, identifying appropriate 'jigsaw pieces', and determining how these may be connected in a chemically consistent manner to help reveal a picture of the molecular structure (Figure 2.1).

In a real laboratory situation, an initial assessment of a spectrum will often involve identifying the presence of peaks arising from species other than the compound of interest. These may include the NMR solvent itself, water in the sample, and common impurities

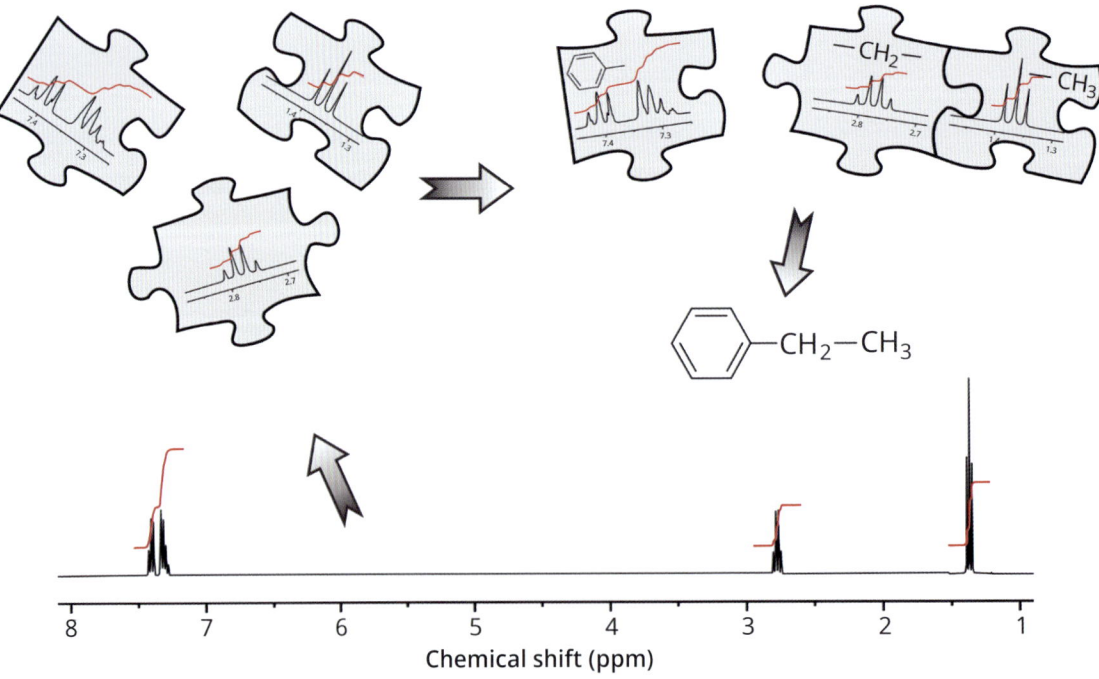

FIGURE 2.1 The linking of ^{1}H NMR spectroscopic jigsaw pieces to reveal a compatible molecular structure.

such as other residual solvents and (silicon) grease. We will consider these practical aspects further in Chapter 6. Answers to the exercises are found at the end of the chapter.

2.1 How NMR spectroscopy works

Nuclear magnetic resonance relies on the nuclei of certain atoms generating a local magnetic field, meaning we may think of them as behaving like microscopic bar magnets; they are said to possess a *nuclear magnetic moment*. This behaviour arises from the fact that all nuclei are positively charged and some exhibit spin (one can imagine nuclei spinning on their axis) such that the moving charge gives rise to a magnetic field. The commonly studied ^{1}H and ^{13}C isotopes both possess nuclear spin (they are said to be spin-½ particles) and can give rise to NMR spectra, whilst the more abundant ^{12}C has (unfortunately) zero spin and is thus NMR inactive.

When spin-½ nuclei are placed in an external magnetic field, they will orient in that field such that the majority will adopt a lower-energy parallel orientation 'with' the field (or 'up') and a minority will orient 'against' the field (or 'down') and possess (slightly) higher energy. This orientational behaviour is much like that of a compass needle orienting itself in the Earth's magnetic field. The application of electromagnetic radiation of appropriate frequency causes the aligned nuclei to invert their orientation due to the absorption of energy (Figure 2.2). The absorption (resonance) frequencies are very sensitive to the external magnetic field (in which the NMR samples are placed) but also, critically, to the very much smaller local magnetic fields each nucleus experiences that are influenced by the local chemical structure around it. These local fields arise from the electron clouds that surround all nuclei, since electrons are themselves magnetic particles. The small differences in absorption frequencies due to localized differences in magnetic fields are reflected in the so-called chemical shifts of the resonance absorption lines (or 'resonances'), as represented on a relative frequency axis of the spectrum (in units of parts per million, or simply ppm). These chemical shifts can be correlated to chemical environments and hence to structural features within a molecule, as exemplified in this and the following chapter.

Magnetically active nuclei can also interact with their neighbouring nuclei in a structure via the electrons in intervening bonds—a process known as scalar- or *J*-coupling.

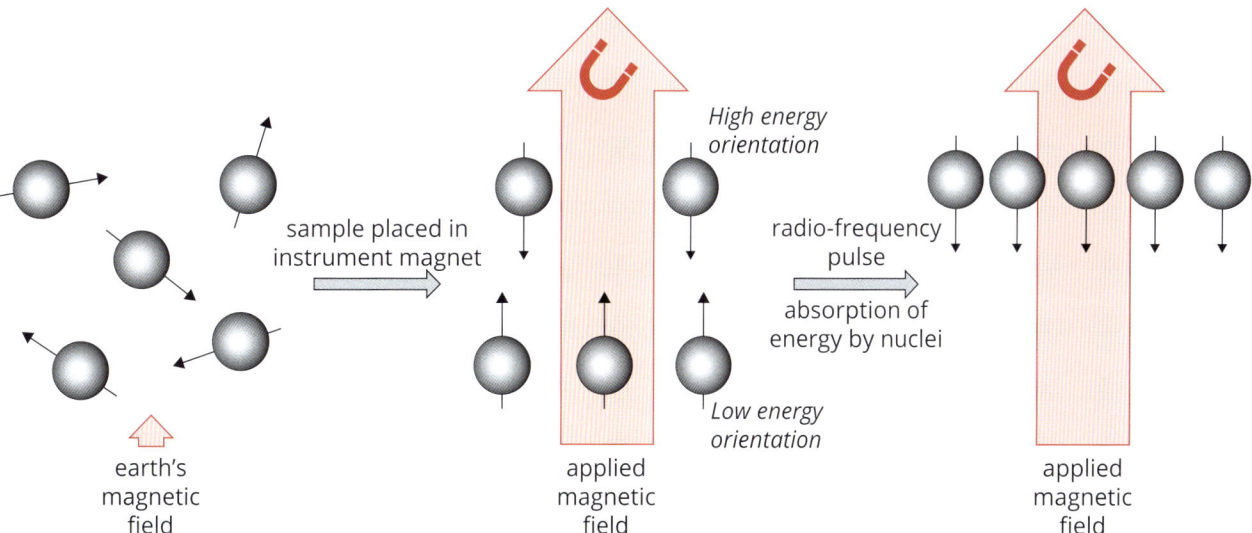

FIGURE 2.2 The process of nuclear magnetic resonance. Nuclei align along an intense applied magnetic field, preferentially in the direction of the field. Application of radio-frequency energy causes nuclei to change orientation from the preferred lower-energy 'up' state to a higher-energy 'down' state. The absorption frequencies can be correlated with the chemical environments the nuclei occupy in a structure.

These interactions cause resonance absorption lines to be split and give rise to multiplet patterns rather than single peaks. This provides evidence of bonds within a molecule, and thus provides additional valuable information on a molecular structure.

When examining organic structures by NMR, we may study analyte samples as neat liquids or, more commonly, as solutions in an appropriate solvent, such as chloroform. The samples are placed within an intense magnetic field (typically at least 200,000 times the Earth's magnetic field) to align the nuclei, and then irradiated with a pulse of electro-magnetic radiation. Since the energy differences between the 'up' and 'down' orientations are rather small, only low-energy radio waves are needed to perturb the nuclear align-ments and yield the required resonance absorption of energy. The absorption profile of the radio waves ultimately provides the NMR spectrum.

Following the resonance absorption (or 'excitation'), the perturbed nuclei will lose energy to their surroundings and return to their preferred orientation, a process known as 'relaxation'. This enables the excitation process to be repeated and recorded data to be co-added to improve detection sensitivity. Sensitivity is inherently low for NMR due to the low energies involved in the transitions and hence low absorption intensities. This en-hancement by signal averaging is key to the success of NMR in structure characterization.

2.2 ^{1}H NMR chemical shifts

The chemical shift of a ^{1}H NMR resonance is the primary indicator of the chemical envi-ronment of that proton.

- The value of the chemical shift gives key information about the type of atom to which the hydrogen is directly attached and whether this is saturated or unsaturated.
- The value of the chemical shift also hints at the nature of atoms one bond further away.

Whilst precise chemical shifts should be recorded, searching directly at this stage through data tables to see what groups appear at a given position on the chemical shift (δ) scale can be a dangerous approach as there will likely be different possibilities for each chem-ical shift value. It is therefore much better to develop a general idea of which resonances appear where on a spectrum and steer away from detailed chemical shift tables as a start-ing point. This may leave some uncertainty at the early stages, but it is worth recognizing that accepting some uncertainty at this point can help prevent mistakes later. Other parts of the spectrum, or information from other analytical techniques, will often reveal the answer at a later stage.

2.2.1 Chemical shift ranges

Figure 2.3 shows the chemical shift (δ) range for most proton environments found in organic molecules (see Appendix 1 for these data in tabulated form and links to related online resources). Generally, the spectrum can be split in half such that:

- The **right-hand side** (smaller ppm values on the δ scale) contains resonances for protons attached to **saturated** carbon atoms.
- The **left-hand side** (larger ppm values) contains resonances for protons attached to **unsaturated** carbon atoms.

An exception to this general consideration is the (rarely encountered) terminal alkyne proton $\equiv$C–H, which appears at around 2.5–3.0 ppm.

The term 'shielded' is often used to describe the environment of a nucleus whose reso-nance displays a relatively small chemical shift, and 'deshielded' for those with a relatively large chemical shift. This is because high electron density surrounding a nucleus serves to protect (or shield) it from the external magnetic field, giving rise to a lower resonance frequency and hence smaller chemical shift (and vice versa).

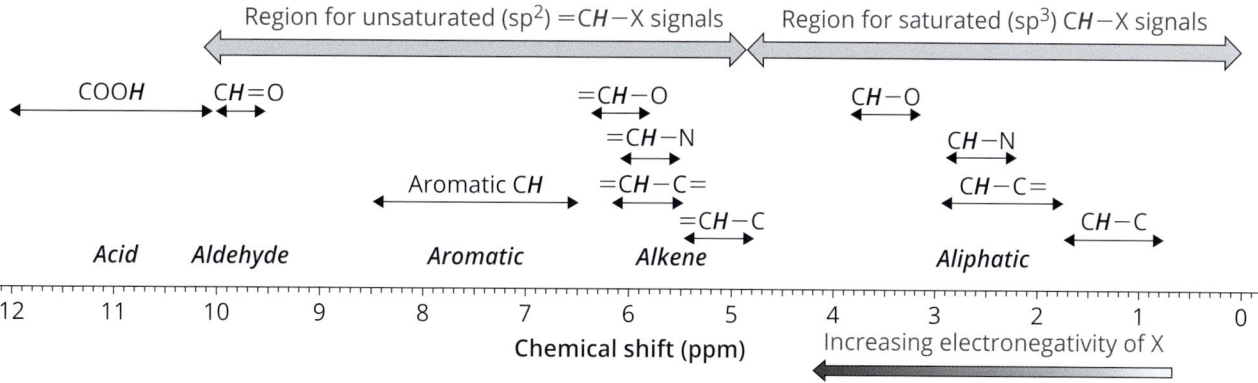

FIGURE 2.3 Typical ^{1}H chemical shift ranges for proton groups. This may be used as a general guide when assessing ^{1}H spectra, but remember that chemical shifts may fall outside these representative boundaries according to structural details.

Considering either half of the spectrum, the figure further shows that:

> for any C**H**–X proton, it is the immediately neighbouring substituent X that most influences where the C**H** resonance appears.

Thus, for a simple alkane, where X is C (i.e. C**H**–C), the H atom has a chemical shift of around 1.2 ppm, and for a simple alkene C**H**=C the H atom has a chemical shift of around 5.5 ppm. These values reflect relatively small chemical shifts within each 'half' of the spectrum. However, if X is oxygen, a highly electron-withdrawing element, the peak appears towards large values of chemical shift in each half of the spectrum. The alcohol C**H**–O has a chemical shift typically around 4.0 ppm, and in the aldehyde C**H**=O, the H atom has a chemical shift of around 9.5 ppm.

> In general, the greater the electron-withdrawing capability (electronegativity) of substituent X, the larger will be the chemical shift value of the adjacent C**H**–X proton.

The influence of electronegative atoms on the chemical shifts of adjacent protons is referred to as the *inductive effect*.

Some regions are often quite characteristic of a particular functionality:

- Peaks from aldehydes will appear around 9–10 ppm.
- Peaks above 10 ppm typically arise for acidic protons involved in hydrogen bonds, such as in carboxylic acids (COO**H**, see section 2.2.2).

Nevertheless, there will often be regions where the exact nature of the adjacent substituent is uncertain. For example, the region around 1.8–3.0 ppm is where resonances from many common groups may appear, including C**H**–C=C, C**H**–Ar (Ar = aromatic ring), C**H**–N and C**H**–C=O. You should keep an open mind as to the exact origin of these peaks until additional supporting evidence enables you to make a more definitive assignment.

> The presence of multiple neighbouring electronegative substituents will tend to have an additive effect on chemical shifts.

Proton shifts in the 4–5 ppm region often arise from additive effects, for example from the combination of neighbouring N and C=O groups in amino acids (NH–C**H**–CO). There may also be aliphatic resonances of saturated centres in the 5–6 ppm window if the

CH carbon bears more than one electronegative element, such as in an acetal O–C**H**–O (~5.5 ppm). A more extreme yet very commonly encountered example of this effect is seen for the widely used solvent chloroform, whose residual C**H**Cl$_3$ signal falls at 7.26 ppm due to the combined influence of three adjacent electronegative chlorine atoms. For simpler molecules, however, these potential areas of confusion are usually not a problem if a systematic approach to the analysis is used.

> When undertaking the initial analysis of a spectrum, it is usually worth considering the left-hand side of the spectrum first and assessing its content. This part of the spectrum is often less crowded, and peaks here will indicate the presence of any aromatic or unsaturated structural elements within the molecule, to which aliphatic groups may be attached. When available, the carbon spectrum will also provide data that may be used to support your initial assessment of the proton chemical shifts (see Chapter 3).

2.2.2 Chemical shifts of acidic protons

The chemical shifts of protons attached directly to electronegative atoms such as O or N deserve special mention as these values can be highly variable and are especially difficult to predict. These protons tend to be acidic in nature.

> Acidic protons are likely to participate in hydrogen-bonding interactions that can greatly influence their chemical shifts.

These H-bond interactions can be with other solute molecules, with the solvent, or with water in the sample. The varying degrees of hydrogen bonding cause correspondingly large variations in proton chemical shifts, which are dependent on the solvent used and on sample concentrations.

> The stronger the hydrogen bonding, the larger the proton shifts tend to be.

For example, the chemical shifts of carboxylic acid protons are often observed above 10 ppm due to 'head-to-head' hydrogen bonding between acid groups (i.e. OH—O=C H-bonds).

> Acidic proton resonances often have a broad appearance which can also lead to any coupling structure being unresolved (see section 2.4.3).

Hence, OH and NH chemical shifts can be difficult to predict but are often broad in their appearance which can help identify their presence. A further practical test for the presence of acidic protons is the 'D$_2$O exchange' experiment described in Chapter 6.

2.3 The integral of the resonances

One of the easiest concepts in ^{1}H NMR to understand is that of integration, that is, measuring the peak area of the ^{1}H NMR resonances.

> Peak integrals (areas) reflect the relative number of hydrogen atoms contributing to each resonance.

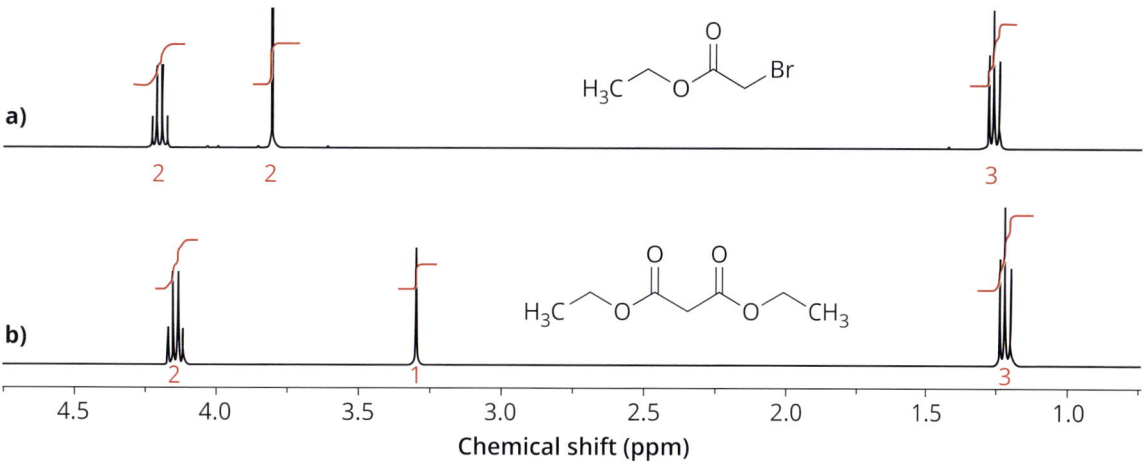

FIGURE 2.4. ^{1}H NMR spectra of a) ethyl bromoacetate ($C_4H_7O_2Br$) and b) diethyl malonate ($C_7H_{12}O_4$) demonstrating the influence of symmetry and *relative* peak integrals (numerical values shown in red below each peak).

These 'integrals' are typically represented as traces drawn above resonances, sometimes accompanied by their relative integrated intensities plotted below the peaks, as in the spectra shown above, where the traces and numerical values are shown in red. If you are new to spectral interpretation, be cautious to avoid assigning resonances arising from solvent residues in the sample; these can often be identified by having integrals that are either very small or much larger than those of the analyte of interest. If you sum the analyte integrals at the very start, this will also give you a rough idea of the size of the molecule under investigation by providing an approximate proton count.

> Peak integrals are relative rather than absolute, and one needs to be aware of symmetry within a structure.

If a structure is symmetrical, then a 'C_xH_y' molecule may in reality be a '$C_{2x}H_{2y}$' molecule, as seen for diethyl malonate in Figure 2.4. Of course, the molecular ion in the mass spectrum is useful to identify such cases, and cross-referencing even a 'guesstimate' of the molecular formula with the molecular ion in the mass spectrum, can help show if this is the case.

EXERCISE 2.1A

Figure 2.5 shows extracts from three ^{1}H NMR spectra of molecules with two CH_3 groups. Using the information about chemical shifts and integrals above, work out to which of the following five structures each spectrum belongs.

EXERCISE 2.1B

For the two molecules from **1–5** below without a spectrum, predict the appearance of the aliphatic region of their ^{1}H NMR spectrum.

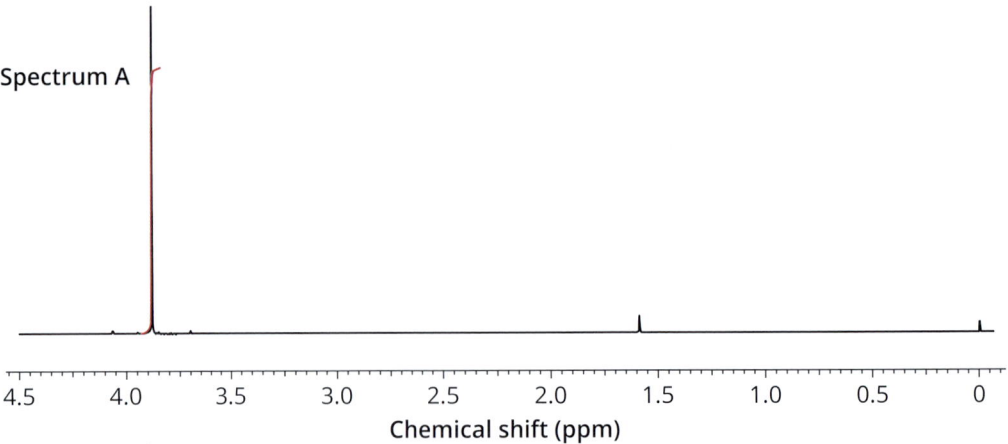

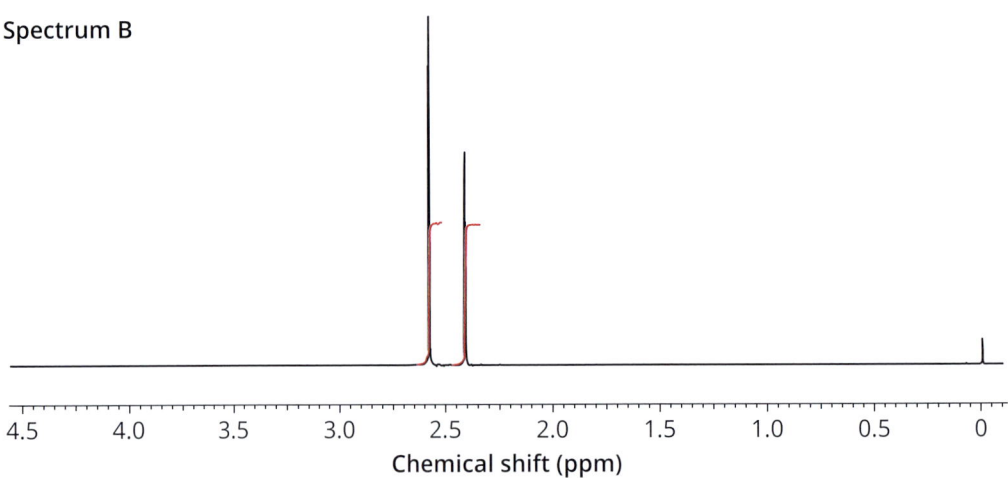

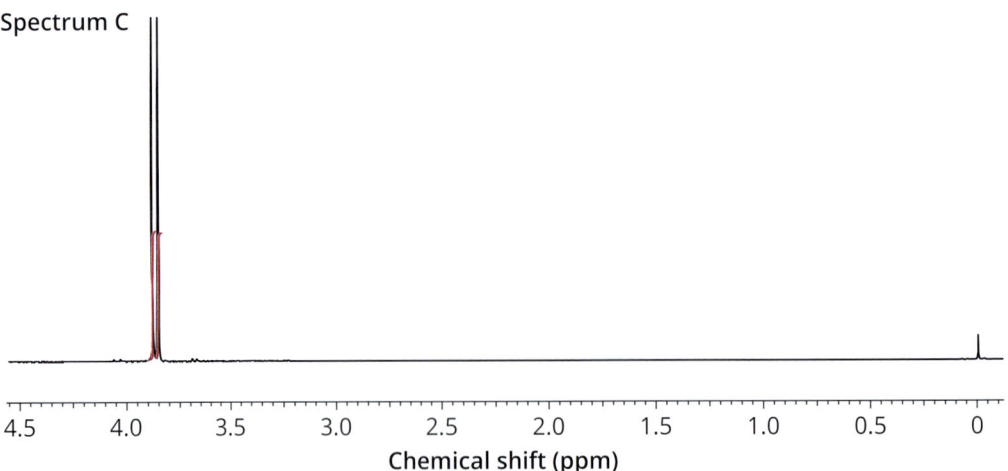

FIGURE 2.5 Extracts from ^{1}H spectra of compounds containing two methyl groups (the chemical reference tetramethylsilane (TMS) is seen at 0.0 ppm).

Finally, we should add some words of caution when examining integral traces on real-world spectra. These are often produced automatically by the processing software, and it is not uncommon for closely separated resonances originating from different proton groups to be shown with only a single integral trace drawn across them. This is simply because the software was unable to recognize that different peaks are present. Thus, you should remain alert to the fact that a single trace *does not necessarily* mean that all the included peaks belong to one proton group. Nevertheless, it is the case that the total integrated area should represent the total for all the groups being integrated. For example, the combined peaks of a CH and CH_2 group should integrate to three protons and the associated integral trace will reflect this sum, but this should not be mistaken for the presence of a methyl group. Consideration of peak multiplet patterns becomes important in such cases. You should also note that integral traces may not always be presented as whole numbers, especially when peaks partially overlap or when impurities are present. Remember that the values only represent the *relative areas of peaks*, and may not directly reflect the proton count in real-world data sets.

2.4 Multiplet patterns

The more complex features to examine in a ^{1}H spectrum tend to be the multiplet structures or 'multiplicities'—the fine detail occurring within peaks when they are split into many lines.

> Resonance splitting arises from through-bond spin–spin coupling interactions between protons (or with other NMR-active nuclei in the molecule).

Multiplet features are usually the last to be considered in any detail. As such, in the worked examples that follow, these are typically interpreted as part of the detailed analysis rather than the initial analysis of spectra.

> Proton multiplet patterns are important because they report on the presence of adjacent protons within a molecule and hence *imply the presence of bonding pathways between these atoms.*

This ability to tell us specifically about *neighbouring* groups is a feature that is unique to NMR spectroscopy.

Although some multiplets have rather complex structures that may not be amenable to complete interpretation, some obvious multiplet patterns can often be readily identified and interpreted. We will highlight some of these in this section as they are typically diagnostic of particular structural elements. Perhaps the most common example of this is the characteristic $-CH_2-CH_3$ quartet-triplet pairing observed for the ethyl group, which is described below. In some circumstances the values of the coupling constants themselves can also provide valuable additional information, and we also present some common examples of these. The coupling constants are given the symbol J and are always measured in hertz (Hz).

> For proton multiplets the observed splittings typically arise for coupling interactions between protons that are separated by only two or three bonds.

These are often referred to as *geminal* and *vicinal* couplings, respectively, and are given the symbols $^2J_{HH}$ and $^3J_{HH}$ to reflect their origins. The term *geminal* comes from *gemini*, meaning 'twins', and *vicinal* derives from *vicinus*, which means 'neighbour' in Latin. Most proton multiplet structures can be interpreted in terms of these short-range interactions.

> Longer-range couplings (mostly across four bonds and involving unsaturation) may also produce splittings, which tend to be small.

2.4.1 Common multiplet structures

The presence of common molecular fragments within molecules can often be recognized from the collection of multiplets they produce. It is therefore helpful to be familiar with simple multiplet patterns:

- The number of lines within simple multiplet structures may be predicted using the '*n*+1' coupling rule: coupling to *n* equivalent nuclei produces *n*+1 multiplet lines.
- Relative line intensities within simple multiplets may be predicted from a Pascal's triangle coupling tree (Figure 2.6).

	n	multiplet
1	0	singlet
1 1	1	doublet
1 2 1	2	triplet
1 3 3 1	3	quartet
1 4 6 4 1	4	quintet
1 5 10 10 5 1	5	sextet
1 6 15 20 15 6 1	6	septet

FIGURE 2.6 Pascal's triangle predicts the relative peak heights within resonance multiplets caused by coupling to **n** identical neighbouring protons.

Pascal's triangle predicts the intensity patterns expected for simple and commonly encountered structures such as doublets, triplets, and quartets (Figure 2.7). An example of a recognizable fragment is that of the $-CH_2-CH_3$ pairing observed for the ethyl group shown in Figure 2.8, which is found commonly in molecules such as ethyl ketones, ethyl ethers, and most frequently ethyl esters. When making such an assignment, it is also important to check that the integrated intensities of the multiplets show the correct ratio of 2:3 and that the coupling constants for each multiplet match. We describe some other common structural elements and their associated multiplet structures in section 2.5.

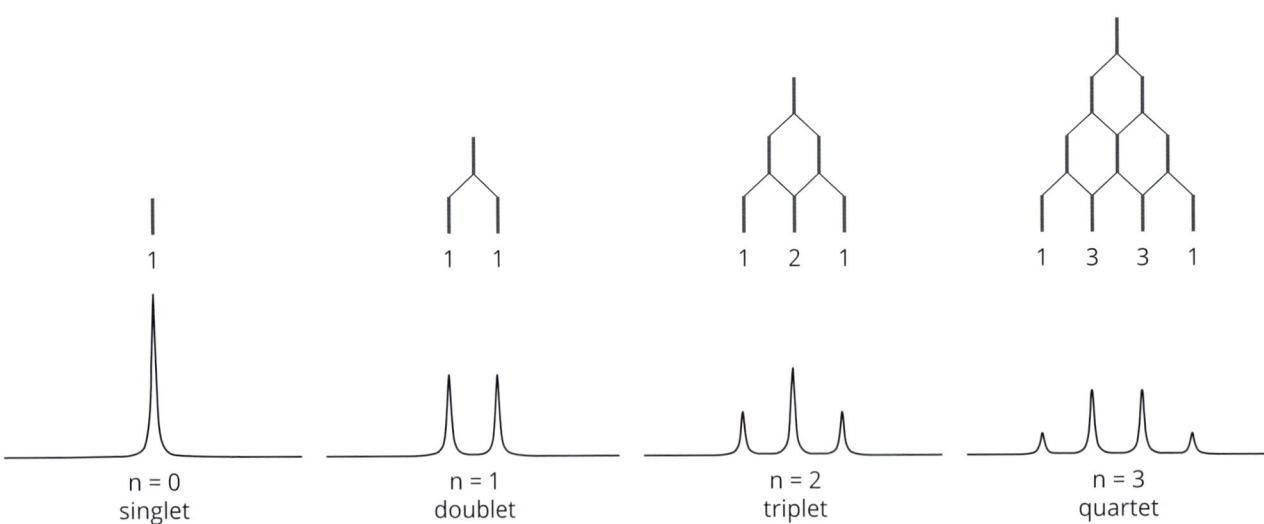

FIGURE 2.7 Common multiplet patterns from coupling to none, 1, 2, and 3 neighbouring protons (with identical coupling constant *J* Hz), as predicted by Pascal's triangle.

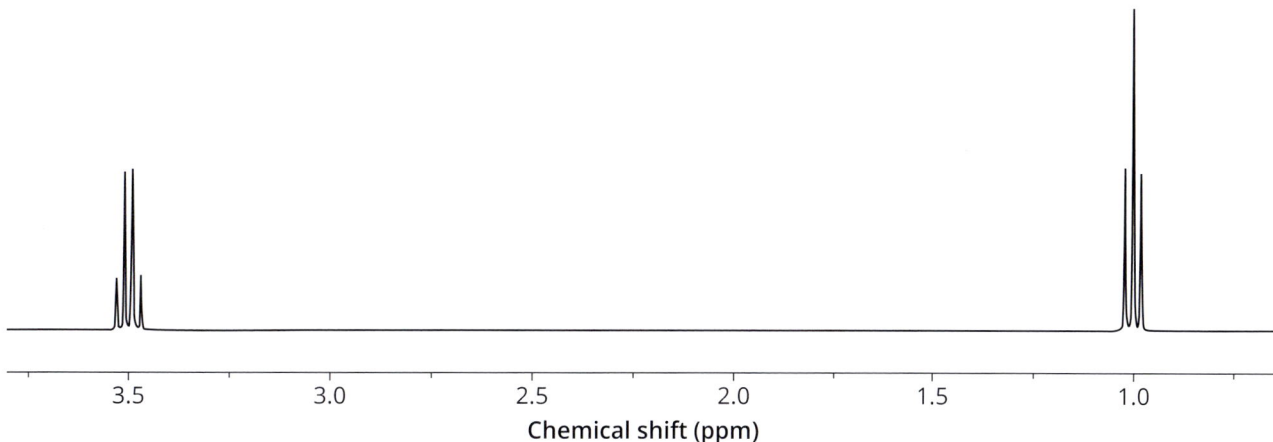

FIGURE 2.8 The commonly seen ethyl group multiplet structure (note: the chemical shift of the CH$_2$ quartet in particular will vary according to the adjacent functionality).

When protons share couplings to multiple protons with differing coupling constants, their multiplet structures can exhibit more elaborate patterns, which can often be understood in terms of 'coupling trees'.

For example, coupling of one proton to two adjacent protons each with a different magnitude (as may occur in a fragment such as X–CH–**CH**–CH–Y) gives rise to a doubling up of a doublet—thus yielding a *double doublet* for the highlighted proton shown as bold (Figure 2.9a). Extending this concept, a fragment such as X–CH$_2$–**CH**–CH–Y would yield either a *triple doublet* (Figure 2.9b) or a *double triplet* (Figure 2.9c), depending on the magnitudes of the couplings involved. When describing multiplets, it is conventional to define the structure of the largest coupling first (here $J^a > J^b$), so in the *triple doublet* the

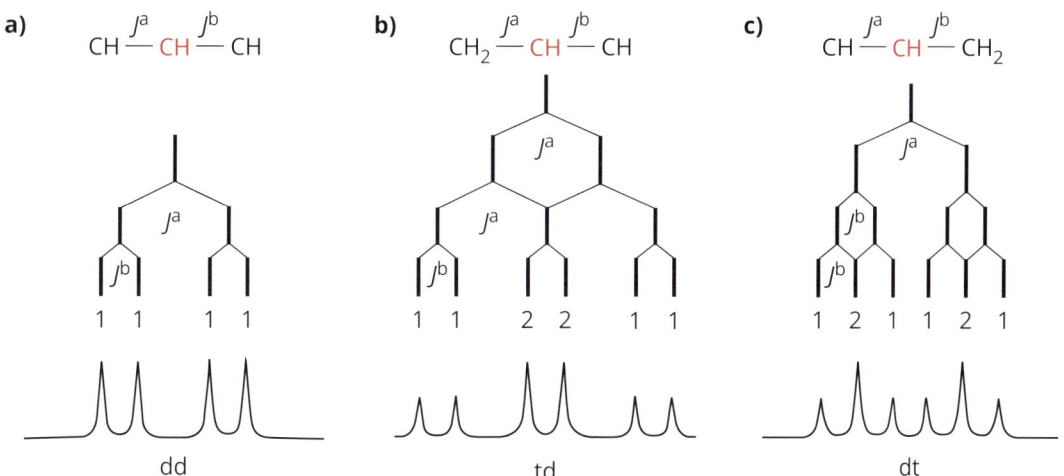

FIGURE 2.9 Commonly observed multiplet structures seen for coupling to multiple neighbouring protons: a) double doublet, b) triple doublet, and c) double triplet. The multiplet trees show the origin of the line intensities within the multiplets, with each branching point reflecting the effect of an additional coupling.

coupling to the CH_2 (yielding the triplet structure) will be larger than that to the CH (the doublet structure). When the coupling to the CH is larger than that to the CH_2, the multiplet would be described as a *double triplet*. The ability to analyse multiplets in this way is useful as it provides information on neighbouring groups of protons in a structure. However, this level of interpretation should usually be reserved only for the detailed analysis of a spectrum and not be part of its initial analysis.

> The complete analysis of more highly complex structures arising from many couplings is often not possible, either because not all lines are resolved or because the peak is overlapped with others, masking its features. The resonance is then described as a *multiplet* (m).

In situations of highly complex multiplet structures, the peak integrals of these multiplets can often be key to revealing the number and nature (CH vs CH_2 etc) of protons present.

2.4.2 Magnitudes of coupling constants

When interpreting proton multiplet structures, it is often helpful to have a feel for the magnitudes of coupling constants as this can help avoid mis-assignments.

> Coupling constant magnitudes are influenced by the bonding environment of the proton and by adjacent substituents; it is more useful to have a general sense of the ranges coupling values may take than to attempt to memorize precise values.

Proton coupling constants usually fall between 0 and 18 Hz, and typical values are summarized in Figure 2.10.

> Coupling constants measured from multiplet line separations must be equal (within measurement error) for mutually coupled protons.

Thus, for the ethyl group, a 7.0 Hz splitting for the CH_2 quartet will be matched with a 7.0 Hz splitting on the CH_3 triplet, and we write J_{HH}= 7.0 Hz for the coupling, with the subscript *HH* denoting the nuclei involved. In most instances, it is sufficient to quote *J* values to 0.5 Hz.

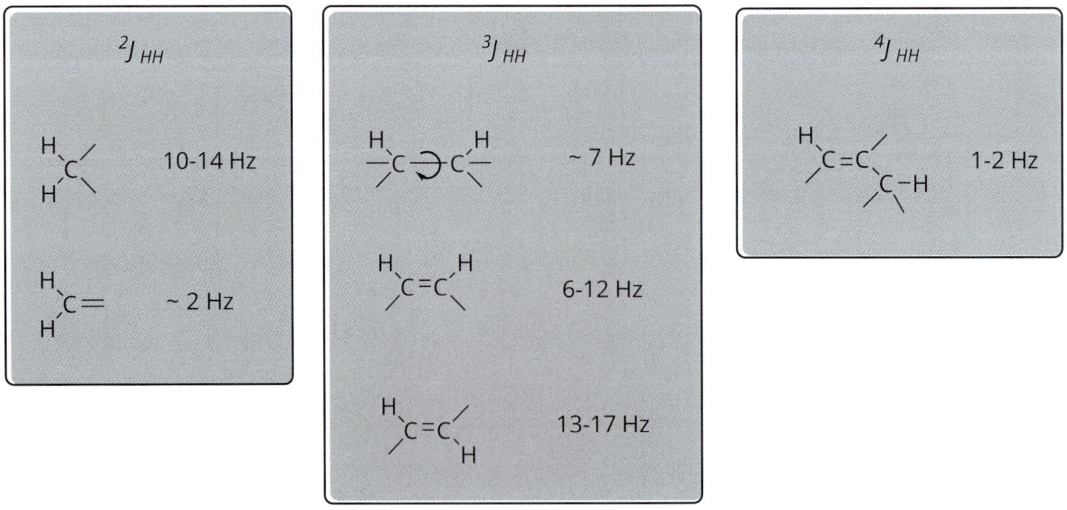

FIGURE 2.10 Typical values for proton–proton coupling constants.

As noted earlier, most observable proton–proton couplings occur over 2 or 3 bonds. Three-bond vicinal couplings ($^3J_{HH}$) in saturated systems can range in magnitude from 0 to 12 Hz and are influenced strongly by the dihedral angle between the bonded protons.

This angular relationship is described by the *Karplus equation* which has the following general form:

$$^3J_{HH} = A + B\cos\phi + C\cos 2\phi$$

where ϕ is the H–H dihedral angle and A, B, and C are coefficients whose values depend on the nature of the neighbouring substituents.

This is represented graphically in Figure 2.11, and although the actual *J* values may vary for different substitution patterns, the general shape of the Karplus curve remains largely similar.

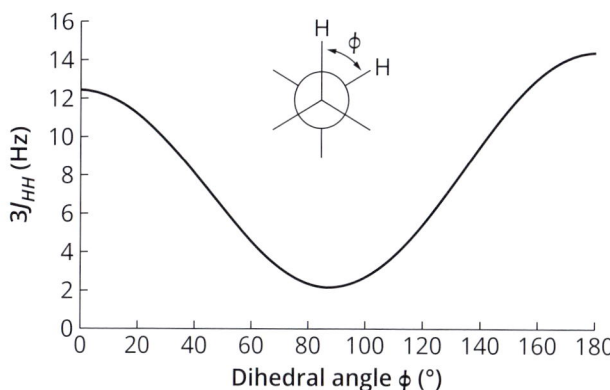

FIGURE 2.11 The Karplus curve: Definition of the dihedral angle and its influence on $^3J_{HH}$.

In aliphatic groups experiencing free rotation about carbon–carbon bonds, vicinal coupling constants will typically average to be ~7 Hz.

In some situations, the magnitudes of vicinal couplings can be diagnostic of geometry within structures. It is, for example, possible to define the double bond stereochemistry in alkenes on the basis of these values (see Figure 2.10).

Cis alkenes have a $^3J_{HH}$ value around 6–12 Hz, whereas in *trans* alkenes these are around 13-17 Hz. (So, there is still possibility for confusion if the *J* value just happens to be around 12-13 Hz!)

1,2-disubstituted alkenes, with no further couplings to neighbouring aliphatic CHs, are especially easy to spot as they give rise to a pair of ^{1}H doublets (Figure 2.12). Even in the presence of additional couplings, *trans* alkenes are usually readily identified due to their rather large *J* values.

Geminal ($^2J_{HH}$) couplings are strongly influenced by carbon hybridization.

- In terminal alkene sp^2 centres, protons share very small geminal $^2J_{HH}$ couplings that may even be unresolved.

- In aliphatic systems where two protons on the same carbon experience different chemical environments and hence have separate chemical shifts (that is, they are inequivalent), these will typically exhibit a large mutual geminal $^2J_{HH}$ coupling of around 10–14 Hz.

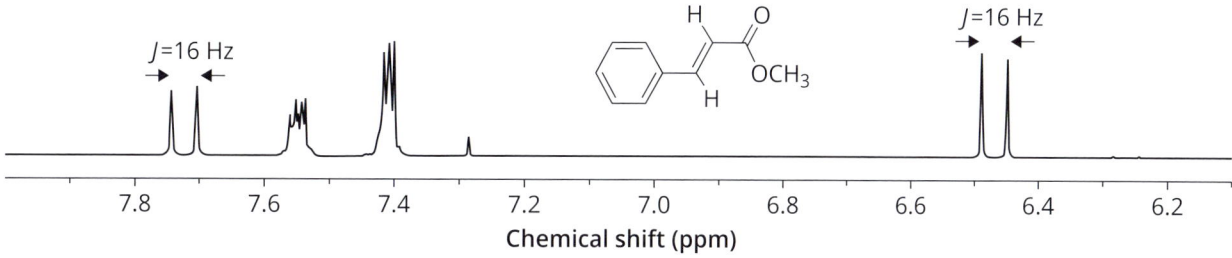

FIGURE 2.12 The doublet pattern of *trans* methyl cinnamate displays a large $^3J_{HH}$ coupling of 16 Hz.

FIGURE 2.13 Common structural features where methylene protons may be inequivalent. In the ring system H^a and H^b have axial and equatorial orientations respectively, whilst in the open system H^a and H^b are in the vicinity of a stereogenic centre (the carbon bearing different groups X, Y, and Z). In both cases, the H^a and H^b will experience different chemical environments, are thus inequivalent, will exhibit different chemical shifts, and will show geminal coupling to each other.

Common examples of this situation arise for methylene CH_2 groups that exist within ring systems or for those that are adjacent to stereogenic centres in a molecule (see Figure 2.13). It is important to realize that these geminal coupling values overlap those expected for some vicinal couplings, and you should watch out for this ambiguity in identifying coupling partners when inequivalent protons in methylene groups may be present. For most examples in this book, methylene groups will have equivalent protons, meaning geminal coupling between them is not observed.

Long-range four-bond couplings ($^4J_{HH}$) are nearly always small in magnitude (<2 Hz) but can sometimes be observed across unsaturation and are most common in aromatic rings.

The presence of long-range couplings gives rise to very fine splittings on multiplets that may only be observed when these are expanded; these will be exemplified in section 2.6.

In the examples and problems in this textbook, we report relevant coupling constants where these impart unique or useful information. However, in many instances, coupling constants themselves need not be fully interpreted for a successful structure definition, and we advise not to consider these in the early phases of spectrum assessment.

2.4.3 Common distortions to multiplet structures

The introductory discussions in previous sections have presented multiplet structures that are symmetrical about their mid-points and have relative line intensities that match those predicted from Pascal's triangle, such as 1:3:3:1 for a quartet. However, in real-world examples, multiplets often display some slight distortions to their relative line intensities, and it is important to recognize their existence and not be confused by these. Once recognized, these perturbations rarely introduce problems with spectrum interpretation, and can even provide additional information that can help with this process.

One distorting effect can be seen in the ethyl group in Figure 2.14, where for both multiplets the outermost lines display slightly differing heights, leading to the appearance of the multiplets as 'roofed'.

- In roofed multiplets, the roofing points towards the resonance peak of the coupled partner, which provides an indication of where in the spectrum this partner peak will appear.

- The extent of the roofing is dictated by how close in the spectrum the peaks of the coupled protons are; the closer together the peaks are, the stronger the tilt.

More formally, it is the ratio of the chemical shift difference $\Delta\delta$ (expressed as the frequency separation in hertz) to the coupling constant J_{HH} (i.e. $\Delta\delta/J_{HH}$) that dictates the extent of roofing, with peak distortions being readily apparent when $\Delta\delta/J_{HH} < 20$ and becoming more pronounced as this ratio diminishes, that is, as peaks become closer. In practice, this means that for ^1H spectra recorded at 400 MHz, mutually coupled protons separated by ~0.5 ppm or less will be expected to show roofed multiplet structures. This dependence on peak separation is apparent in the roofed doublet peaks for a CH–CH group (see Figure 2.15).

When coupled multiplets approach each other very closely, more severe multiplet distortions occur. This happens, as a rule-of-thumb, when $\Delta\delta/J_{HH} \ll 10$, and the

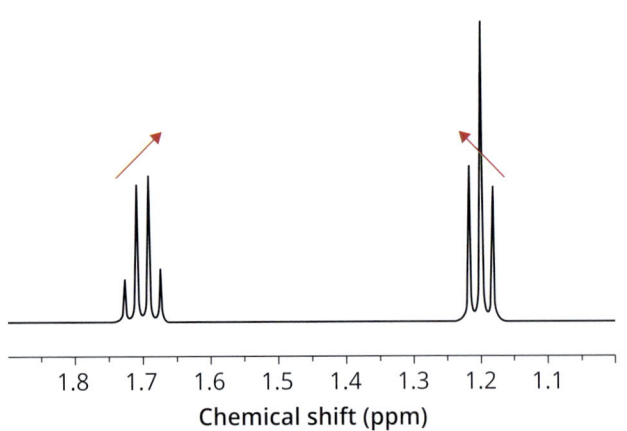

Chemical shift (ppm)

FIGURE 2.14 Multiplet distortions seen for the resonances of an ethyl group. The arrows show the direction of roofing towards the coupled partner.

features are then considered to be 'second-order' and are more difficult to interpret using the simple 'first-order' coupling rules usually employed. Note, however, that spectra in this book have largely been recorded to avoid such severe distortions that complicate analysis.

It is also helpful to realize that roofing effects (and other second-order distortions) are more apparent in spectra recorded on lower field magnets and less so on higher field spectrometers. This is because chemical shift differences ($\Delta\delta$), when expressed in hertz, are smaller for lower magnetic field strengths, whilst the magnitude of J is field independent. This means that $\Delta\delta/J_{HH}$ ratios become relatively smaller at lower fields. Fewer multiplet second-order distortions are one of the advantages of using higher magnetic fields in NMR.

> In cases of multiplet roofing, the gross structures of the multiplet shapes are retained, remaining recognizable as doublets, triplets, et cetera, and the integrated area of each multiplet is unchanged by the roofing.

It is therefore only necessary to recognize and be comfortable with the influence of roofing on peak structures when interpreting spectra, as the overall analysis process remains unchanged. Many of the multiplets in the spectra in this book will in fact display some degree of roofing, so it is worth being alert to this.

Another 'distortion' to multiplet shapes, not predicted by the n+1 coupling rule, is commonly observed for symmetrically substituted aromatic rings.

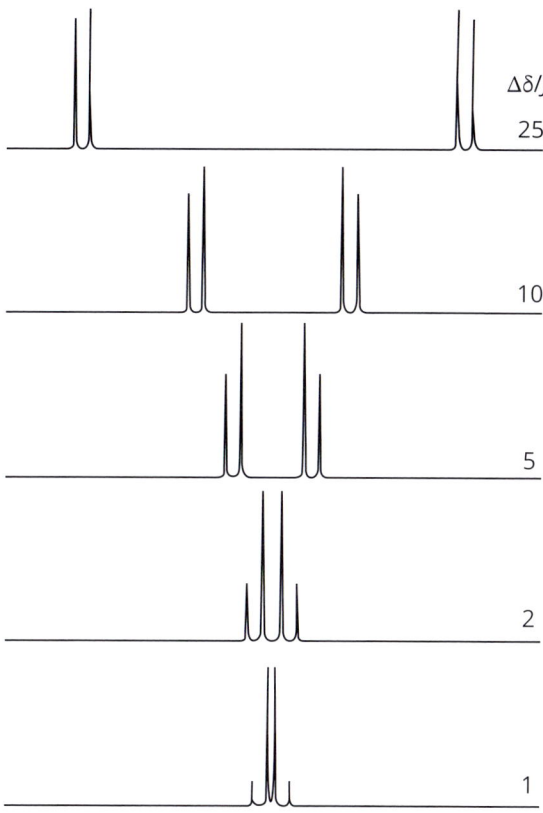

FIGURE 2.15 The increased roofing effect on the doublets of a coupled CH–CH group as the shift difference between them ($\Delta\delta$) reduces whilst J_{HH} remains constant.

> 1,4 disubstituted benzene rings display additional fine structure around the base of their multiplets, which predominantly exhibit doublet structures (see Figure 2.16).

This arises due to the symmetrical nature of this molecular fragment and the presence of long-range (4-bond) couplings that exist between symmetrically identical protons, for example Ha to H'a. This fine structure is diagnostic of the 1,4-substitution pattern and can often be easily recognized in the spectrum of molecules containing these fragments. It is frequently referred to in shorthand notation as an AA'BB' pattern (if roofing is also present) or an AA'XX' pattern (if roofing is not apparent).

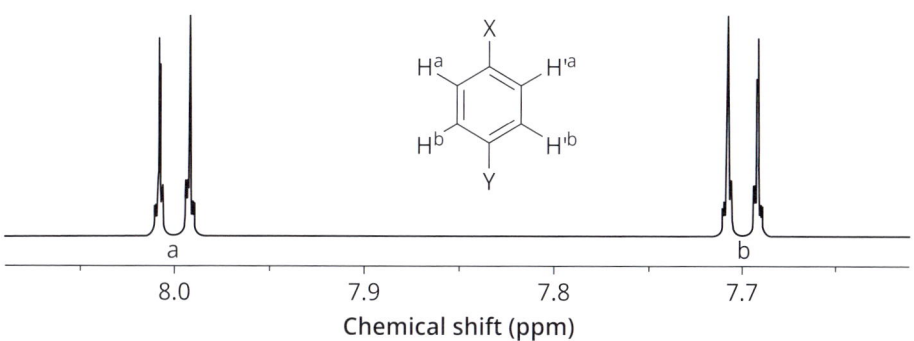

FIGURE 2.16 Multiplet patterns for 1,4-disubstituted benzene rings (the AA'BB' pattern).

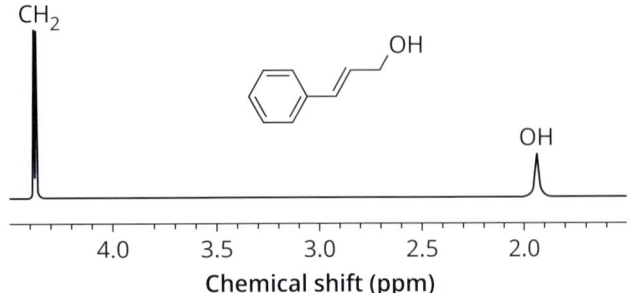

FIGURE 2.17 The broadened resonance of the acidic OH proton in cinnamyl alcohol. The proton exchange also masks its coupling to the adjacent CH_2 (which appears as a doublet due to its coupling to the adjacent alkene proton).

The final distortion we consider is the disappearance of multiplet splittings altogether (Figure 2.17).

> The resonances of acidic protons (e.g. alcohols, amines, and amides) often show characteristically broadened resonances which mask their splitting patterns.

This broadening arises from their acidic nature and their ability to exchange with another proton, coming either from another molecule or, quite frequently, from water in the solvent. Their appearance will vary greatly depending on the hydrogen-bonding environment and rate of proton exchange, which in turn will be affected by the choice of NMR solvent and solute concentrations. Given that the chemical shift of these protons can also be quite variable (see section 2.2.2), spotting a broadened resonance can be a useful indicator of an exchangeable acidic proton. Further experimental proof can be obtained from the D_2O exchange experiment described in Chapter 6.

2.5 Common features of aliphatic molecular fragments

Being comfortable with structure determination, whether using ^{1}H NMR or other spectra, often depends on competence in pattern recognition, which largely comes from practice and experience. Common structural elements give rise to easily identifiable 'spin systems' which represent groups of coupled protons that are separated from others in the molecule such that they share no further couplings. As already noted, a 3H triplet together with a 2H quartet gives evidence for the $CH_3CH_2–$ spin system of the ethyl fragment. Extending this idea, we see that ethyl butyrate can be recognized as comprising two separate spin systems: the $–CH_2CH_3$ group and the $–CH_2CH_2CH_3$ group (Figure 2.18).

> The ability to recognize spin systems from multiplet structures and to correlate these to structural elements is one of the key skills to develop for efficient spectrum interpretation.

Examples of simple spin systems and their multiplet patterns are shown in Figure 2.19. Common structural elements frequently give rise to characteristic peak patterns that enable their identification (Figure 2.20). The chemical shift of the proton(s) at the connection

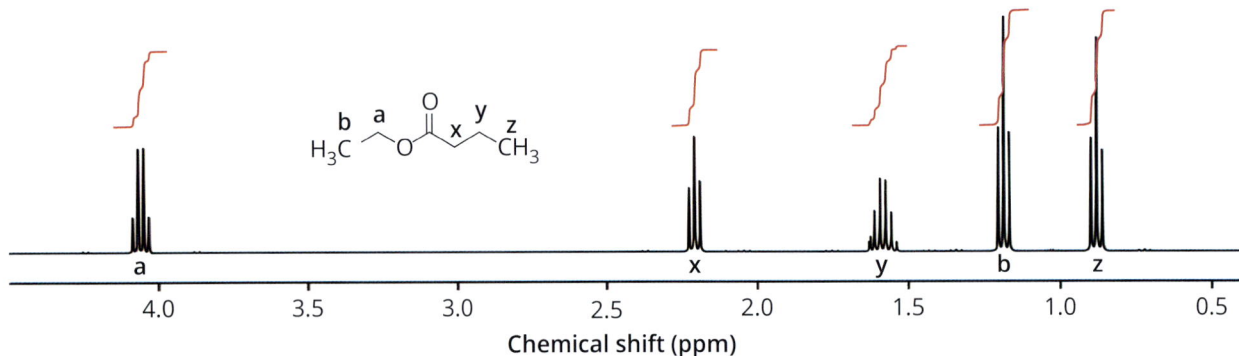

FIGURE 2.18 Ethyl butyrate comprises two isolated proton spin systems.

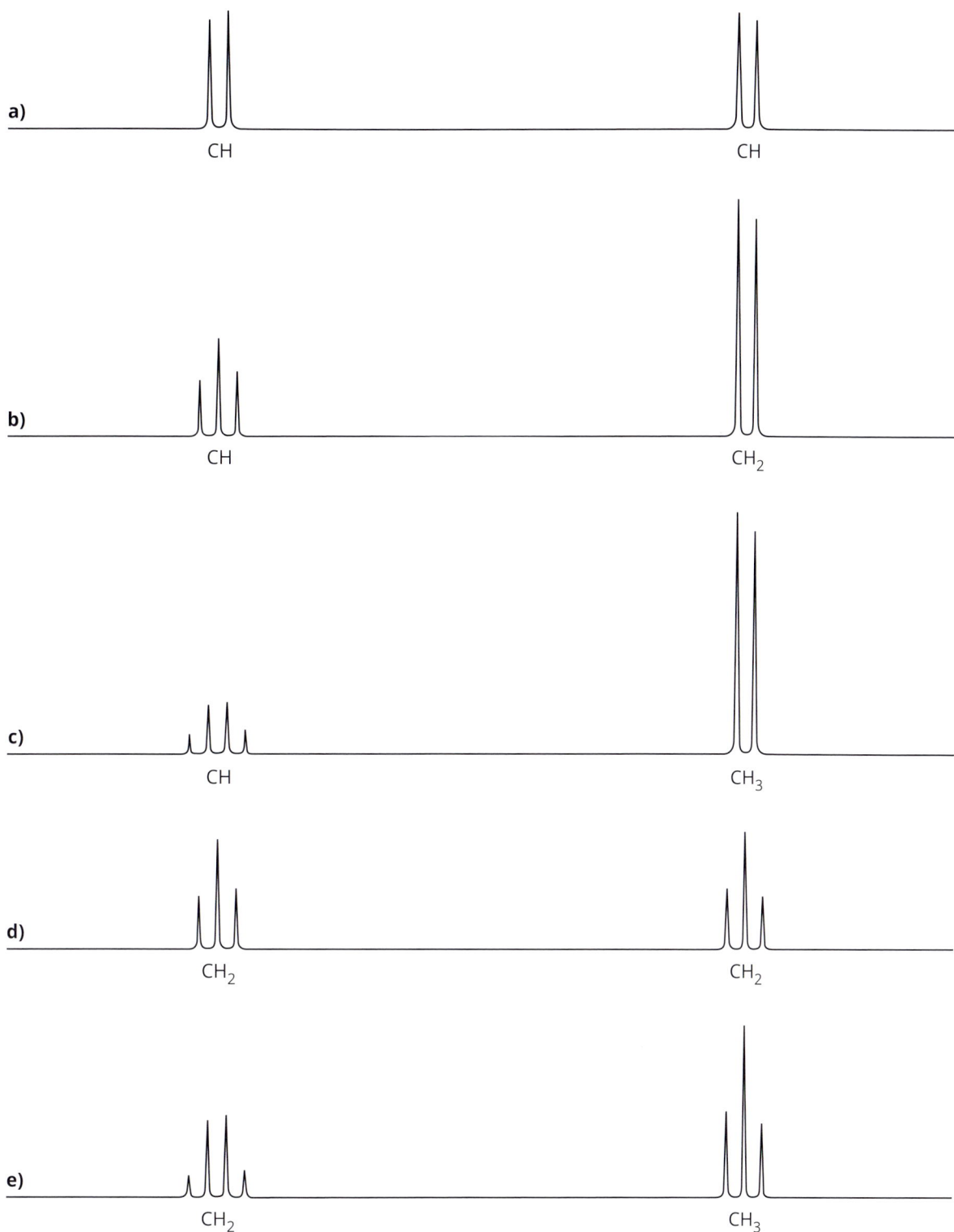

FIGURE 2.19 Common aliphatic spin systems and their ^{1}H multiplet structures a) CH–CH, b) CH–CH$_2$, c) CH–CH$_3$ d) CH$_2$–CH$_2$, and e) CH$_2$–CH$_3$.

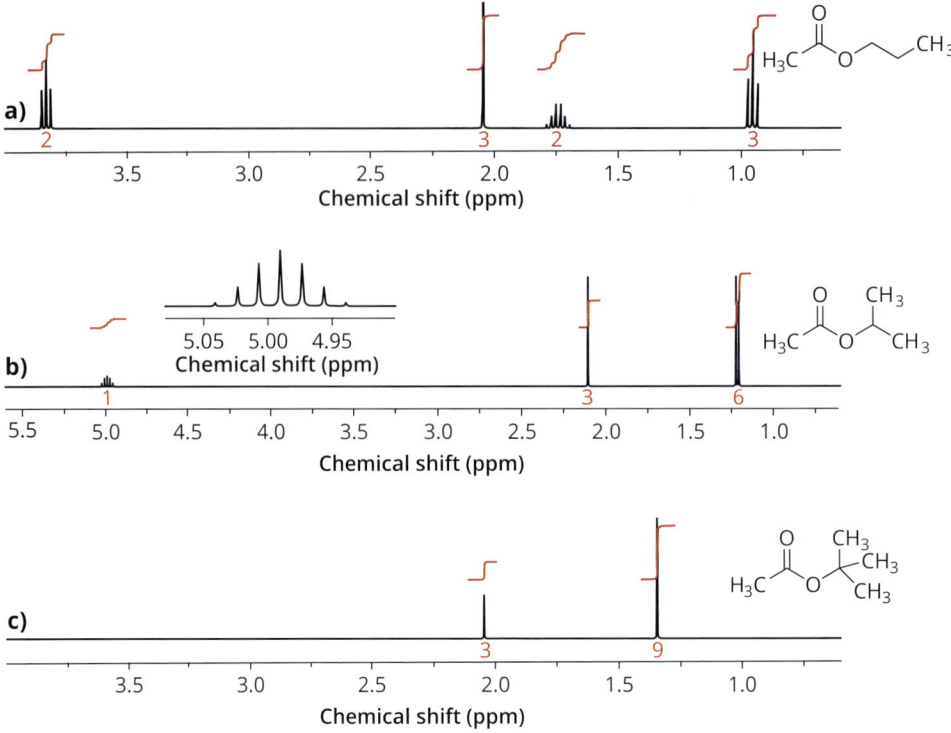

FIGURE 2.20 ¹H NMR signals of (a) *n*-propyl, (b) *iso*-propyl, and (c) *tert*-butyl ester groups. In each case the acetyl methyl group also produces a singlet at ~2.1 ppm.

point of the fragment will vary, and this may then give you some indication as to the neighbouring functionality to which the fragment is attached (O, N, C=O, Aryl, etc). In the examples of Figure 2.20, the adjacent ester functionality gives rise to rather large proton chemical shifts for the connecting methine (CH) or methylene (CH₂) groups. Despite its lack of coupling structure, the *tert*-butyl group is readily identified as an intense 9-proton singlet. Similarly, isolated methyl groups lacking neighbouring protons appear as distinct 3-proton singlets standing above other multiplets, and again their chemical shifts may give evidence as to the nature of the adjacent group. For example, OMe groups are often seen as distinctive singlets around 3.5–4.0 ppm, as already illustrated in Exercise 2.1A above.

EXERCISE 2.2

Match the aliphatic regions of the three ¹H NMR spectra shown in Figure 2.21 to the following molecules: (**1**) diethyl ether; (**2**) ethyl benzoate; (**3**) triethyl amine.

1   **2**   **3**

EXERCISE 2.3

Identify the proton spin system, and hence molecular fragment, associated with the proton spectrum in Figure 2.22.

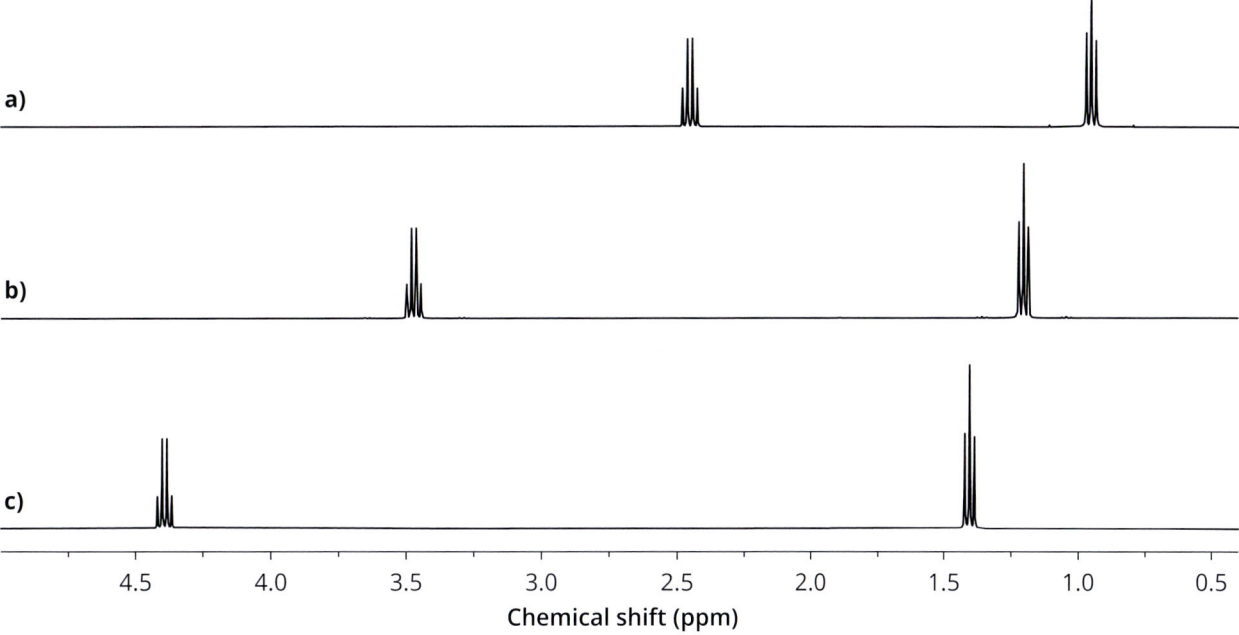

FIGURE 2.21 The aliphatic regions of three ^{1}H spectra to be correlated with the three structures **1** to **3**.

2.6 Common features of unsaturated molecular fragments

Alkenic =CH resonances occur typically at chemical shifts between 5.5 and 7.5 ppm. The splitting patterns arising from such alkene resonances can be simple, such as doublet, but can also be complicated due to coupling to several neighbouring nuclei with quite different magnitudes.

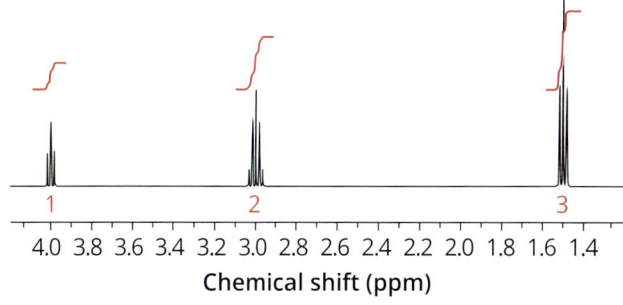

FIGURE 2.22 The ^{1}H spectrum of an unidentified fragment, showing integral traces and their numerical ratios.

> 1,2-Disubstituted alkenes, with no further couplings to neighbouring aliphatic CHs, are easy to recognize as they give rise to a pair of ^{1}H doublets.

The magnitude of the coupling constant, $^3J_{HH}$, can be used to work out the stereochemistry of the double bond since in most cases $J_{trans} > J_{cis}$, as noted in section 2.4.2. The large reciprocal couplings of *trans* alkenes in particular are easy to spot, as there are not many other systems which give rise to such large couplings (13–17 Hz) in this region of the spectrum. A useful way to link alkene doublets, which is especially useful for *cis* alkenes where the J value is much smaller, is to look for evidence for roofing of the doublets (section 2.4.3) since each will show a similar degree of asymmetry in their line intensities (Figure 2.23).

Aromatic =CH resonances occur typically at chemical shifts between 6.5 and 8.5 ppm.

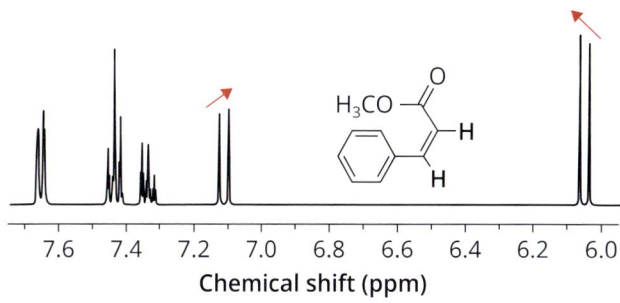

FIGURE 2.23 Multiplet roofing for the *cis* alkene coupling of *cis* methyl cinnamate ($^3J_{HH}$ = 10 Hz).

Aromatic resonances often overlap with those of alkenes, which may lead to ambiguous assignment—so you should be cautious in the initial stages of assessment.

Ultimately, inspection of multiplet structures can play a key role in recognizing aromatic molecular fragments and will often allow the nature of benzene ring substitution to be identified.

A critical part of interpreting multiplet patterns of substituted benzene rings is recognizing the symmetry that arises with different substitution patterns.

The different types of multiplet patterns anticipated for various ring substitutions are illustrated in Figure 2.24, noting that in reality the chemical shifts of each multiplet will be influenced by the groups on the aromatic ring system.

In the case of the monosubstituted benzene ring (Figure 2.24a), the axis of symmetry means that only three resonances are expected in a 2:2:1 integral ratio of *ortho:meta:para* protons. However, it is very often the case that the *meta* and *para* protons are coincident, or almost so, meaning that their multiplet structures are often unclear. The dominant splittings within each multiplet arise from 3-bond *vicinal* couplings to adjacent protons (~8 Hz), whilst the finer splittings arise from longer 4-bond *meta* couplings (~1–2 Hz). Hence, $H^{a/a'}$ have a dominant doublet structure from the neighbouring $H^{b/b'}$ protons, whereas H^b and H^c are primarily (roofed) triplets due to them each having two immediate neighbours.

The 1,4-disubstituted system (Figure 2.24b) has a distinctive 'distorted' doublet-like structure to its multiplets (as described in section 2.4.3), each of which integrates to two protons, making this a fragment that is often readily identified.

The 1,3-disubstituted system (Figure 2.24c) lacks internal symmetry, provided X and Y are different groups, and so is predicted to display four peaks, each representing a single proton. Again, the dominant multiplet structures are governed by the adjacent protons (one neighbour for H^b and H^d leading to their doublet appearances, and two for H^c giving it a triplet structure), with finer splittings from 4-bond couplings for both H^b and H^d due to their interactions with H^a. Similarly, H^a appears as a fine triplet due to its long-range $^4J_{HH}$ couplings to both H^b and H^d.

The 1,2,4-trisubstituted system (Figure 2.24d) is characterized by one proton displaying only a long range $^4J_{HH}$ coupling, which may be used to identify its coupled partner that displays one large and one small splitting.

Finally, the 1,3,5-trisubstituted system (Figure 2.24e) will show only small $^4J_{HH}$ couplings for all protons which, if unresolved, can give peaks the appearance of slightly broadened singlets. The importance of recognizing symmetry within benzene rings is also highlighted in Chapter 3 on carbon-13 NMR.

EXERCISE 2.4

Predict the multiplet patterns you would expect for a 1,2,3-trisubstituted benzene ring when the 1- and 3- substituents X are identical.

We finish this chapter with a reminder:

Whilst chemical shift data are extremely useful, remember not to draw firm conclusions from these too early on with regard to structural assignments. It is better to interpret the chemical shifts with a degree of caution until you have obtained further evidence.

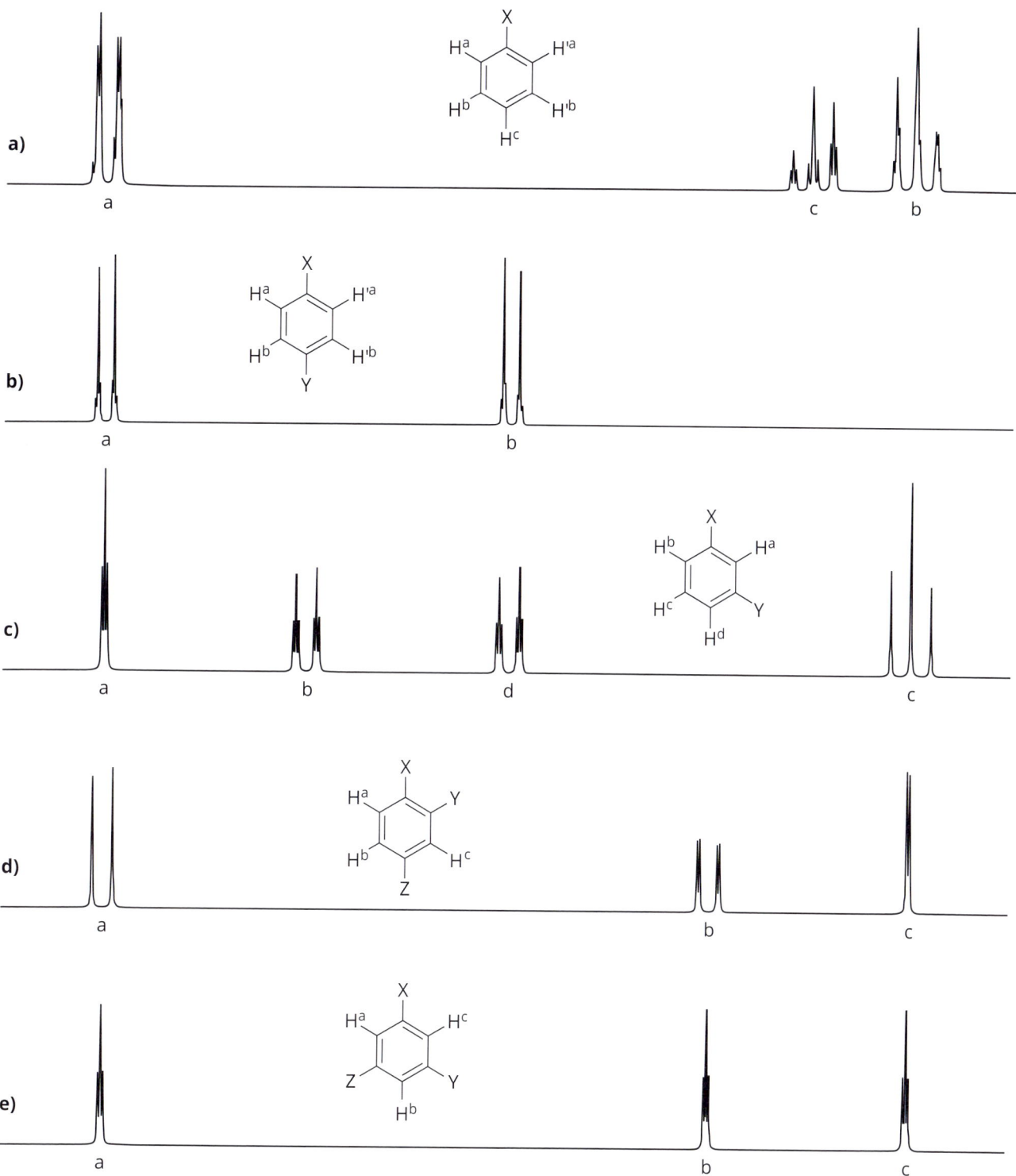

FIGURE 2.24 Representative multiplet structures for differing benzene ring substitution patterns: (a) monosubstituted, (b) 1,4-disubstituted, (c) 1,3-disubstituted, (d) 1,2,4-trisubstituted, and (e) 1,3,5-trisubstituted. In all spectra, $^3J_{HH}$ = 8 Hz and $^4J_{HH}$ = 1 Hz. In real cases, the relative positions of peaks will depend on the ring substituents X, Y, and Z.

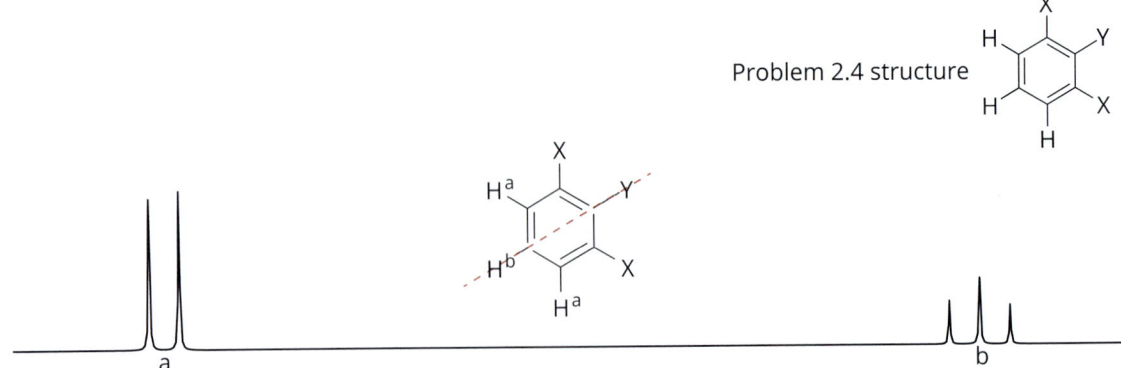

Problem 2.4 structure

FIGURE 2.25 The expected multiplet structures for the 1,2,3-trisubstituted benzene ring when the 1- and 3- substituents X are identical.

The worked examples in Chapter 8 of this book describe a *general approach* to solving spectroscopic problems in which the solution develops progressively and ultimately considers all available data. Even though this process typically begins with a consideration of chemical shift data, the identification of structural fragments through recognition of their spin systems often plays a significant role in this, although the recognition of more complex spin systems typically comes together only towards the end of structural interpretation process.

Solutions to Exercises

Solution 2.1A

The signal at 1.6 ppm in Spectrum A is not a methyl singlet due to the small relative integral. This signal is actually due to a trace of water in the NMR solvent, and similar signals most often have integrals with non-integer ratios compared to other resonances and so can be easily discounted. This spectrum therefore shows only one signal, at 3.8 ppm, and so is assigned to Structure **5** in which both OMe groups are equivalent. In Spectrum B the integrals can be used to determine the relative number of hydrogen atoms for each signal. Do not be misled by the peak height. In this example the peak heights are unequal as the resonance at 2.8 ppm is slightly broader due to very small long-range couplings (which you cannot see without an expansion). The integrals, however, reveal that each signal arises from the same number of hydrogen atoms. Chemical shift values around 2.0–2.5 ppm indicate typically aliphatic CH groups neighbouring an unsaturated functional group, such as a C=O or C=C. Spectrum B is therefore assigned to Structure **1**. In Spectrum C the chemical shift values around 3.5–4.5 ppm indicate aliphatic CH_3 groups neighbouring the electronegative oxygen atom. Spectrum C is therefore assigned to Structure **2**.

Solution 2.1B

The ^1H NMR spectra for compounds **3** and **4** will be similar in the aliphatic regions, and hard to differentiate by looking at the methyl resonances alone. Both have aryl methyl ether groups, which would appear at around 3.8 ppm (as in **2** and **5**). The chemical shifts of a methyl group attached directly to an aromatic ring and of one in a methyl ketone are, however, both expected to appear in the same region, at around 2.0–2.5 ppm. So, without further details (such as considering the aromatic region of the proton spectra), these could not be differentiated.

Solution 2.2

If you see a triplet and quartet in a spectrum, you can think of this as having identified a 'CH$_3$CH$_2$–X' jigsaw piece, and you will have a fair idea of what it will be attached to, based on the chemical shift of the methylene group. Here the three spectra are assigned as (a) = **3**, (b) = **1**, and (c) = **2** due to the electron-withdrawing (deshielding) effects of the adjacent groups being C(=O)O– > O– > N–. Remember, however, that the observed chemical shifts in any example will vary somewhat, so don't try to learn exact values.

Solution 2.3

We see three discrete resonances with an intensity ratio 1:2:3. The first at 4.0 ppm integrates as a single proton and its triplet structure suggests that it is adjacent to a CH$_2$ group. The second resonance at 3.0 ppm integrates to two and has a more complex multiplicity comprising five lines whose relative intensities match those of the Pascal's triangle (Figure 2.6) for four neighbouring protons with similar coupling constants (1:4:6:4:1). Feasibly, this may be due to either two adjacent CH$_2$ groups or an adjacent CH and CH$_3$ group since either scenario would equate to four neighbouring protons. The final multiplet at 1.5 ppm is a three-line triplet indicating it to have an adjacent CH$_2$ group. Since we have identified a CH group at 4 ppm and a CH$_3$ at 1.5 ppm, this is sufficient to explain the 5-line multiplicity at 3.0 ppm, meaning this CH$_2$ sits between the CH and CH$_3$ groups. Thus, the molecular fragment is CH–CH$_2$–CH$_3$.

Solution 2.4

This molecule has an axis of symmetry, so only two unique proton environments exist, giving two peaks with integrals in the ratio 2:1. The middle proton will have two neighbours with identical coupling constants due to symmetry and so will appear as a triplet while the two equivalent protons will have a single neighbour and will therefore have a doublet structure. The two protons labelled 'a' will show no coupling to each other since they are fully equivalent (Figure 2.25).

CHAPTER 3

Analysis of ^{13}C NMR spectra

The carbon NMR spectrum of an organic molecule provides a 'fingerprint' for the core of the molecule and is used routinely alongside ^{1}H NMR spectroscopy in the identification of chemical structure. Since carbon-12 is not NMR active (nuclear spin $I = 0$), we observe only carbon-13 nuclei ($I = ½$) when recording a carbon NMR spectrum, and the fact that this isotope has only 1.1% natural abundance has important consequences. At a practical level, it means that we require more material to record a carbon spectrum than a proton one since only ~1 in every 100 carbon atoms will contribute to the detected NMR signal (meaning that carbon spectra invariably also have more noise in the baseline than proton spectra). It also means that coupling to neighbouring ^{13}C atoms can be ignored since it is very unlikely that two ^{13}C atoms will exist as neighbours (unless, of course, the molecule is specifically labelled with ^{13}C isotopes).

When recording a carbon spectrum, the spectrometer is normally set up to remove the coupling with neighbouring protons, a process known as 'decoupling' (Figure 3.1).

Proton decoupling greatly simplifies a ^{13}C spectrum, making interpretation easier since each carbon environment (type) gives rise to a unique resonance, typically without any multiplet structure—that is, carbon peaks usually appear as singlets.

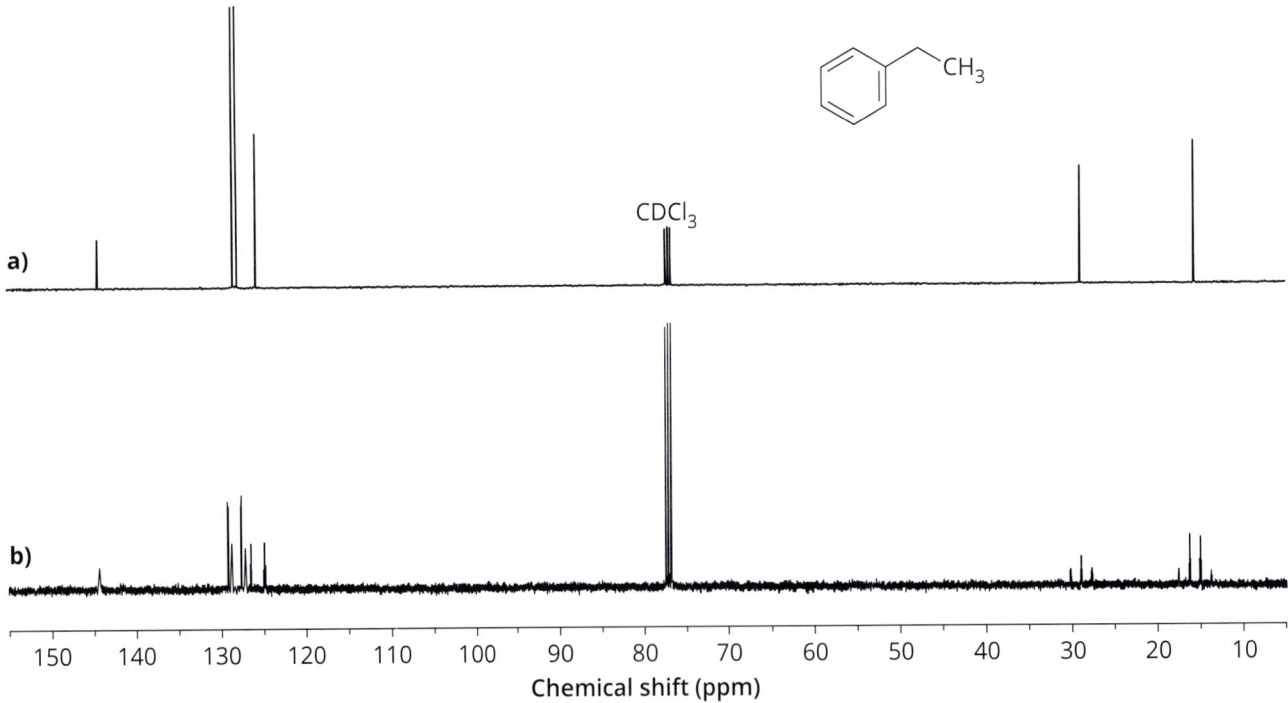

FIGURE 3.1 Carbon spectra are simplified with proton decoupling. a) The more usual proton-decoupled carbon spectrum and b) a fully proton-coupled carbon spectrum of ethyl benzene in CDCl$_3$. In (a) each peak is simpler and concentrated into a single line so that it is also more intense and clearer to see.

It is important to realize, however, that proton decoupling in ^{13}C NMR *does not* remove coupling to any other NMR-active nuclei, so these may still cause observable splittings. In organic chemistry the most common nuclei which couple to carbon-13 resonances are fluorine-19 and phosphorus-31 (both with $I = \frac{1}{2}$ and 100% natural abundance). Note that no examples of fluorine- and phosphorus-containing molecules are included in this book. What is very commonly observed is coupling of the solvent carbon resonance with attached deuterons ($I = 1$) since NMR solvents are invariably deuterated (see Chapter 6). This means that the carbon resonance of chloroform appears as a characteristic 1:1:1 multiplet at 77 ppm, as can be seen in Figure 3.1 and more clearly in the inset of Figure 3.2.

In the absence of couplings, the number of peaks in a carbon spectrum can be correlated with the number of unique chemical *environments*, which often matches the number of carbon atoms in a molecule.

The important exception to this is when symmetry exists within the structure.

Molecular symmetry causes the number of peaks observed to be lower than the number of carbon atoms in the molecule.

For example, the carbon spectrum of diethyl ether ($CH_3CH_2OCH_2CH_3$) displays only two resonances (at 65.9 and 15.3 ppm), and not four, due to the equivalent environments either side of the central oxygen atom (Figure 3.2). Symmetry is also frequently observed for aromatic ring systems, so a monosubstituted benzene ring will display only four peaks rather than six, as seen above 100 ppm in Figure 3.1. We will discuss this further in section 3.4.

Even with the simplification arising from proton decoupling, however, it may be that resonances are so close that they can only be differentiated by looking at an expansion of the spectrum (as with the $CDCl_3$ example below). This is often the case in aromatic molecules where carbon resonances may only differ in a few tenths of a ppm. If an expansion is not available, then a peak list will normally reveal if resonances are very close to one another. Peak listings are usually given either below each resonance and the δ scale, or 'tagged' to the top of each peak, as in Figure 3.2 (and as for the spectra contained in this book).

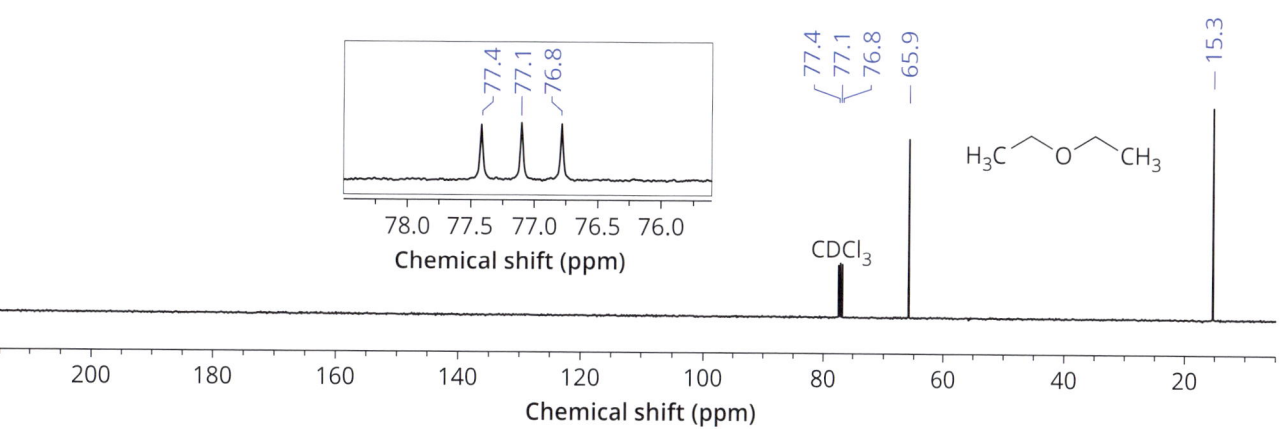

FIGURE 3.2 The carbon spectrum of diethyl ether shows only two peaks due to internal symmetry. The cluster of three peaks around 77 ppm arises from the solvent resonance ($CDCl_3$) and shows characteristic coupling to the attached deuterium atom (see inset expansion).

3.1 ¹³C NMR chemical shifts

Carbon chemical shifts are especially useful indicators of chemical environments, typically covering 0–220 ppm, a range significantly greater than the mere 10 ppm occupied by protons. The greater shift range, and the singlet structure of carbon resonances, means that peak overlap is also rarely encountered. As with ¹H spectra, the data provided in chemical shift tables for ¹³C resonances may be broadly useful for their identification (Appendix 1), yet also potentially misleading as there are still regions that may coincide for differing chemical environments. Thus, we again advise splitting the spectrum into regions and recommend that you develop a general idea of which types of resonances appear where on the spectrum (Figure 3.3). Whilst Appendix 1 tabulates key regions, it also lists resources for chemical shift predictors that are nowadays preferable to lengthy data tables.

> There are three main regions of the ¹³C NMR spectrum which reflect the following environments:
>
> - saturated (aliphatic) environments (typically 0–100 ppm)
> - unsaturated (alkenic and aromatic) environments (typically 100–160 ppm)
> - carbonyl environments (typically 165–220 ppm).

The carbonyl resonances often reflect carbon type with typical aldehyde and ketone resonances appearing at larger chemical shifts (185–220 ppm) compared to esters, amides, and acids (typically 165–180 ppm). Being familiar with what types of resonances appear in these three regions is useful and means, as with the ¹H NMR, that there is no need to turn to data tables to make a start at spectrum interpretation. For completeness, we should also note that resonances of alkyne groups fall (perhaps unexpectedly) between 70 and 90 ppm, although these occur rather less commonly in small molecules and are not included in the problems we cover in this book.

> For the interpretation of carbon spectra, it is advisable to start from the left-hand side of the spectrum, beginning with an assessment of the larger chemical shifts and working towards smaller chemical shifts further to the right.

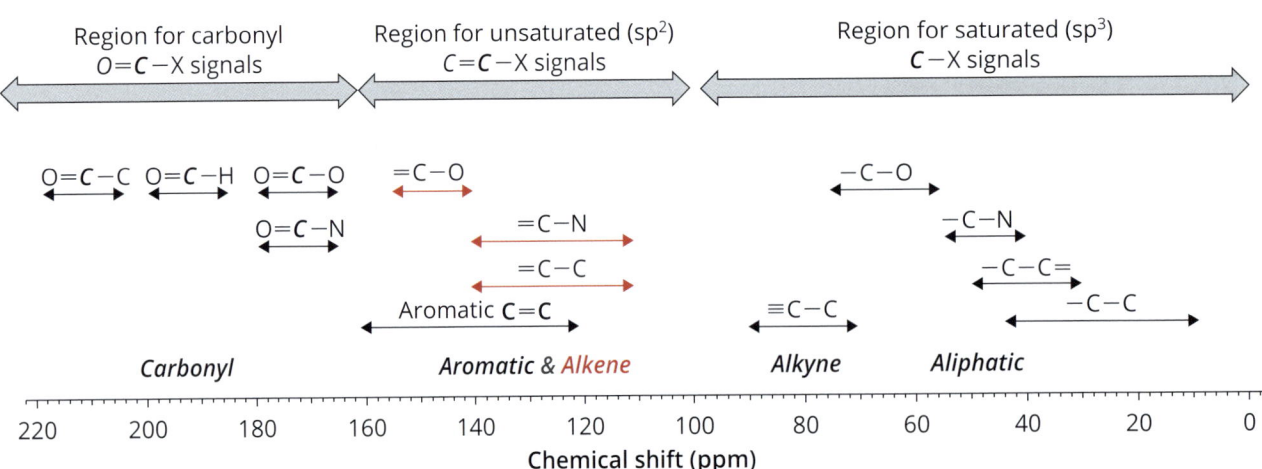

FIGURE 3.3 Typical ¹³C chemical shift ranges for carbon atoms. These may be used as a general guide when analysing spectra, but remember that chemical shifts may fall outside these boundaries according to structural details. The broad arrows at the top highlight the three main regions for interpretation and the fine red arrows distinguish alkene groups from aromatic groups shown below these in black.

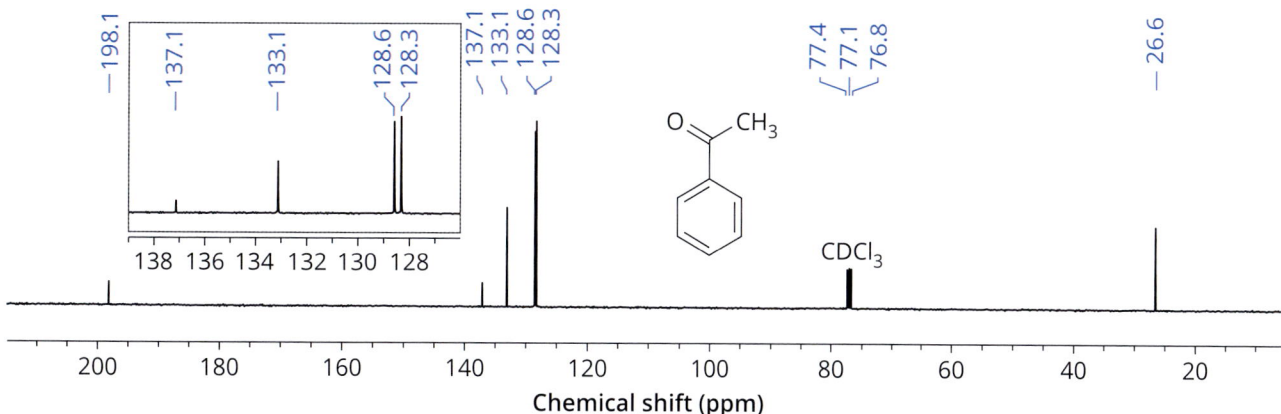

FIGURE 3.4 The ^{13}C NMR spectrum of acetophenone. The inset shows an expansion of the aromatic peaks whilst the cluster of three peaks at 77 ppm is again characteristic of the CDCl$_3$ solvent peak.

The region of larger chemical shifts is often least crowded and is indicative of some particular environments, especially carbonyl groups, that are not typically reflected in ^{1}H spectra (with the exception of aldehydes). These observations can also be usefully correlated with evidence for carbonyl groups in infrared spectra. Below the carbonyl region, we find evidence of C=C unsaturation in resonances above 100 ppm, with the region below 100 ppm dominated by peaks from saturated carbon environments.

> Chemical shifts in the range 80–40 ppm are often suggestive of electronegative heteroatoms in aliphatic systems, commonly O or N.

Resonances below 40 ppm arise typically for carbons remote from heteroatoms. The additive nature of electronegative effects means that peaks in the 80–100 ppm window are often indicative of multiple electron-withdrawing atoms attached to a carbon. For example, acetals (O–**C**H–O) typically appear at around 100 ppm, and (as already seen) chloroform (**C**HCl$_3$) at 77 ppm; notice the similar influence as for the ^{1}H shifts of these groups highlighted in section 2.2.1.

The ^{13}C spectrum of acetophenone in Figure 3.4 illustrates some of these points. The largest shift is indicative of the carbonyl resonance, here a ketone with a shift of 198.1 ppm. Although we do not observe the oxygen in this molecule by NMR, its presence is suggested by this carbon peak and would be something to seek additional evidence for in supporting infrared and mass spectrometry data. The group of peaks between 140 and 120 ppm indicates the presence of C=C double bonds, in this case a substituted benzene ring containing four unique chemical environments (as shown by the peak picking and in the inset expansion) arising from monosubstitution. The peak at 26.6 ppm tells us of the presence of an alkyl group, here the methyl group.

3.2 **Resonance intensities**

When you are working on determining a structure, simply adding up the number of resonances gives you a good, if sometimes rough, idea of the size of the molecule and will indicate the number of different chemical environments it contains.

- A ^{13}C spectrum does not yield meaningful integral values, unless the spectrometer is set up in a special way.
- Despite a lack of meaningful integrals, resonances due to carbon environments with more than one atom (mostly from symmetry within a structure) *tend* to be more intense than resonances due to just one atom.

- Resonances arising from carbon atoms *without* attached hydrogens (quaternary centres or carbonyl groups) are often of much lower intensity than those bearing a hydrogen atom (CH_n).

Whilst a complete explanation for these effects is outside the scope of this particular book, the observation of lower intensity peaks for non-protonated carbons can nevertheless be helpful in signal assignment.

By way of example, the typical ^{13}C spectrum of a monosubstituted aromatic ring (see the inset in Figure 3.4) shows two large, one medium, and one small resonance for the benzene ring. The larger peaks arise from the overlap of the two *ortho* carbons and likewise for the two *meta* carbons due to symmetry within the ring. The weakest signal at 137.1 ppm is from the substituted carbon which has no attached hydrogen. The order of the peaks on the δ scale will depend on the nature of the substituent, but as the resonances are all aromatic carbons, in the same ring and with similar environments, the peak height can be useful in their assignment.

EXERCISE 3.1

Correlate the three ^{13}C spectra in Figure 3.5 with the three C_4H_8O isomeric structures **1–3**.

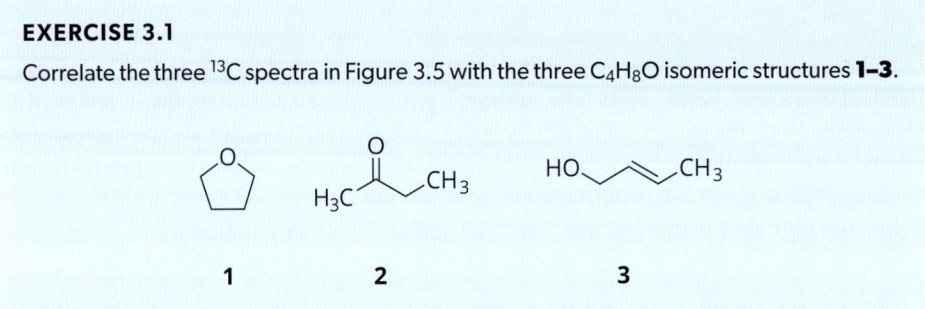

1   2   3

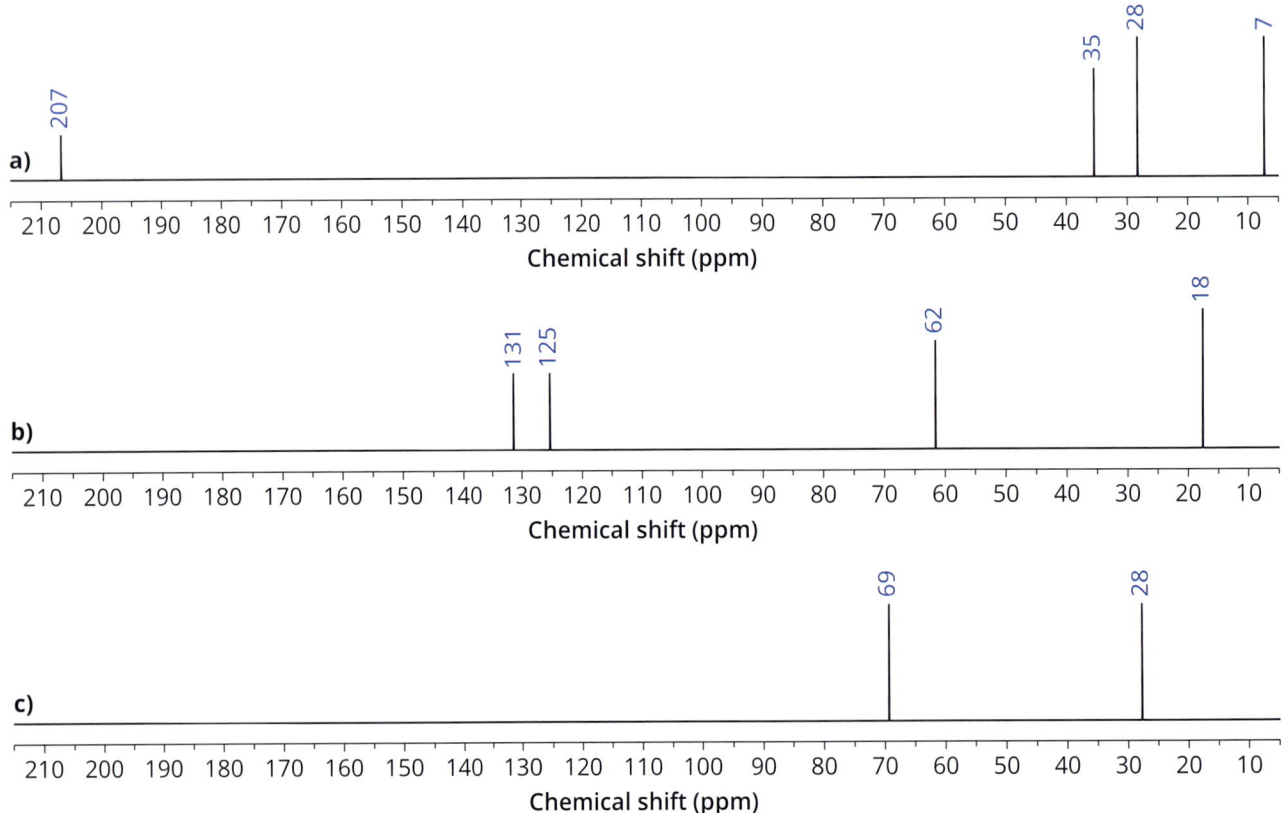

FIGURE 3.5 The carbon-13 spectra of structures **1** to **3**.

3.3 **Common features of aliphatic molecular fragments**

The chemical shift region below 100 ppm will often allow you to determine the number of carbon atoms present in the structure of an aliphatic molecule (or aliphatic fragment) simply by counting the peaks. Coincidental overlap in chemical shifts tends to be rare (although not impossible) even for relatively large molecules, allowing all peaks to be distinguished. However, there are two principal confounding features that can limit this process.

> Firstly, the presence of symmetry in a structure will reduce the number of carbon peaks observed.

The example of diethyl ether noted previously illustrates this point as it displays only two resonances despite containing four carbon atoms (Figure 3.2). Similarly, groups with internal symmetry may also show this behaviour so that, for instance, an isopropyl group will often display only two resonances since the methyl carbons usually have equivalent chemical environments. Likewise, dimethylamino groups will display only a single carbon methyl peak and diethylamino groups only peaks for a single ethyl group (Figure 3.6).

> Secondly, one should keep in mind that there is always the chance that a peak is hidden under that of the solvent resonance (most of which fall in the aliphatic region).

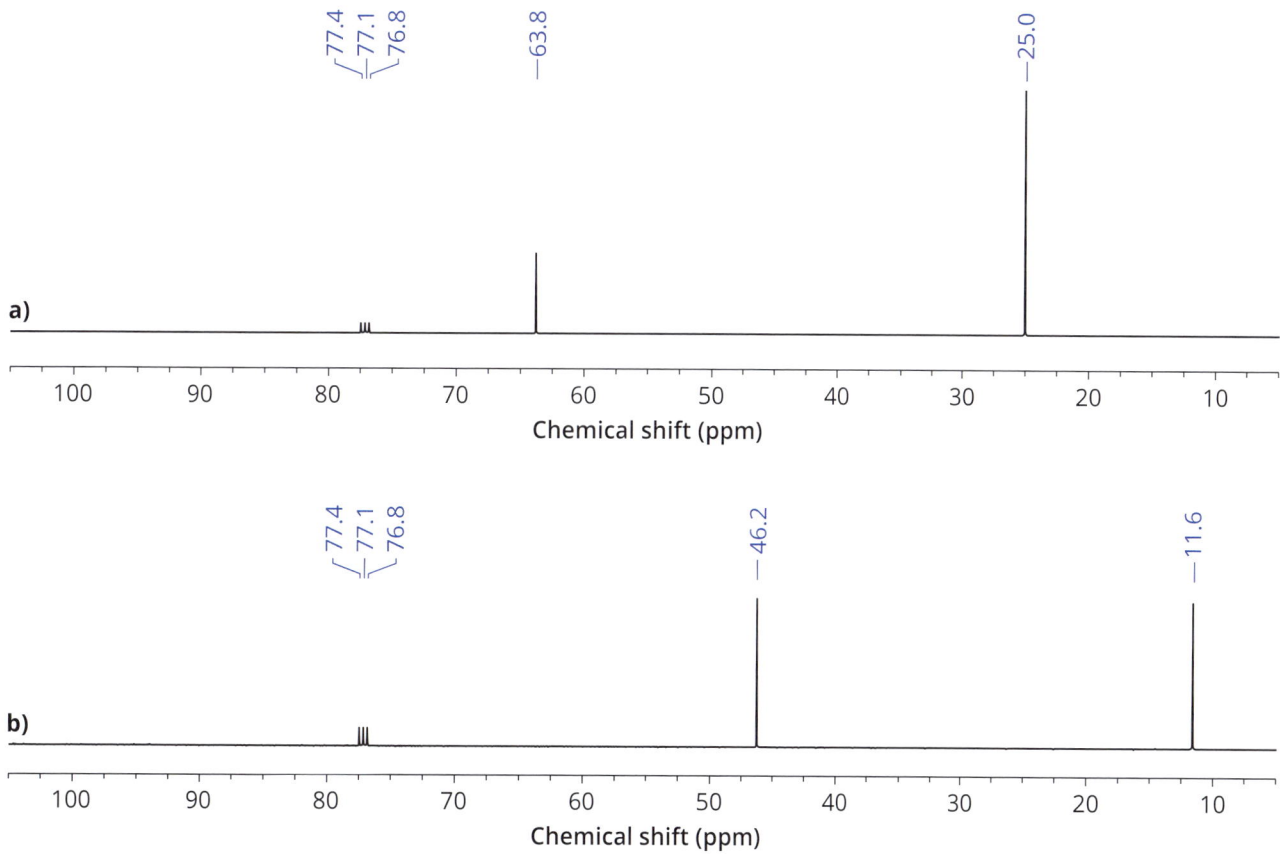

FIGURE 3.6 ^{13}C NMR spectra of a) isopropyl (R–O–CH(CH$_3$)$_2$) and b) diethylamino (R–N(CH$_2$CH$_3$)$_2$) groups.

Since deuterated solvents are used (recall that deuterium is an NMR active nucleus with $I=1$), the intense solvent resonance will appear with a wide multiplet structure due to its coupling to attached deuterons, and can easily mask a smaller solute peak. The most commonly used solvent, chloroform, falls at ~77 ppm and is therefore conveniently far away from many peaks from aliphatic molecules. But most other solvents have lower shifts and can be more problematic in this regard (see Chapter 6).

> In the absence of unsaturation, the chemical shifts of aliphatic groups are influenced significantly by the electronegativity of atoms or groups attached directly to the carbon centres.

This inductive influence from electronegative atoms means that carbon shifts can be indicative of the locations of heteroatoms within aliphatic molecules, as suggested by Figure 3.3. However, as previously noted, you should not be overly concerned with trying to interpret *exact* chemical shift values, especially in the early stages of spectrum assignment.

3.4 Common features of unsaturated molecular fragments

A primary factor to be aware of when dealing with unsaturated molecules is the potential for similar chemical shifts to arise from both alkene and aromatic groups, as highlighted in Figure 3.3. Because of this possibility of coincidence, one cannot always be confident as to which functionality is present, or indeed if both are present in your structure.

> The presence of many peaks in the unsaturated region is likely to be associated with the presence of an aromatic ring; substituted benzene rings are a very commonly encountered feature of small molecules. For substituted benzene rings, the number of peaks seen will be defined by the number and nature of substituents on the ring, which dictate its symmetry and number of unique carbon environments.

The influence of symmetry is illustrated in Figure 3.7 for a range of common substitution patterns. It is important to be comfortable with identifying these when assessing the spectra of aromatic molecules.

Substitution pattern

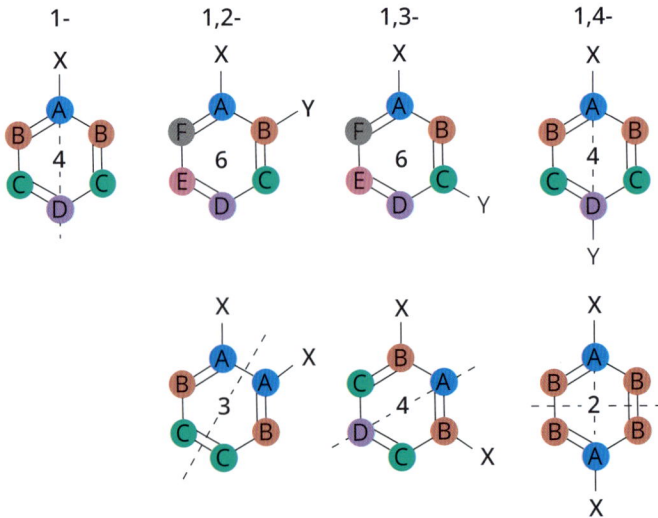

FIGURE 3.7 Common substitution patterns for benzene rings and the number of associated carbon resonances (shown in the centre of each ring). X and Y indicate differing substituents, whilst the dashed lines indicate symmetry axes and the lettered balls indicate the individual carbon environments.

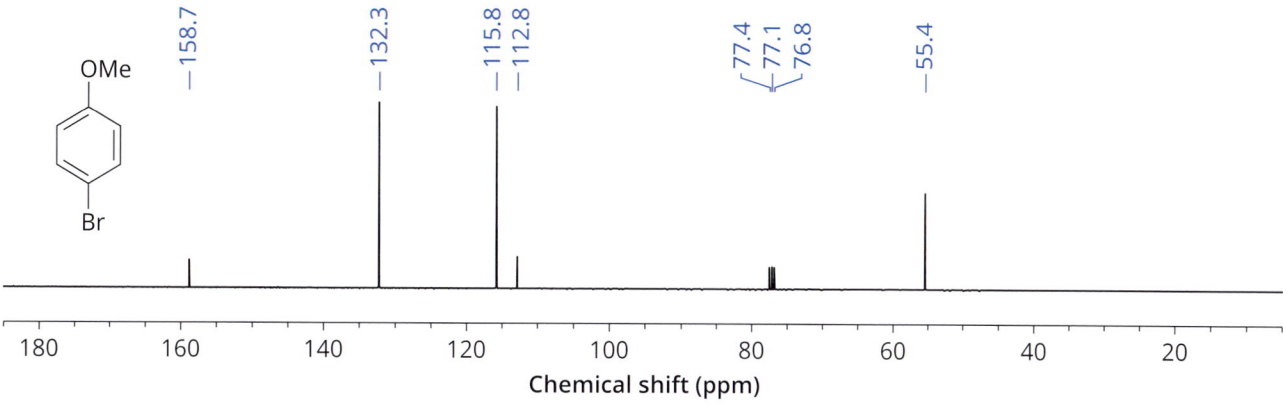

FIGURE 3.8 The ^{13}C NMR spectrum of 4-bromo anisole in CDCl$_3$.

The chemical shifts in the spectra of substituted aromatic rings will vary considerably according to the attached groups, and the precise assignment of each peak may not always be possible.

> Electronegative groups on an aromatic ring will tend to increase the shifts of the carbons to which they are attached, making these stand away from 'the crowd'.

Recall that low peak intensities can often be associated with non-protonated carbons and will therefore also indicate positions bearing substituents. This is illustrated for 4-bromo anisole in Figure 3.8, where the O substitution leads to a rather large carbon shift (158.7 ppm) and both the O and Br additions give reduced carbon peak intensities (at 112.8 ppm in the case of Br).

EXERCISE 3.2

Predict the number of aromatic carbon resonances for the differing benzene ring substitution patterns in the following structures **1–3**, in which X, Y, and Z represent different groups.

1   **2**   **3**

Solutions to Exercises

Solution 3.1

We start the analysis for each spectrum at the left-hand side and move down towards the right looking for characteristic resonances. In Spectrum (a) we quickly come to the peak at 207 ppm, which is indicative of a carbonyl functional group and more precisely in this case a ketone due to the large shift value. This matches only structure **2**. All other resonances fall below 40 ppm and are consistent with the remaining carbons being aliphatic and remote from heteroatoms. Notice also the relatively lower signal intensity for the ketone carbon due to the lack of an attached proton. In Spectrum (b) the first two peaks we come to fall around 125–130 ppm and tell us that we have unsaturated centres in our structure. The fact that we see only two resonances suggests the presence of an (unsymmetrical) alkene, which matches structure **3**. The next peak at 62 ppm arises from the carbon attached to the hydroxyl group, with the methyl carbon falling at a characteristically small chemical shift. The final Spectrum (c) displays only two resonances, both within the aliphatic window. Structure **1** has a plane of symmetry and only two unique carbon environments and is therefore consistent with this, with the larger shift of the two peaks being the carbon adjacent to the oxygen atom.

Solution 3.2

There is no symmetry within structures **1** and **2** since the three substituents are different, and they will both be expected to exhibit six carbon resonances. For structure **3**, a symmetry axis that divides the molecule in two between the X groups can be recognized, and so we would expect only four carbon peaks for this ring in the structure.

CHAPTER 4

Analysis of infrared spectra

In the identification of organic structures, infrared (IR) spectroscopy is used primarily in an empirical manner to seek evidence of functional groups within a structure. The IR spectrum rarely provides a complete picture of atom connectivity within a structure (unless you are matching data to a known compound) and so is unlikely to be used in isolation. Instead, it provides information that complements that from NMR and mass spectrometry and helps with identifying a chemical structure. Evidence for a functional group in the IR spectrum can often help confirm what has already been suggested by other techniques and thus allow you to be more confident that a proposed structure is correct. Unlike NMR spectroscopy, many of the absorption bands seen in an IR spectrum remain unassigned, even when a structure has been defined, and it is more often the case that we interpret only specific, and perhaps dominant, features in the IR spectrum. Thus, IR interpretation tends to focus on specific regions within the whole spectrum, and in this chapter, we will focus on the analysis of these key regions.

4.1 How infrared spectroscopy works

The absorption of infrared radiation by a molecule corresponds to the excitation of bond vibrations within the structure in the form of bond stretching (a change in bond length) or bending (a change in bond angle) (Figure 4.1). The IR frequencies of these absorptions can be characteristic of the presence of certain structural features or functional groups within the molecule. The infrared spectrum displays a plot of the absorption intensity (or, more commonly, the extent of transmission of radiation, as percentage transmittance) versus infrared frequency. For convenience, this frequency is represented in units of wavenumber, which is simply the reciprocal wavelength, in cm^{-1}, with higher wavenumbers corresponding to higher frequency vibrations. For characterizing organic structures, the wavenumber range of interest is ~600–3600 cm^{-1}.

The vibration frequencies of bonds are influenced primarily by the masses of the atoms involved and the strength of the bond between them; a common analogy employed for this is the vibration of two masses (atoms) connected by a spring (bond). Hence, stronger bonds and/or lighter atoms will give rise to higher frequency vibrations and thus higher wavenumber absorptions. For example, as we shall see, a carbonyl C=O will absorb at higher wavenumber than C–O due to the stronger double bond, and H–O will be higher than C–O due to the lighter atom masses involved.

The strength of an IR absorption correlates with the extent of change in the molecular dipole moment associated with the bond vibration. Thus, bonds associated with a large charge separation, such as C–O and C=O, tend to yield strong absorptions which are readily observed, whereas bonds with a more symmetrical charge distribution, such as C=C, typically give rise to weaker absorptions. Absorption intensities tend to be loosely referred to as strong, medium, and weak within a spectrum.

Samples are typically studied in either the liquid or solid state. Analytes that are themselves liquids will usually be held as a thin film between salt plates that are transparent

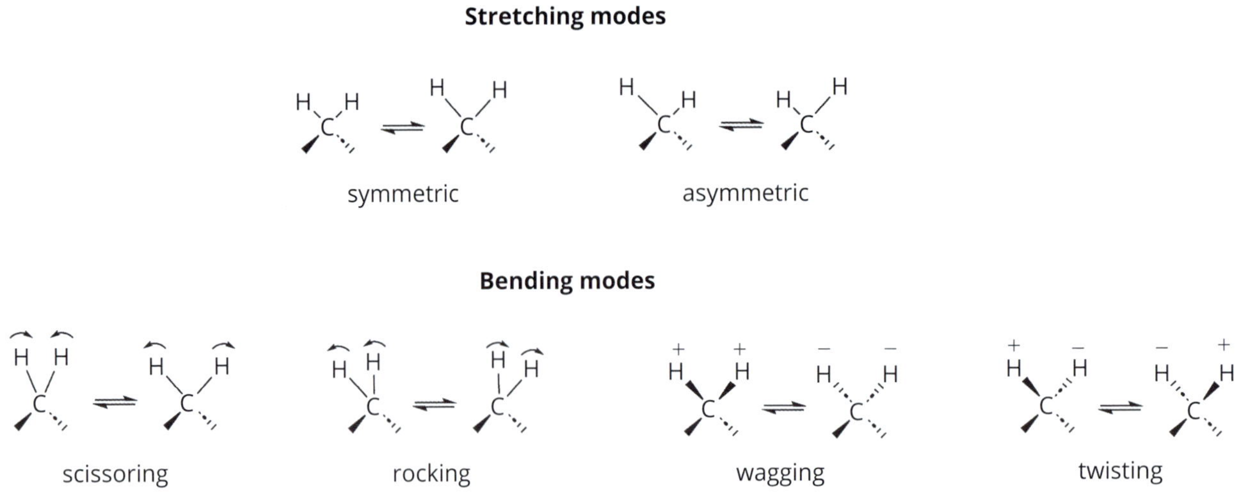

Stretching modes

symmetric   asymmetric

Bending modes

scissoring   rocking   wagging   twisting

FIGURE 4.1 Bond stretching and bending modes illustrated for a CH_2 group. The changes shown are highly exaggerated to make these more apparent.

to infrared and commonly made of sodium chloride. Solid samples can be finely ground with a salt—most commonly potassium bromide—and then compressed into a pellet that can be irradiated, yielding 'KBr discs' for analysis. Alternatively, solids can be mixed with a mineral oil (usually nujol, a paraffin oil) to form a thick suspension known as a mull that can be placed between salt plates for irradiation; in this case a reference spectrum of the nujol is subtracted to remove interference from the hydrocarbon peaks. Modern instruments that employ a method known as attenuated total reflectance (ATR) can directly analyse both solid and liquid samples without the need for sample preparation. In this case, the sample is placed onto a crystal surface with a high refractive index and IR absorption occurs at the analyte–crystal interface. The flexibility and simplicity associated with ATR IR spectroscopy means that it is increasingly widespread. It has also been used for all IR spectra included in this book.

4.2 Using infrared spectroscopy

In organic structure determination, the IR spectrum in isolation is probably the least useful for providing detailed information about an unknown structure. Many absorption bands from molecular vibrations as a whole fall in the region between ~600 to 1500 cm^{-1}, making this region often crowded and complex. Whilst this so-called 'fingerprint' region of a spectrum may be highly characteristic of the molecule in question, it is only useful if you have a reference spectrum with which to compare the unknown—that is, some knowledge or guess as to what the compound is likely to be. However, as stated above, the IR spectrum can provide highly useful evidence for corroborating an interpretation suggested by NMR and MS spectra.

> The IR spectrum is typically used to provide information on the presence or absence of a few key absorption frequencies that represent functional groups in a structure.

The key regions in an IR spectrum for analysis are summarized in Figure 4.2 and presented in tabulated form in Appendix 1.

For initial stages of structure determination, the main feature to look out for in an IR spectrum is the presence or absence of a carbonyl group (C=O) (Figure 4.3).

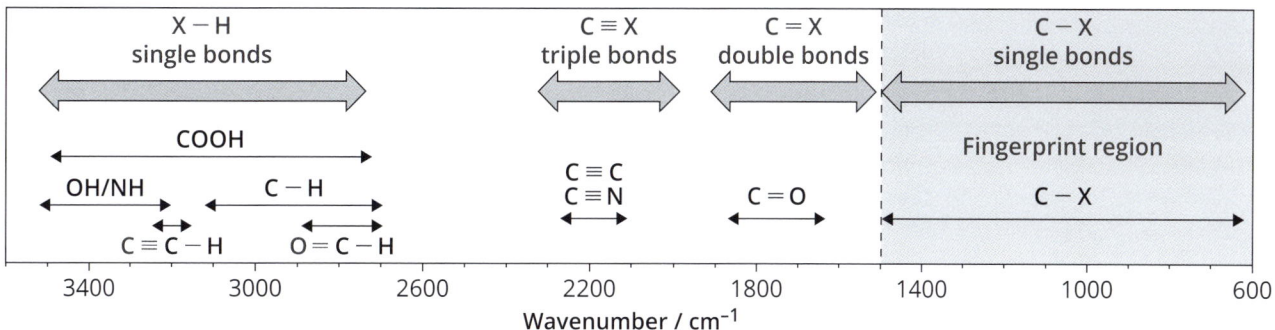

FIGURE 4.2 Typical IR absorption regions for common functional groups.

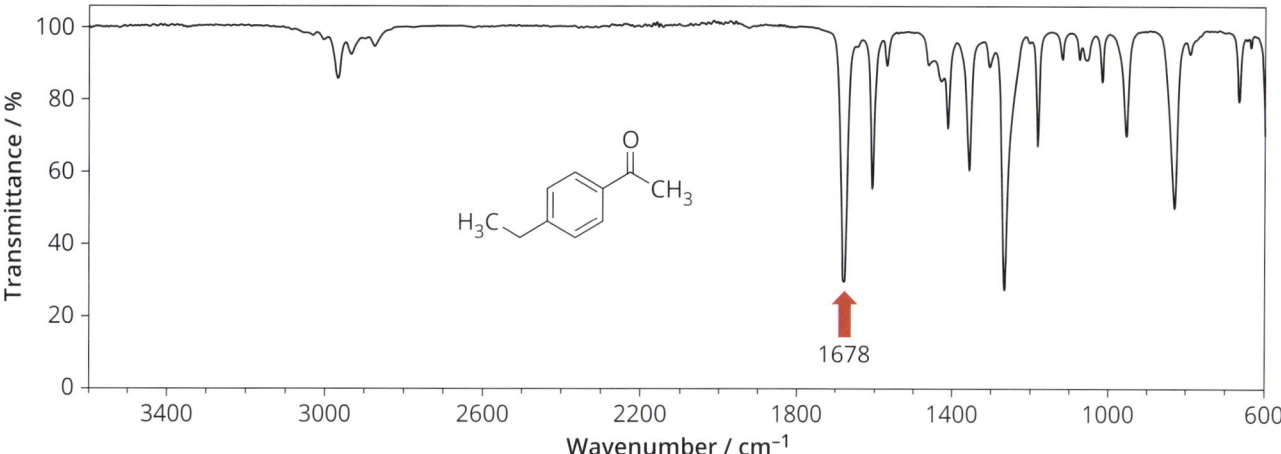

FIGURE 4.3 The intense absorption at 1678 cm^{-1} is characteristic of the presence of the carbonyl group in 4-ethyl acetophenone.

A carbonyl functional group shows a strong absorption in the spectrum within the frequency range 1650–1850 cm^{-1}.

Data tables (Table A.3 in the Appendix) may help you decide if you have an acid chloride (1750–1815 cm^{-1}) or amide (1630–1700 cm^{-1}), but esters, ketones, and aldehydes often have absorptions in similar regions of the spectrum (typically 1660–1750 cm^{-1}).

Other structural features can cause variation in absorption frequencies for these functional groups. These include the presence of conjugation (specifically α,β-unsaturation), which reduces frequencies by 20–40 cm^{-1} due to a reduction of the C=O double-bond character, and the influence of strain within cyclic systems. For example, the compression of bond angles as ring sizes reduce across the cyclic ketones shown in Figure 4.4 leads to a progressive increase in their C=O absorption frequencies, shifting the cyclobutanone to unusually high values. Ring sizes of six or more atoms tend to absorb at similar frequencies to their linear counterparts. Lastly, carbonyl groups involved in hydrogen bonds display absorption frequencies lowered by 40–60 cm^{-1}. These various effects tend to be approximately additive, meaning that a precise interpretation of

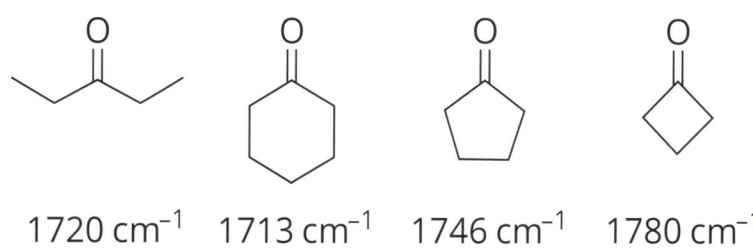

1720 cm^{-1} 1713 cm^{-1} 1746 cm^{-1} 1780 cm^{-1}

FIGURE 4.4 Greater ring strain causes absorption frequencies to increase, as seen for the cyclic ketones.

a carbonyl stretching frequency may not be easily made in the early stages of structure determination.

> In an initial analysis, it is often best to simply focus on whether or not a carbonyl group is present. If there is one, you may consider its frequency at a later stage, when you evaluate a candidate structure or fragment.

The other main region of interest in the IR spectrum falls above ~2500 cm^{-1} and reveals single bond stretching frequencies. CH bonds are ubiquitous, and their vibrations are rarely diagnostic and often rather weak. Whilst it is often possible to distinguish unsaturated from saturated CH stretches (at 3000–3100 and ~2900 cm^{-1} respectively), the most diagnostic stretches in this region occur for aldehydes, which often display two bands in the region 2700–2900 cm^{-1}, and for terminal alkyne groups, which absorb around 3300 cm^{-1}. The presence of an aldehyde should be corroborated by the corresponding carbonyl stretch.

> C–H stretches absorb around 2700–3100 cm^{-1} and are commonly seen but generally not diagnostic, the main exceptions being aldehyde and alkyne stretches.

Most useful in the region above 2500 cm^{-1} are the stretches of OH and NH groups, which are present typically in amines, alcohols, amides, or acids (Figure 4.5).

- Above ~3200 cm^{-1} alcohols and phenols have generally broad, yet smooth absorption bands, whereas the NH stretching absorptions of amines and amides tend to be less intense and sharper.
- Carboxylic acids give typically very broad absorptions in the region around 2500–3500 cm^{-1} and will always have a matching absorption corresponding to the carbonyl group.

It is normally straightforward to distinguish acid OH stretching bands from those of alcohol and phenol OH stretches, as the acid OH absorption band is influenced by hydrogen bonding (commonly the OH of one acid group hydrogen bonds to the carbonyl group of another). This means that the bands are typically jagged in appearance, and not smooth like that for an alcohol. H-bonding also causes OH bonds to lengthen, leading them to absorb across a wider range of lower frequencies, yielding broader acid absorption bands (compare Figure 4.5 (c) and (a)).

Due to the possibility of synchronous (symmetric) and asynchronous (asymmetric) stretching of primary amine and amide NH bonds (Figure 4.6), these may be distinguished by the presence of two distinct absorptions in the region of 3200–3500 cm^{-1} (Figure 4.7). Secondary amines and amides, on the other hand, will typically show just one absorption band in this region of the spectrum, as seen in Figure 4.5 (b).

The method of sample preparation of an acid, alcohol, or amine may also affect the appearance of the OH and NH bond absorptions. For example, in traditional absorption IR spectra where the analyte may be either in solution, in a nujol mull, or in a potassium bromide pellet, the different hydrogen bonding opportunities in these preparations can affect the appearance of the absorption. As the more recent technique of ATR IR spectroscopy uses a neat solid or liquid sample, the resulting spectrum may again vary. Thus, whilst the absorption of a carbonyl group C=O stretch may appear quite consistent, with just the frequency being different from compound to compound, one needs to be careful and be prepared for greater variation in the appearance of the NH and OH stretches in the spectra of acids, alcohols, and amines.

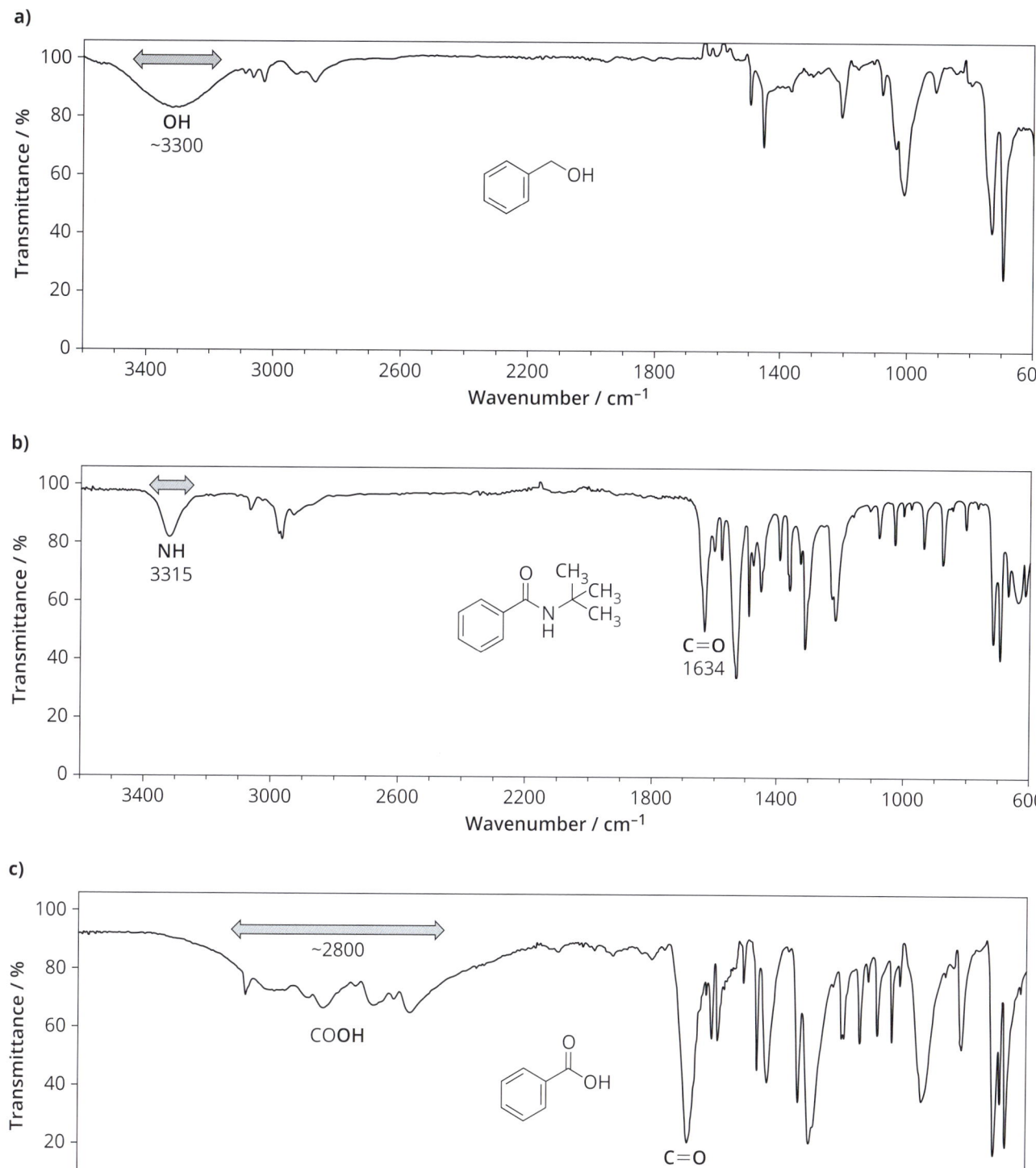

FIGURE 4.5 The distinctive IR absorptions of the a) OH, b) NH, and c) COOH functional groups. Where appropriate, the carbonyl C=O stretches are also indicated.

synchronous   asynchronous

FIGURE 4.6 The synchronous and asynchronous stretches of an NH_2 group. The grey arrows indicate the relative motions of the NH bond stretching in each case.

The final region worth noting is that representing the presence of triple bonds.

> Although triple bonds tend to be less common, their absorptions appear in the otherwise uncluttered region around 2200 cm^{-1}, which often makes them easy to recognize.

Nitrile groups and terminal alkynes give rise to distinctive peaks of moderate intensity (Figure 4.8) whilst disubstituted alkynes can often be weaker due to their higher symmetry.

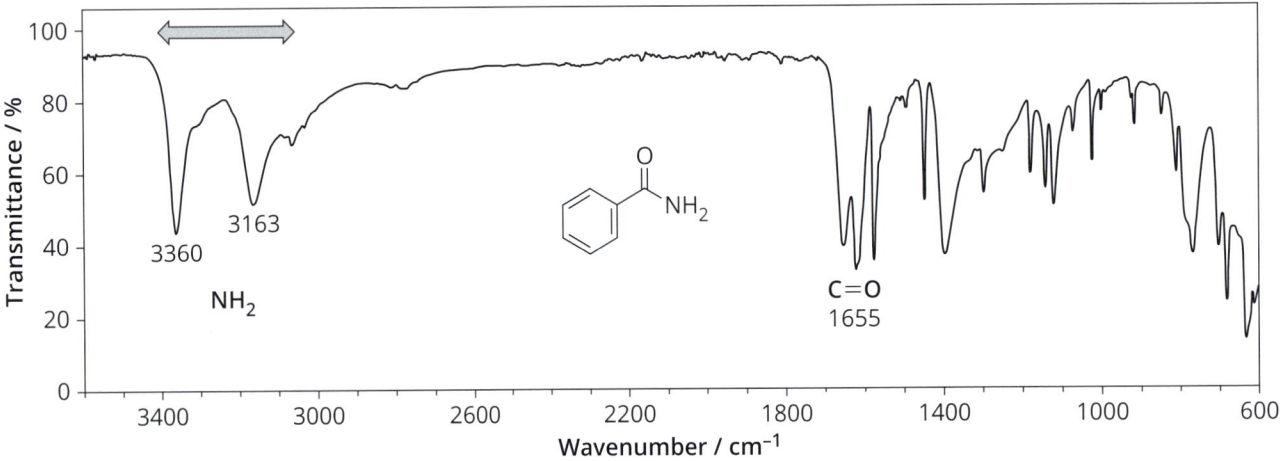

FIGURE 4.7 The NH absorptions of a primary NH_2 group are seen above 3000 cm^{-1}.

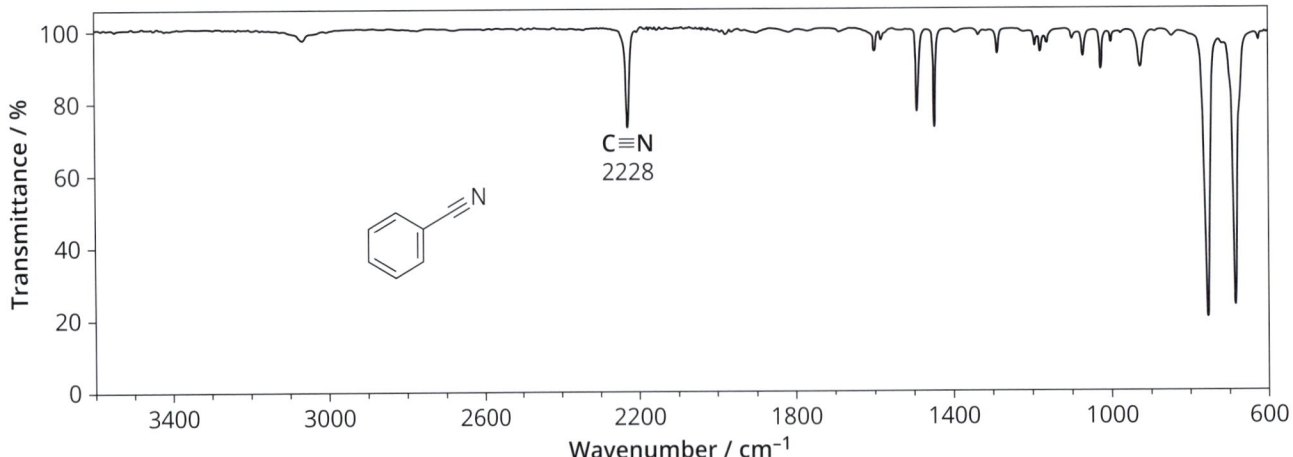

FIGURE 4.8 The triple-bond absorption of benzonitrile.

CHAPTER 5

Analysis of electron ionization (EI) mass spectra

The mass spectrum of an organic molecule contains a wealth of information about its identity and structure but, as with other spectroscopic data, this usually exists in the form of multiple clues—or pieces of the structural jigsaw—rather than as a single piece of definitive evidence. The skill of identifying unknown compounds from their mass spectra therefore lies in knowing what information may be present and how best to assemble this information to reveal chemical structure.

In this chapter we focus exclusively on mass spectral interpretation to elucidate chemical structure using electron ionization mass spectrometry (EI-MS). Once the pre-eminent mass spectrometry technique, EI-MS is now one of a number of mass spectrometry approaches available but remains particularly useful for the analysis of thermally labile, stable organic small molecules. Its highly reproducible, structurally sensitive, fragmentation patterns make EI-MS a preferred approach for many structural analysis studies. We will focus on 'low-resolution' EI mass spectra that provide integer monoisotopic m/z values, although the same approaches discussed here can also be applied to the interpretation of high-resolution EI mass spectra where 'accurate mass' measurement provides chemical formula prediction capabilities. Information about other mass spectrometry techniques, and the interpretation of mass spectra derived from 'high-resolution' instruments using soft ionization approaches, is discussed in Chapter 12.

5.1 How mass spectrometry works

All mass spectrometers measure the mass-to-charge ratio (m/z) of ions in the gas phase. These ions are usually generated from neutral analyte molecules in the ion source. During ionization an electron is removed from a neutral molecule to form a positively charged ion (discussed further in section 5.3.1). Once formed, the ion is referred to as a 'radical cation' because it has an unpaired electron and is positively charged. Many ions are formed and these are transferred from the ion source to a mass analyser using voltage potentials or 'lenses' to focus the ions into a beam and accelerate or deaccelerate them as required during their analysis. The instrument is maintained under high vacuum to provide a mean-free path for the ions, so they do not encounter neutral gas molecules. Ions of different m/z are subsequently separated by a mass analyser, of which there are two basic types: those which separate ions by 'time' (e.g. how long it takes an ion to reach the detector under certain conditions) or in 'space' (e.g. separating ions according to their deflection in a magnetic field). Once the ions have been separated, a detector counts the number of ions of each m/z to provide an abundance value. Figure 5.1 provides a schematic of a mass spectrometer illustrating each of the main parts involved in these processes.

Mass spectrometry can be a highly accurate and selective analytical technique. Some instruments are capable of measuring the mass difference a single electron makes to the overall mass of a molecule, and they can measure many different m/z values from a sample in the same experiment. m/z is a unit-less value, and it is not a physical quantity. However, when the charge is 1 (which is common for small molecules), the m/z value is equivalent to the

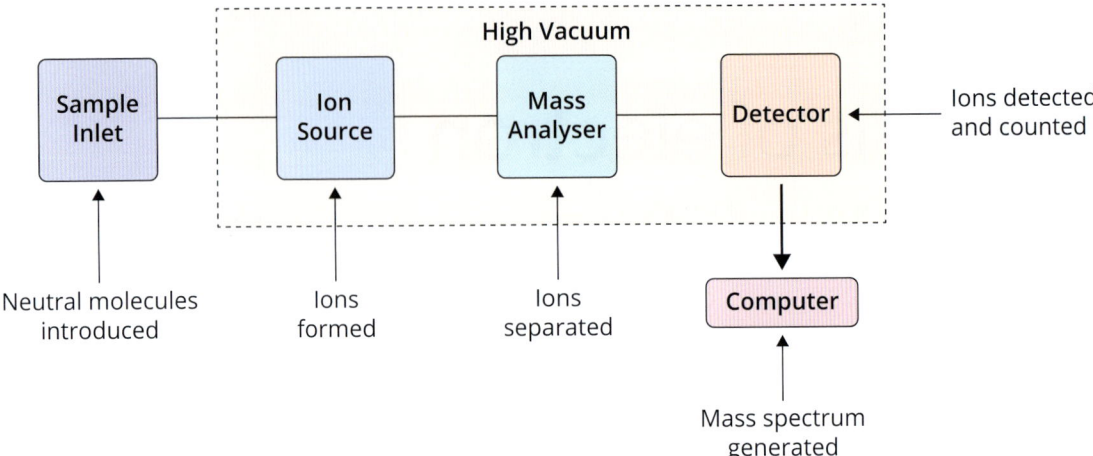

FIGURE 5.1 A schematic of a generic mass spectrometer. The sample inlet and ion source enable a sample to be introduced into the instrument and ionized. Ions are then transmitted electrostatically to a mass analyser which separates ions based on their mass-to-charge ratio (*m/z*). Each *m/z* value is then counted using an ion detector. The information is digitized and sent to a computer where a graph of the *m/z* value (*x*-axis) and ion abundance (*y*-axis) is usually created. This is called a 'mass spectrum'.

mass of the ion, which makes it possible to determine molecular mass directly. The official unit of mass measurement at the atomic and molecular scale is the unified atomic mass 'u', but the unofficial unit dalton (Da) is commonly used synonymously. These terms are used interchangeably in this book, as this is also the case in the general scientific literature.

5.2 **Measurement of the mass spectrum**

A mass spectrum is the basic visual output from a mass spectrometry experiment and is essentially a bar graph displaying ion abundance on the vertical axis and *m/z* on the horizontal axis. An example of an EI mass spectrum is shown in Figure 5.2. The *m/z* values provide information about the molecular weight of the compound being analysed and its fragments, whilst the associated isotope peaks and their relative abundances provide information about elemental composition. It is important to point out that the relative abundance of the different peaks, indicated by the *y*-axis scale, cannot be used for quantification purposes directly but does indicate the relative stability of the ions formed in the mass spectrometer. This provides some indirect information about chemical structure.

Some fundamental characteristics of EI mass spectra:

- The most intense peak in the mass spectrum is known as the *base peak*.
- The base peak does not usually represent the *m/z* value of the analyte directly (although this is possible, especially for aromatic compounds) and is more commonly a stable fragment ion resulting from fragmentation of the analyte ion in the ion source.
- The radical cation does not always appear in the mass spectrum at all; sometimes only its fragment peaks are observed.
- The molecular ion will be the highest *m/z* value in the spectrum if present (except for associated isotope peaks).
- Fragment ions will always have a lower *m/z* value than the singly charged radical cation.

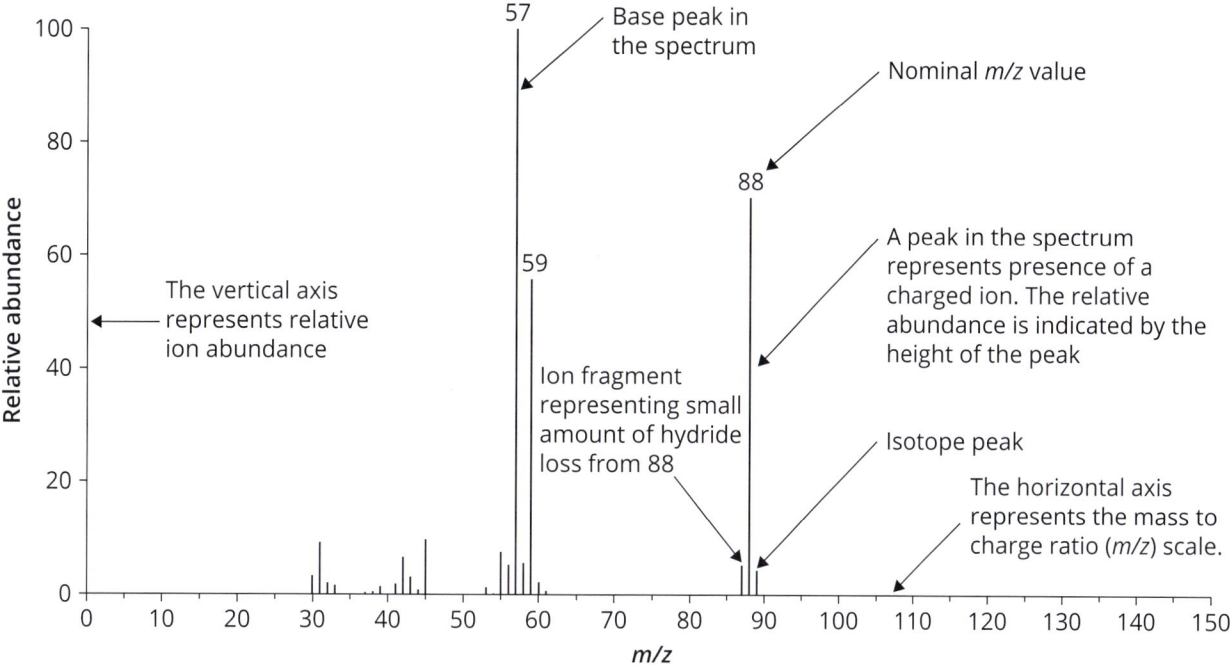

FIGURE 5.2 The EI mass spectrum of methyl propionate ($C_2H_8O_2$).

The accuracy of the reported m/z value is determined largely by the type of mass analyser in the mass spectrometer, as there are various different types. When a relatively low-resolution analyser (such as a quadrupole, for example) is used, the resulting m/z value may be reported as an integer value (that is, with no decimal places). This means that only ions with at least one mass unit difference can be differentiated from one another. Note that even for low-resolution mass spectrometers, the isotope peaks of a singly charged ion will be distinguished as a series of separate peaks in this case. Higher-resolution mass spectrometers (such as those incorporating Time of Flight or Orbitrap mass analysers) can produce individual m/z values with up to 4 or even 5 decimal place mass accuracy. With such accuracy it is often possible to calculate the chemical formula of small organic molecular ions and fragments directly from the m/z value (see Chapter 12). At unit mass resolution, however, the chemical formula cannot be calculated directly and needs to be deduced from additional information in the mass spectrum.

5.2.1 Isotopes

Isotope peaks are ubiquitous in EI mass spectra. Although they may appear to make the spectra complicated, they can be a very useful source of information about chemical composition because they represent the chemical elements present in the radical cation and its fragment ions. The isotope peak intensity reflects the sum of the naturally occurring isotopes of the elements present and is therefore unique to the chemical formula.

Elements can be categorized according to their isotopic composition. For example, elements that do not have multiple natural abundance isotopes are referred to as 'A' elements—these include fluorine, phosphorus, and iodine. Those which have a natural abundance isotope differing by 1 mass unit, such as carbon and nitrogen, are referred to as A+1 elements, and those where the next most abundant isotope differs by two mass units are referred to as 'A+2' elements. Common A+2 elements include, chlorine, bromine, sulfur, silicon, and oxygen. These isotope spacings, and their relative abundances, can be used to predict the number of carbon atoms and often identify which A+2 element is present.

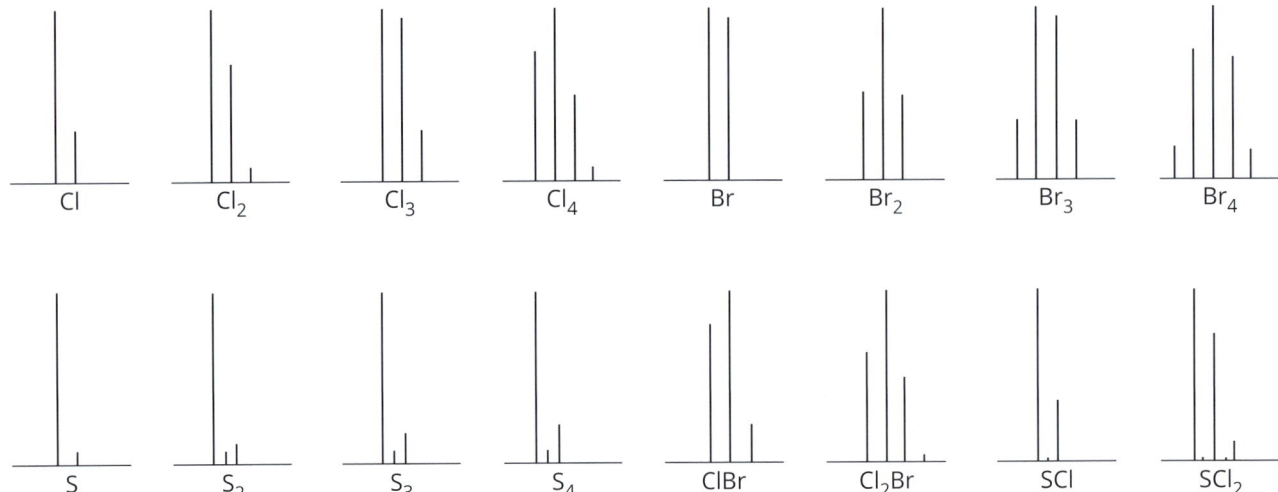

FIGURE 5.3 Relative isotope peak intensities for selected Cl, Br, and S element combinations.

The uniqueness of chlorine and bromine isotope abundances (see Table 5.1) means that analytes containing these elements can readily be identified from the relative heights of the A+2 peaks in a mass spectrum.

As the number of carbon atoms and A+2 elements increase in a molecule, the isotope pattern becomes more complicated. When more than six carbon atoms are present, for example, the height of the A+2 peak (2× ^{13}C) starts to become significant, making it difficult to distinguish whether oxygen or nitrogen is present. The A+2 elements generally have large A+2 abundances such as bromine and chlorine, sulfur, and silicon, making their presence relatively easy to identify. However, note that when more than one A+2 element is present, the relative heights are modified. Figure 5.3 illustrates the effect this has on the isotope peak abundances in the mass spectrum for chlorine, bromine, sulfur, and their combinations.

TABLE 5.1 Isotopic abundances of selected common elements.

Element	Element type	A relative isotope abundance (%)	A+1 relative isotope abundance (%)	A+2 relative isotope abundance (%)
Carbon	A+1	^{12}C: 98.9	^{13}C: 1.1	n/a
Hydrogen	A+1	^{1}H: 99.9	^{2}H: 0.01	n/a
Nitrogen	A+1	^{14}N: 99.6	^{15}N: 0.4	n/a
Oxygen	A+2	^{16}O: 99.7	^{17}O: 0.04	^{18}O: 0.2
Sulfur	A+2	^{32}S: 95.02	^{33}S: 0.75	^{34}S: 4.21
Phosphorus	A	^{31}P: 100	n/a	n/a
Fluorine	A	^{19}F: 100	n/a	n/a
Chlorine	A+2	^{35}Cl: 75.77	n/a	^{37}Cl: 24.23
Bromine	A+2	^{79}Br: 50.69	n/a	^{81}Br: 49.31
Iodine	A	^{127}I: 100	n/a	n/a

5.3 Mechanisms of ion formation and dissociation

The mass-to-charge ratio of the radical cation is the primary source of information closest to the mass of the intact analyte and correlates directly with its relative molecular mass. To understand what we see in the EI mass spectrum, it is worth thinking about how the radical cation is formed and subsequent mechanisms of dissociation.

5.3.1 Radical cation formation

The process of electron ionization takes place in the ion source of the mass spectrometer via the interaction between high-energy electrons and uncharged gaseous molecules of the analyte. Electrons are typically accelerated to 70 electron volts (eV) and enter a reaction chamber in the ion source which has a low pressure of analyte molecules present. A rapid transfer of energy takes place between the high-energy electrons and analyte molecules as they pass close to each other, leading to an increase in the internal molecular energy of the neutral analyte molecule. This increase in the internal molecular energy promotes electrons in the molecule into higher energy states, eventually leading to the loss of a single electron from the molecule. This loss of an electron leads to the formation of a positively charged radical cation. The m/z of the newly formed ion represents the mass of the neutral molecule minus the mass of a single electron, that is, a very small mass change from the original molecule—much less than a single mass unit.

> Generally, the electron with the lowest ionization energy will be ejected from the molecule, transforming the neutral analyte into a radical cation. A radical cation is indicated with a superscript plus sign representing the single charge and a dot next to it representing the unpaired electron: e.g. $M^{+\cdot}$

Figure 5.4 illustrates the process of ion formation accompanied by indicative changes to internal energies.

For the purposes of interpreting molecular structure, the location of the electron in the molecule that is removed during ionization is important as it often initiates subsequent fragmentation mechanisms. Knowing which electron is removed can therefore help with

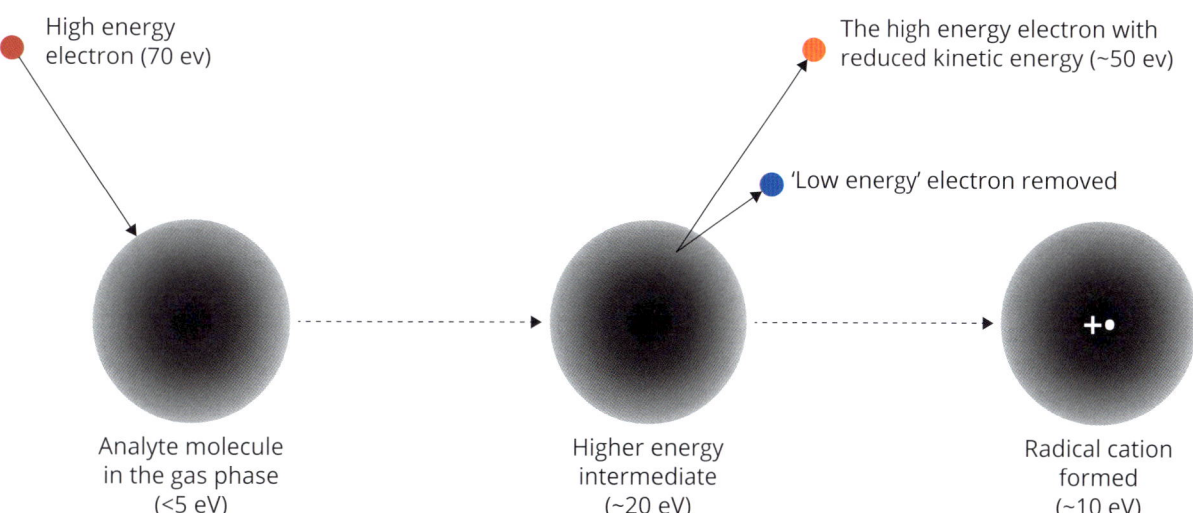

FIGURE 5.4 Schematic representing the formation of a radical cation via electron ionization of a molecule with indicative energy changes.

interpreting chemical structure. Although the amount of energy absorbed during the ionization process is often sufficient to eject electrons from many different positions in the molecule, in practice the one from the highest (energy) occupied molecular orbital (HOMO) is usually favoured. Non-bonding electrons associated with heteroatoms are the most likely to be removed (if present). The next most likely are pi-bond electrons and then sigma-bond electrons (non-bonding $> \pi > \sigma$). This order reflects relative bond energies (from weakest to strongest). Because one electron is always removed from a pair, the radical cations formed are always odd-electron species (OE). Which particular electron is removed within these categories for representative atoms is often guided by the electronegativity of the atom.

> In EI the electron that is removed from a neutral molecule is usually predictable. The likelihood of an electron being removed is: heteroatom non-bonding electrons > pi-bond electrons > sigma-bond electrons (non-bonding $> \pi > \sigma$).

5.3.2 Radical cation dissociation

As a result of energy transfer from the high-energy electron, the newly formed radical cation often has an excess of internal energy. This internal energy is dissipated by the emission of a photon but also by rapid redistribution of internal energy into the vibrational and rotational motion of the newly formed ion. When the redistributed energy exceeds internal bond energies, this leads to unimolecular fragmentation resulting in the formation of molecular fragments.

> Fragmentation of the radical cation often occurs in EI, and the degree to which it occurs depends upon the types of bonds present in the analyte molecule.

Sometimes fragmentation is so extensive that no radical cation appears in the mass spectrum because it has been completely converted into fragments. More commonly, however, the radical cation is present but at low or medium abundance compared to other fragment peaks. The extent to which the radical cation fragments is usually dependent on the functional groups present. When the radical cation is at higher abundance in the mass spectrum, this indicates that the structure is better able to stabilize the positive charge.

5.3.3 Odd- and even-electron fragmentation products

Figure 5.5 represents the process of radical cation dissociation to form fragment ions. Unlike the radical cation, which is always an odd electron species because it contains an unpaired electron as a result of ion formation, charged fragments can be either odd- or even-electron (OE or EE) species. Both types will be visible as peaks in the spectrum because they have a positive charge. Odd-electron cation fragments arise due to the loss of a neutral molecule, which leaves an odd-electron charged fragment. Even-electron cation fragments are accompanied by a neutral radical (an OE species). The neutral

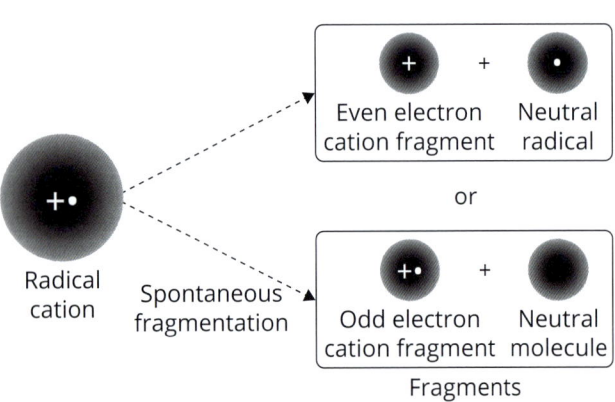

FIGURE 5.5 Schematic showing the main types of fragments that can be formed as a result of the unimolecular dissociation of radical cations. Often fragmentation of the radical cation leads to formation of a cation (EE+, will be seen in the mass spectrum) and loss of a neutral radical (top box). Rearrangements can lead to a new radical cation (OE+·, seen in the mass spectrum) with or without loss of a neutral molecule (bottom box). OE+· = odd-electron species; EE+ = even-electron species.

molecules or radicals do not appear as peaks in the mass spectrum (as they possess no charge) but their mass can be inferred from the mass difference between the radical cation and fragment peak. We will see that it is an important distinction whether a fragment is an odd- or even-electron species because it determines, among other things, whether the fragment has an odd or even integer mass. When combined with information about the radical cation, this can provide useful chemical and structural information.

5.3.4 Types of bond cleavage

Two basic types of bond cleavage can occur when a single bond is broken to create two fragments. The two electrons can be divided equally between the two fragment products (homolytic fission), or two electrons can go to one fragment and none to the other (heterolytic fission). This is illustrated in general terms in Figure 5.6. Both types of bond cleavage can occur during fragmentation of radical cations and determine whether a fragment is an odd- or even-electron species. It is important to follow the movement of individual electrons and pairs of electrons when illustrating fragmentation mechanisms. The movement of a single electron is shown by convention using a single barbed fishhook arrow. When a pair of electrons moves together, this is indicated by a double-barbed arrow, as also shown in Figure 5.6.

5.3.5 Mechanisms of fragmentation

The unimolecular decomposition of radical cations to form fragment ions and neutral species is reproducible under the same conditions, and the mechanisms involved are often predictable, which is useful when trying to interpret chemical structure. Whilst we provide a short overview of the main mechanisms, we do not cover this aspect in exhaustive detail; you can find detailed information on fragmentation mechanisms in other textbooks. Our aim is to provide an overview of the principles in sufficient detail to enable you to start working on the examples in this book. There are four basic fragmentation mechanisms that predominate in the decomposition of radical cations:

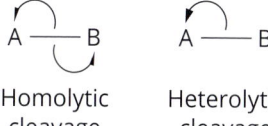

FIGURE 5.6 Illustration of homolytic and heterolytic bond fission with the movement of a single electron and a pair of electrons indicated by single and double fishhook arrows respectively.

- **Radical site-driven cleavage** (also known as **α-cleavage**) is driven by the tendency of the unpaired electron at the radical site to form a new bond with an adjacent carbon atom (α-carbon). This often occurs when the charge is on a nitrogen, oxygen, phosphorus, or sulfur atom (illustrated in Figure 5.7a).

- **Charge site-driven cleavage** (also known as **inductive or i-cleavage**) is initiated by the charge site attracting both electrons from an adjacent bond, which leads to its heterolytic cleavage and subsequent migration of the charge site to an adjacent electron-deficient atom. This often occurs at bonds next to highly electronegative elements, such as the halogens, from which non-bonding electrons are often preferentially removed during ionization (illustrated in Figure 5.7b).

- **Sigma-bond cleavage** (**σ-bond cleavage**) takes place when the electron that is lost during ionization comes from a single bond, occurring predominantly between C–C and C–H bonds when more electronegative atoms are not present. This type of fragmentation occurs, for instance, in hydrocarbons (illustrated in Figure 5.7c).

- **Rearrangements** also commonly take place where the analyte does not simply lose part of its structure through bond cleavage, but new bond formations also occur, reconfiguring the molecular structure. Rearrangements can be complex and varied. One common rearrangement is known as the *McLafferty rearrangement* and tends to be favoured when certain functional groups are present, notably carbonyl groups (illustrated in Figure 5.7d).

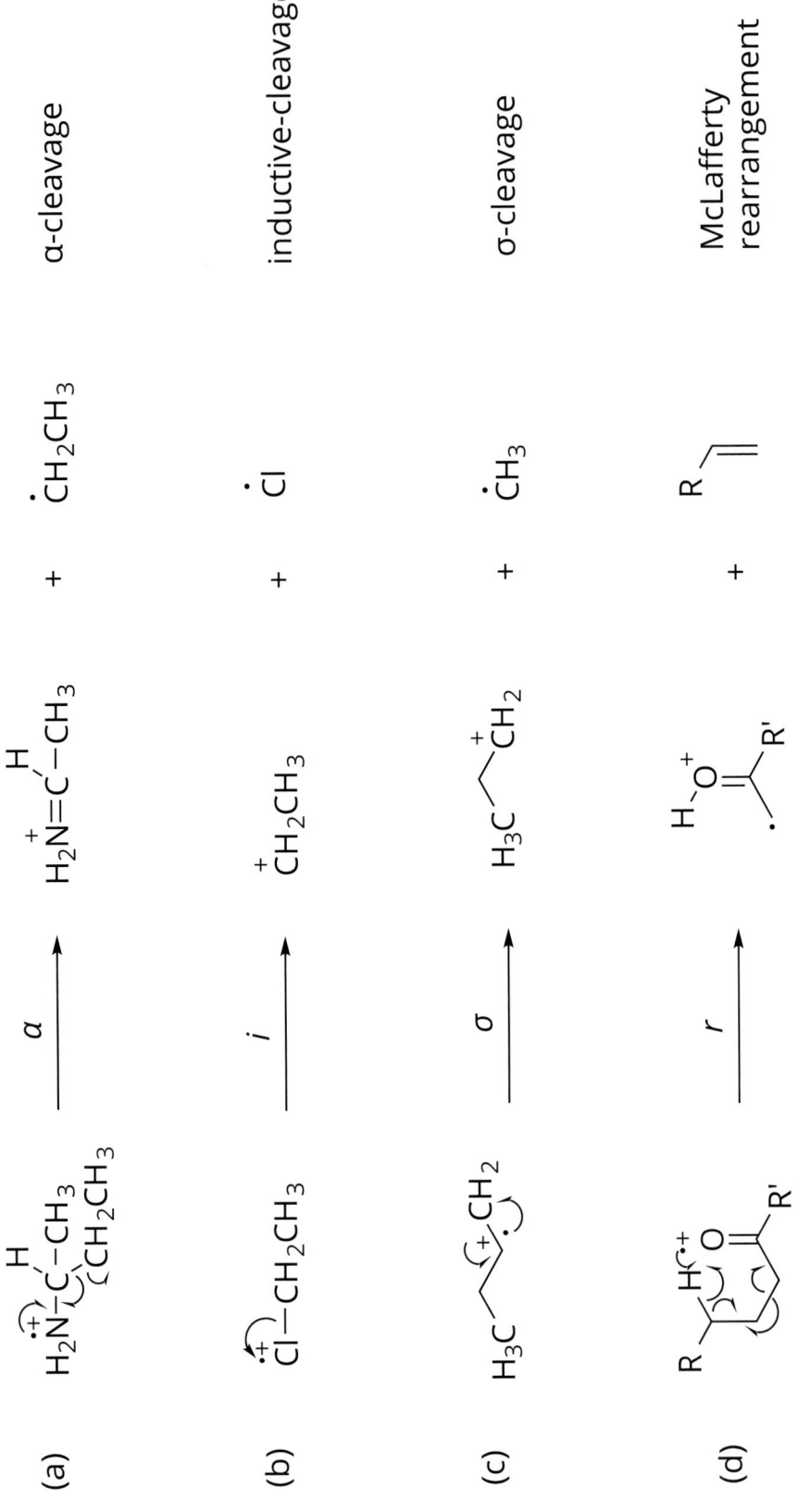

FIGURE 5.7 The most common EI fragmentation mechanisms for aliphatic molecules: (a) radical site-driven cleavage (also known as α-cleavage), (b) charge site-driven cleavage (also known as i-cleavage), (c) sigma-bond cleavage, and (d) McLafferty rearrangement.

5.4 **Interpretation of mass spectra**

The interpretation of mass spectra is a process of piecing together multiple chemical and structural clues derived from the spectra (and any additional information from other spectroscopic techniques that might be available). This includes information about the elements present, the stability of the radical cation, the predominant mechanism of the fragmentation process, and the actual fragments formed. It is usually not necessary to identify every peak in the mass spectrum to elucidate chemical identity. Indeed, even when the compound has been successfully identified, it can be difficult to rationalize every peak, particularly as compounds become more structurally complex.

5.4.1 **Identification of the molecular ion**

Identifying whether the molecular ion is represented in a mass spectrum is an important first step in the interpretation of EI mass spectra. To help with this process, there are a number of criteria that a molecular ion must meet:

1) If present, the molecular ion must be represented by the highest m/z value in the mass spectrum of the analyte (including its isotope peaks).
2) It will always be an odd-electron species and therefore usually have an even m/z value, unless it contains an odd number of nitrogen atoms (see section 5.4.2).
3) Its fragments must represent logical neutral losses.
4) Its isotope peak distributions provide information about elemental composition.

In addition, whether it has an odd or even integer value, the radical cation usually dissociates into fragments which are the opposite in terms of their odd or even integer value. For example, if the radical cation has an even integer m/z value, its fragments will usually have an odd m/z value and vice versa. In Figure 5.2 the highest mass ion has an even integer value whilst all the lower mass ions of significant abundance have odd integer values. This knowledge can be used to help determine if the radical cation is present in the mass spectrum. The reason for the change in m/z from odd to even or vice versa is that the original odd-electron radical cation commonly forms an even-electron cation fragment (see sigma-bond cleavage, radical α-cleavage, and inductive cleavage mechanisms in section 5.3.4), changing from odd to even or even to odd depending on the number of nitrogen atoms present (see section 5.4.2). However, there are exceptions to this general situation that are useful to know about. For example, rearrangements such as the McLafferty rearrangement produce a radical cation fragment (that is, it is still an odd-electron species just like the original radical cation), which means that there is no change from even to odd m/z, or vice versa, when the radical cation's structure is rearranged.

The characteristics of molecular ions and their fragments can be summarized as follows:

• The mass and elemental composition of the molecular ion, if present and identifiable, provide some of the most useful information for interpreting chemical identity.
• The radical cations can possess odd or even m/z depending on whether nitrogen atoms are present and, if so, indicate how many.
• Fragments of an even integer radical cation will have odd m/z, and fragments of an odd integer radical cation will have even m/z unless a rearrangement takes place.
• Note that neutral molecules rather than radicals must have been lost when a fragment remains an odd-electron species.

5.4.2 The nitrogen rule

In general, if an analyte contains the elements C, H, O, N, P, S, Si, F, I, Cl, or Br, the radical cation will have an even integer value unless the ion contains an odd number of nitrogen atoms. This is known as the 'nitrogen rule' and can be useful in the interpretation of mass spectra as it indicates whether a nitrogen atom is present.

> When applying the nitrogen rule to EI mass spectra, the *nominal monoisotopic mass* should be used (that is, the mass rounded to the nearest integer value).

In cases when the analyte contains no nitrogen atoms, the radical cation has an even *m/z* value and fragments will have an odd integer value (as long as they are even-electron fragments). Similarly, analyte ions which contain an even number of nitrogen atoms will produce radical cations with an even *m/z* value, and analytes containing an odd number of nitrogen atoms will produce radical cations with odd *m/z* values. Care must be taken when the radical cation contains one or more nitrogen atoms because the fragments may contain all, some, or none of the nitrogen atoms (as none, one, or more nitrogen atoms may be lost in the neutral fragment). Note that this lack of certainty arises only when dealing with integer masses, for instance when using low-resolution mass spectra. High-resolution mass analysers can often provide sufficient *m/z* accuracy to directly determine the elemental composition, including the number of nitrogen atoms in the radical cation and fragments (see Chapter 12).

There are a number of different permutations as to whether the radical cation or fragment has an odd or even *m/z* value, and these are summarized in Table 5.2. In general, an odd integer radical cation will give even fragment ions and vice versa, unless 1) there has been a change in the number of nitrogen atoms between the radical cation and its fragments; 2) an even-electron neutral species has been lost (e.g. H_2O, CO, etc) or another type of rearrangement has occurred, such as the McLafferty rearrangement, leading to what is termed a distonic radical cation fragment. A distonic radical cation is where the unpaired electron (radical site) and the charge are located on different atoms within the molecule.

5.4.3 Working out the chemical formula

Predicting the number of carbon atoms in the formula of an unknown analyte can be invaluable when trying to differentiate possible chemical formulae for an unknown analyte. A useful rule of thumb is derived from the theoretical abundance of the ^{13}C isotope

TABLE 5.2 Summary of possibilities for odd/even integer values for radical cation and fragment pairs and their formation.

Radical cation	Fragment	Possibilities
Even	Odd	No nitrogen is present (or an even number of nitrogen atoms; 2,4,6, etc) in the radical cation and this remains the same for a charged fragment.
Odd	Even	An odd number of nitrogen atoms are present in the radical cation and remain on the charged fragment.
Even	Even	No nitrogen is present (or even number 2,4,6, etc). The fragment either contains a different number of nitrogen atoms, or it contains the same number but a non-radical neutral was lost, or the fragment is a rearrangement product.
Odd	Odd	1,3, or 5 nitrogen atoms are present in the radical cation. The fragment has lost a nitrogen, or a non-radical neutral was lost, or the fragment is a rearrangement product.

of carbon which is found at approximately 1.1% of the abundance of the ^{12}C isotope in almost all carbon-containing compounds on Earth.

> The number of carbon atoms contained in the formula for an analyte can be estimated by comparing the relative heights of the mass spectral peaks representing the ^{12}C and ^{13}C carbon versions of the radical cation or fragment ion:
>
> The number of carbon atoms (for a molecular ion or fragment ion) = (relative abundance of the M+1 isotope peak / relative abundance of the M peak) / 0.011.

A prediction of the number of carbon atoms in an analyte molecule can be made by taking the height of the isotope peak (M+1) in the mass spectrum and dividing it by the height of the M peak before dividing the result by the average abundance of ^{13}C atoms (1.1%). Figure 5.8 illustrates this process. In this example the M+1 isotope peak height relative to the M isotope peak is 15%, so the calculation is: (15/100)/0.011 = 13.6 (or 14 rounded to the nearest integer). Performing this calculation often results in a fractional value because of slight variations in peak height measurement accuracy. Rounding to the nearest integer and applying a nominal +/−1 error to this value is a useful rule of thumb. Although we usually use this calculation to predict the number of carbon atoms, in theory the isotope abundances of other common elements can influence the M+1 abundance such as ^{2}H, ^{15}N, and ^{17}O, which are all very close to the M+1 m/z value. The relative abundance of the M+1 isotope peak predominantly represents ^{13}C, however, as it is by far the most abundant of these common stable isotopes. In practice it therefore usually makes up the majority of the M+1 signal, making the M+1 isotope peak height in the mass spectrum a useful predictor of the number of carbon atoms in the molecule it represents.

5.4.4 Fragmentation mechanisms for common organic compounds

Different compound classes, and the presence of particular functional groups, often lead to characteristic fragmentation mechanisms and the formation of often diagnostic fragments. Whilst there are usually some exceptions, it is still useful to keep these in mind when interpreting fragmentation patterns. Table 5.3 provides a list of compound classes with information about typical molecular ion intensities, mechanisms of fragmentation, and common fragment masses.

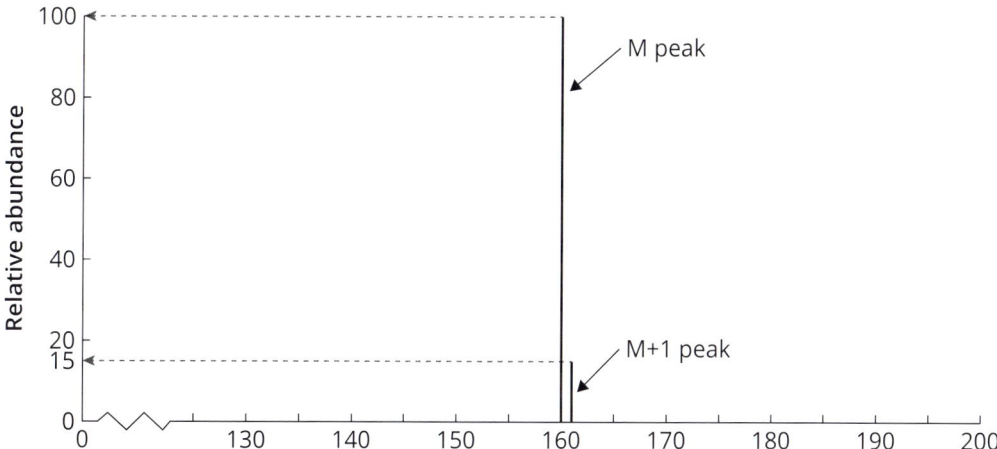

FIGURE 5.8 The height of the M+1 peak relative to the M peak in a mass spectrum can be used to predict the number of carbon atoms in the analyte molecule with an approximate error of +/− 1.

TABLE 5.3 Typical observations and mechanisms for EI fragmentation of common compound classes.

Compound class	M+ intensity	Typical mechanisms	Typical Fragments (m/z)	Notes
Alkanes	weak	σ-bond	29,43,57,71,85 etc	Fragments separated by 14 mass units of type $[C_nH_{2n+1}]^+$.
Cyclic alkanes	medium	σ-bond/ α-cleavage	98,70,56	Loss of ethene.
Alkenes	weak/ medium	α-cleavage	41,42,55,56,69, etc	Fragments separated by 14 mass units by $[C_nH_{2n-1}]^+$, $[C_nH_{2n}]^+$ and $[C_nH_{2n+1}]^+$.
Cyclic alkenes	medium	σ-bond/ α-cleavage	2 units lower than for cyclic alkanes	Determining the position of double bond difficult.
Alkynes	varies	α-cleavage	$[C_nH_{2n-3}]^+$ and $[C_nH_{2n-1}]^+$.	Molecular ion intensity depends on triple bond location.
Aromatics	strong	α-cleavage/ i-cleavage	M−1, 91, 92, 65, 77 78, 51	m/z 91 is highly stable tropylium ion (can rearrange to form tolyl ion).
Alcohols	weak/ absent	α-cleavage/ rearrange	$[M-H_2O]^{+\cdot}$, $[M-H]^{+\cdot}$, 31, 57 (cyclic), 77 (phenols)	Phenols/aromatics show more intense molecular ions.
Ethers	Weak (strong for aromatic)	α-cleavage/ β-cleavage/ i-cleavage	CO loss (aromatic ethers), 105, 77	Rearrangements when α-carbon is substituted. Aromatic often fragments at O–Ar bond leading to subsequent loss of CO and ring opening.
Thiols	weak/ medium	α-cleavage/ i-cleavage	Similar to alcohols and ethers	Presence of sulfur isotopes observable.
Amines	varies	α-cleavage	$[M-H]^{+\cdot}$, 30	Presence of nitrogen often identified by odd molecular weight and even fragments.
Amides	varies	α-cleavage/ McLafferty	86 (primary), 59 (McLafferty)	Similar fragmentation to carboxylic acids.
Aldehydes	weak/ medium	α-cleavage/ rearrange	Loss of H·, M−18, M−28, M−43, 44, 58 (McLafferty)	McLafferty can occur, i-cleavage can be present.
Ketones	weak/ medium	α-cleavage/ rearrange	105 (aromatic), 55 (cyclic)	McLafferty can occur, acylium ion formation, i-cleavage can be present.
Carboxylic acids	weak	α-cleavage/ McLafferty	M−OH, M−CO_2H, M−H_2O (aromatic)	Multiple functional groups lead to complex fragments.

Compound class	M⁺ intensity	Typical mechanisms	Typical Fragments (m/z)	Notes
Esters	Weak (strong for aromatic)	α-cleavage/ McLafferty	$[C_nH_{2n-1}O_2]^+$ 14 m/z units apart. CH_3O^+ (aromatic), 105, 77	Both acid and alcohol groups can initiate fragmentation. Loss of methoxy and ethoxy radicals from these esters are diagnostic.
Nitriles	weak/missing	α-cleavage/ McLafferty	M–H, 41 (McLafferty), 97 (straight chain >7)	Odd molecular weight nitrogen, molecular ion usually absent.
Organic halides	varies	i-cleavage	Loss of halogen radical, e.g. loss of 19 or 127 for F and I respectively. H⁻halogen loss.	Cl and Br show characteristic isotope patterns. F and I have no additional naturally occurring isotopes (monoisotopic) but also often no molecular ion.

5.4.5 Fragmentation of aromatic compounds

In contrast with many aliphatic compounds, aromatic compounds often produce highly stable molecular ions. The delocalized electrons in the aromatic ring are usually better able to stabilize the positive charge of cationic fragment ions. Loss of a hydrogen radical from the aromatic ring directly, or from alkyl substituents (e.g. aldehydes, amines, and nitriles), can also occur. Many alkyl-substituted aromatic compounds form a tropylium ion represented at m/z 91, after fragmentation of the alkyl group (the figure shows this formation). This is a highly stable 7-membered ring structure often represented by a prominent or base peak in the mass spectrum. It can decompose through loss of acetylene to form a 5-membered ring at m/z 65. m/z 91 and 65 in a mass spectrum are therefore highly indicative of an alkyl-substituted aromatic structure.

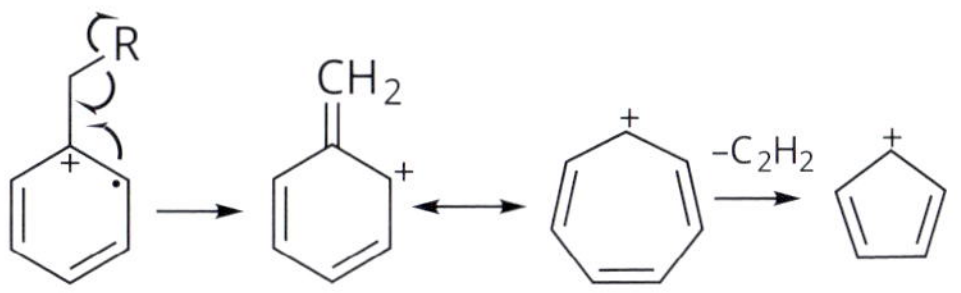

m/z 91   m/z 91   m/z 65
            (Tropylium cation)

When a carbonyl group is an aromatic substituent (e.g. aldehyde, ketone, ester etc) m/z 105 (Ar≡CO) and m/z 77 (phenyl cation) are typical fragments.

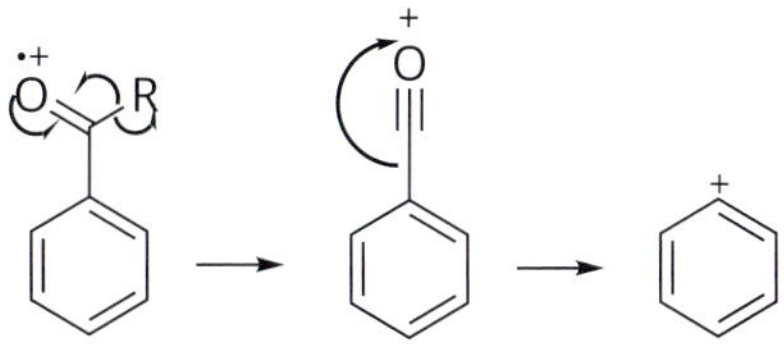

m/z 105   m/z 77
              (Phenyl cation)

5.5 A step-by-step approach to the interpretation of mass spectra

So far in this chapter we have addressed the characteristics of EI mass spectra, the mechanisms underlying radical cation formation, and fragmentation processes, along with various tools for interpreting chemical and structural identity. In this final section we will look at how to go about analysing EI mass spectra in practice: where to start and what questions to ask of experimental results. To do this, we provide a step-by-step approach to interpreting the mass spectrum of an unknown compound. We use ethyl chloride as an example (Figure 5.9 and Table 5.4). In this case information about the analyte is provided in addition to the mass spectrum so that you can see how mass spectral information relates to chemical structure as we work though the problem.

Analyte information	Formula/structure
Ethyl Chloride	C_2H_5Cl
Monoisotopic mass	64.01
Structural formula	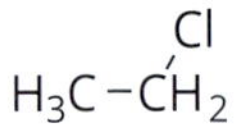 $H_3C-\overset{\displaystyle Cl}{\underset{}{CH_2}}$

a) The first step in the interpretation of an EI mass spectrum is to identify whether there is evidence that the molecular ion is represented in the mass spectrum. What are the possibilities in this example? The molecular ion could be i) the peak at m/z 66 or ii) the base peak at m/z 64 with 65,66,67 as isotopes. Another option is iii) the molecular ion is not present in the spectrum. These options narrow down the possibilities but do not identify a specific m/z value for the radical cation yet.

b) It is also useful to consider the relative abundance of the suspected molecular ion peak and the base peak (if different). This provides qualitative information about the stability of a molecular ion and main fragments. If the suspected radical cation is also the base peak, this would imply a chemical structure that is able to stabilize the positive charge of the radical cation well. Look for evidence of an aromatic group in fragments, e.g. m/z 77 (phenyl), m/z 91, or m/z 92 (e.g. a tropylium ion which can be confirmed by presence also of m/z 51, which is a tolyl ion fragment). In this example we can rule out the presence of an aromatic group because the mass is too small.

c) Consider which characteristic isotope peaks can be identified. This can help reveal the molecular ion and whether the molecule contains (or does not contain) characteristic elements. In the spectrum there is an isotope profile representing the A+2 element

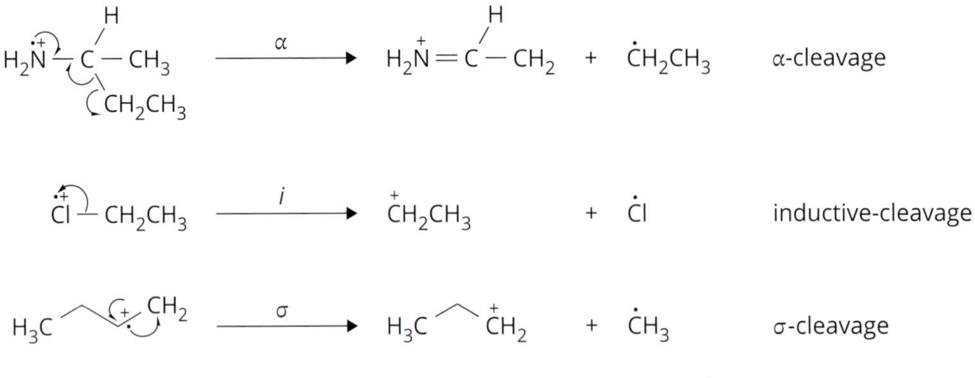

FIGURE 5.9 Example fragmentation processes for aliphatic molecules: a) radical site driven cleavage, b) charge site driven cleavage and c) sigma bond cleavage.

TABLE 5.4 Peak list and relative intensities from the EI mass spectrum of ethyl chloride

MS Peak list	Rel. Intensity	MS Peak list	Rel. Intensity
m/z 15	1.4	m/z 50	1.0
m/z 24	0.9	m/z 51	8.0
m/z 25	4.0	m/z 52	0.1
m/z 26	22	m/z 59	2.0
m/z 27	61	m/z 60	3.8
m/z 28	82	m/z 61	4.0
m/z 29	76	m/z 62	3.9
m/z 30	1.6	m/z 63	0.83
m/z 35	2.6	m/z 64	100
m/z 36	2.0	m/z 65	2.4
m/z 37	1.5	m/z 66	33
m/z 38	1.1	m/z 67	0.70
m/z 47	3.1		
m/z 48	4.0		
m/z 49	28		

chlorine at m/z 64 and 66 which corresponds to the ^{35}Cl and ^{37}Cl isotopes respectively (see (a) in Figure 5.10). The presence of chlorine isotopes can also be recognized by this motif in the fragment series m/z 49–51 (see (b) in Figure 5.10). This provides evidence that the molecular ion is present with the main isotope peak at m/z 64. It also provides evidence that the analyte contains a single chlorine atom. Additional evidence that m/z 64 and 66 represent the molecular ion is provided by the fact that they have even-numbered integer masses, suggesting an odd-electron radical cation, whilst the fragment m/z pair 49–51 is odd, suggesting even-electron ions (fragments).

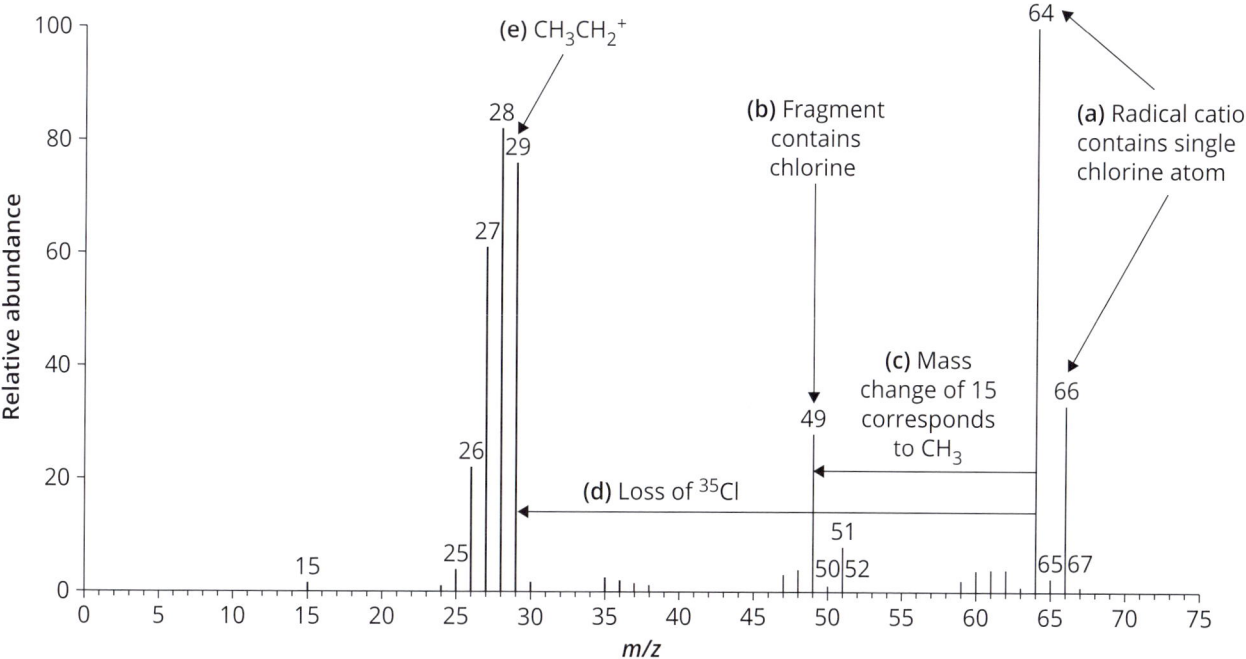

FIGURE 5.10 Interpreting the mass spectrum of ethyl chloride (CH_3CH_2Cl).

d) Having established that there is evidence for the presence of the radical cation, the next step is to predict the number of carbon atoms in the analyte. From the peak list we can see that the intensity of the m/z 65 isotope peak is ~2.4% of the m/z 64 peak: $(2.4/100)/0.011 = 2.4/1.1 = 2.1$. We can therefore predict that the molecule contains 2 (+/−1) carbon atoms.

e) Next it is useful to apply the nitrogen rule to estimate how many nitrogen atoms may be present, if any. As the proposed molecular ion is an even integer value it must either contain no nitrogen atoms or an even number of nitrogen atoms (2,4, etc). Given our prediction that the number of carbon atoms is 2, and knowing that the molecule contains a chlorine atom, we can rule out 2 or more nitrogen atoms because they would increase the molecular weight above 64. At this stage we would therefore predict that the analyte contains approximately 2 carbon atoms, a single chlorine atom, and no nitrogen atoms.

f) If we consider the cluster of fragment ions at m/z 49,50,51,52 we can see that these indicate the presence of chlorine based on the isotopic pattern with the loss of a non-charged fragment from the radical cation of 15 mass units (see (c) in Figure 5.10). The loss of 15 is likely to represent loss of a neutral methyl radical as this is the most plausible explanation for a mass change of 15—not many other changes could cause this. This implies that three hydrogen atoms are bonded to a carbon atom in the molecule. The m/z 49 value therefore corresponds to CH_2Cl and this is confirmed by the presence of a fragment at m/z 29 which corresponds to CH_3CH_2 via loss of a Cl radical from the molecular ion (see (d & e) in Figure 5.10). Having deduced the various pieces of information about the molecule in sections a–f above, we can now arrive at a logical conclusion that its chemical formula must be C_2H_5Cl representing ethyl chloride.

This example illustrates the systematic steps to take when interpreting the EI mass spectrum. To conclude this chapter, it is worth reiterating that the EI mass spectrum in isolation may not always provide enough information for unambiguous structural identification. This is particularly common as the analyte structure increases in size and chemical complexity. One of the aims of this book is to illustrate that, although it is possible to interpret some structures by mass spectrometry alone, it is usually better to combine multiple analytical techniques, most commonly NMR, MS, and IR. Together these techniques offer much greater structural specificity, as each provides complementary and confirmatory information that helps us accurately piece together the puzzle of a chemical structure.

CHAPTER 6

Dealing with spectra from real samples

When dealing with samples derived from real laboratory synthetic procedures, there can be additional features to consider that may not be present in textbook examples or that may cause differences in spectra related to the way the samples were prepared.

> NMR spectra of real samples often contain resonances arising from impurities, most commonly the solvents used in the production of the sample.

Common impurity solvents are ethyl acetate or ether, and you should always consider how likely they are to be present in your spectra. These types of impurities are often present in improperly dried samples after recrystallization. However, even for purified samples extra peaks ('solvent impurities') may just arise from the NMR solvent itself.

> Although in NMR we use deuterated solvents, these will always contain traces of undeuterated (or partially deuterated) solvent that will appear in proton spectra.

Common examples are $CHCl_3$ for chloroform or CHD_2OH for methanol. These will typically also display a resonance from traces of dissolved water, the shift of which will vary significantly according to the solvent itself (Figure 6.1).

> Solvents that contain carbon atoms will display strong resonances that often dominate the carbon spectrum and are split by coupling to the directly attached deuterium atom, which has a spin of 1.

Deuterium coupling is why the carbon spectrum of $CDCl_3$ always contains a characteristic 1:1:1 multiplet at 77 ppm, and the carbon resonances of other deuterated solvents are also split into multiple lines (Figure 6.2). Resonances arising from some NMR solvents in their undeuterated or partially deuterated form are summarized in Table 6.1.

> The recognition of the solvent chemical shifts in spectra provides a useful means of checking that the spectrum itself is correctly referenced.

Referencing errors can occur, for example, if an incorrect solvent is selected on the spectrometer, which will cause all chemical shifts in the resulting spectrum to be erroneous. Chemical shifts arising from traces of common laboratory solvents are shown in Table 6.2; it should be noted that the exact chemical shifts of these can vary depending on which NMR solvent is being used.

We noted in section 2.2.2 that the identification of proton resonances from 'acidic' protons can be difficult to achieve reliably due to their highly variable chemical shifts

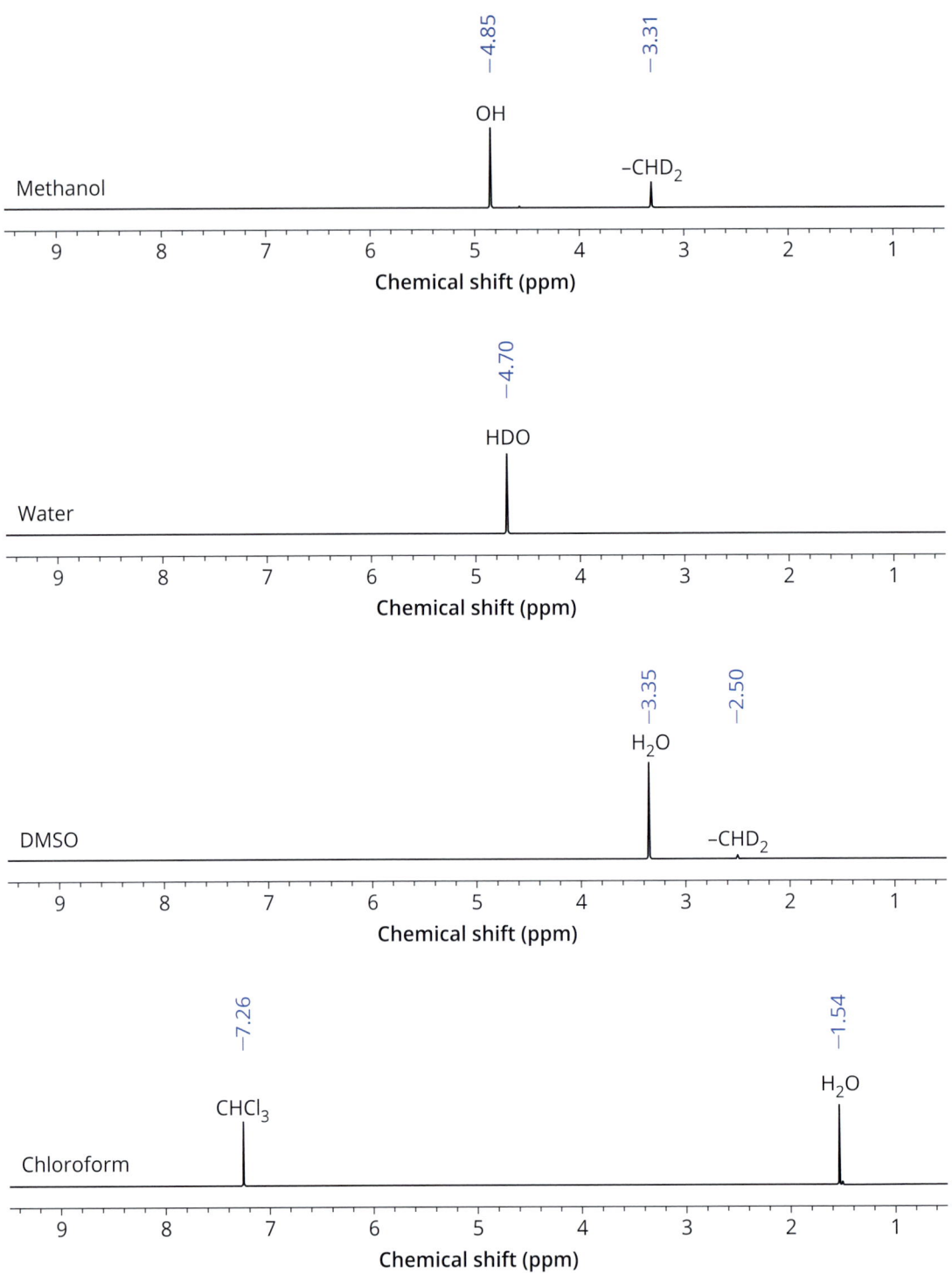

FIGURE 6.1 Proton NMR spectra of some common deuterated solvents showing the shifts of the residual protonated solvent peak and that of the common contaminant water. Dimethyl sulfoxide (DMSO) is very hygroscopic and often contains relatively high levels of water, as seen here.

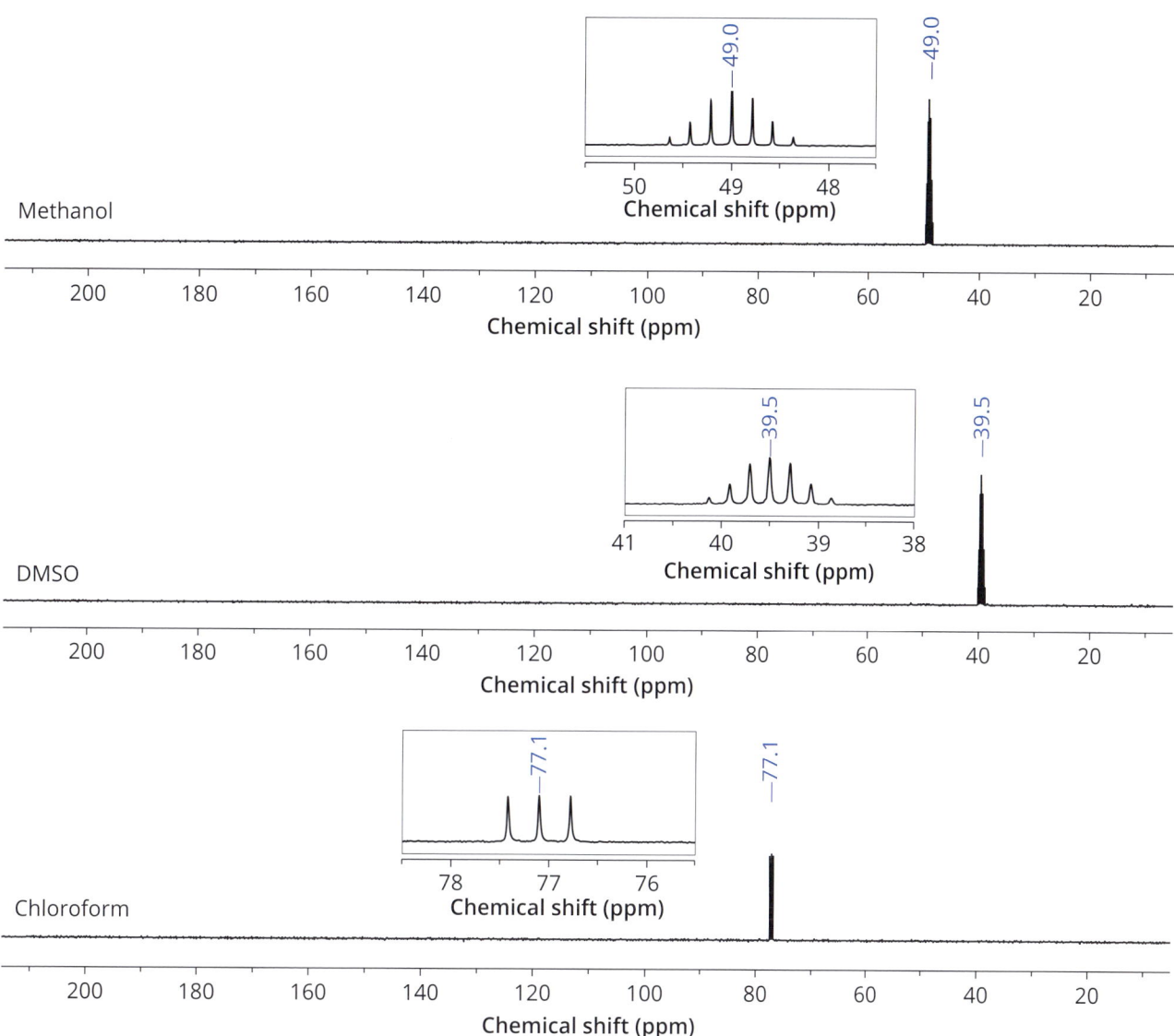

FIGURE 6.2 Carbon NMR spectra of common deuterated solvents. Expansions show the multiplet structures arising from coupling to the attached deuterium atom(s). The centre of each multiplet represents the chemical shift.

TABLE 6.1 Chemical shifts of resonances arising from common deuterated NMR solvents, and of water dissolved in the solvent δ_H (H_2O) which is always apparent. The ^1H solvent shifts are for the *residual protonated* forms of each solvent.

NMR solvent	Formula	δ_H	δ_H (H_2O)	δ_c
Chloroform	$CDCl_3$	7.26	~1.6	77.1
Dimethylsulfoxide (d_6-DMSO)	$(CD_3)_2SO$	2.50	~3.3	39.5
Water	D_2O	4.70	–	–
Methanol (d_4-MeOH)	CD_3OD	3.31, 4.87	~4.9	49.0

TABLE 6.2 Chemical shift of resonances arising from common solvent 'impurities' in an NMR sample dissolved in CDCl3. Data source: J. Org. Chem. 1997, 62, 7512–7515; *OH resonances not always observed.

Solvent	Formula	δ_H	δ_C
Chloroform (Trichloromethane)	$CHCl_3$	7.26	77.4
Acetone (Propanone)	$Me_2C{=}O$	2.17	30.9, 207.1
Methylene chloride (Dichloromethane)	CH_2Cl_2	5.30	53.5
Ethanol	EtOH	1.25 (t), 1.32* (s), 3.72 (q)	18.4, 58.3
Ethyl acetate (Ethyl ethanoate)	EtOAc	1.26 (t), 2.05 (s), 4.12 (q)	14.2, 21.0, 60.5, 171.4
Diethyl ether	EtOEt	1.21, 3.48	15.2, 65.9
Methanol	MeOH	1.09* (s), 3.49 (s)	50.4
Water	H_2O	1.56	–

and often lack of resolved multiplet structure. This behaviour arises from the fact that the acidic proton is able to leave the molecular structure and transit to another environment. For example, an alcohol OH proton can swap with another proton from water (often found in solvents); the protons are then said to have *exchanged*. Note that *acidic* is a general term widely used to describe a proton that can undergo exchange, and does not have to relate only to acids; alcohols and amines are also common examples.

An experimental trick that can be used to identify acidic protons is the 'D₂O exchange' or 'D₂O shake' method which leads to the disappearance of exchangeable proton peaks in a spectrum. This process starts with the collection of a standard ¹H NMR spectrum of your sample in an organic solvent such as chloroform. After this, one drop of D₂O is added to the NMR tube and the sample is capped securely and (carefully) shaken to mix the solvents. The emulsion is then left to stand and allowed to separate, leaving the D₂O above the organic solvent phase. The proton spectrum is then recorded once more, by which time exchangeable protons will have been replaced with deuterons and their resonances removed from the ¹H spectrum, as revealed by comparison with the initial reference spectrum (Figure 6.3). It should be noted that acidic protons will not be seen when the NMR

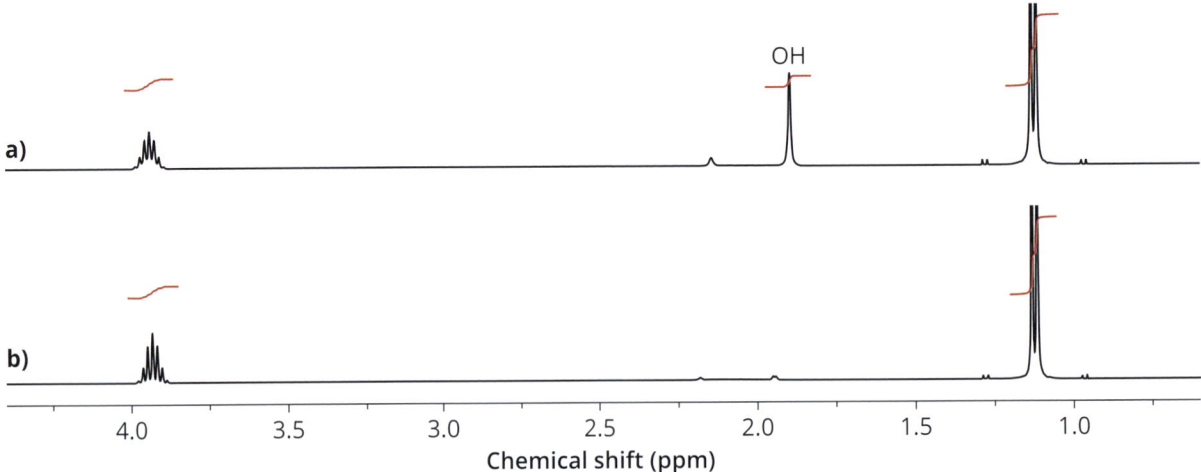

FIGURE 6.3 The D₂O shake experiment performed on isopropanol identifies the exchangeable (acidic) hydroxyl proton resonance. The conventional ¹H spectrum in CDCl₃ (a) shows a broadened OH resonance that is lost following D₂O exchange (b).

solvent itself possesses exchangeable deuterons as these will always be replaced through exchange, as is the case for the deuterated forms of water (D_2O) and methanol (CD_3OD).

Depending on the particular method of analysis, infrared spectra of real samples may also contain evidence of 'impurities' as described above. Most commonly seen is water, which is readily detected in poorly dried samples and often appears from compressed salt discs. Traces of water will be apparent in the region >3200 cm^{-1}. For samples that have been measured as solutions (most commonly in carbon tetrachloride or chloroform) or as mulls (with Nujol), there may also be interference from the IR absorption peaks of the solvent itself, even though instruments attempt to eliminate this by subtracting a solvent reference spectrum. This means certain regions in the spectra may be considered 'dead zones' from which no useful data can be obtained. The increasing use of the ATR methodology, employing neat samples (as in this book), means that this complication is now less often encountered in the laboratory.

> Mass spectra can also contain extraneous peaks and altered analyte peak intensities related to experimental and sample preparation conditions which can confound interpretation of spectra.

EI spectra are affected by, for example, ion source conditions such as source pressure, ionization energy, and contamination from previous samples in the sample injection port. These can both alter the *m/z* spectrum by providing additional peaks resulting from contamination and alter the relative abundances of analyte peaks. In cases in which the molecular ion is particularly unstable, adjusting the source conditions can lead to the presence or absence of a molecular ion peak, which can have a significant impact on how the mass spectrum is interpreted. Ionization energy can also affect the relative abundance of peaks; this can become apparent when comparing spectra from the same compound between two instruments and may affect how closely a spectrum matches a reference spectrum, which is relevant, for example, when matching a spectrum you collected to spectral databases.

> Sample contamination or contamination in the instrument itself—such as a dirty injection port—is commonly identified by mass spectrometry and is often observed via additional peaks in the mass spectrum. These have the potential to confound or confuse interpretation of an otherwise 'purified' sample.

Note that where gas chromatography (GC) systems can be coupled to the mass spectrometer, the resulting GC-MS analysis can be a useful way to eliminate sample and injection port contamination by separating the analyte from the contaminants. However, it is also possible for this approach to lead to additional peaks from, for example, chromatography column bleed. Note also that chromatographic retention times can be highly laboratory-specific, and the same compound may elute at different places in the chromatogram (and lead to different contaminating peaks), especially if slightly different temperature gradients are used.

CHAPTER 7

Analysis checklist

Before attempting to determine the structure of a molecule from spectroscopic and spectrometric data, it is worthwhile considering first what can be determined with a high degree of certainty from each technique. Other information can often be inferred early on, but it is important not to jump to a definite assignment without corroborating evidence to support it. Even then, it may only be at the end of the data analysis process that all the pieces fall into place and the structural picture can be completed. This is one of the reasons why it's important to utilize information from multiple analytical techniques when possible.

The checklist presented below provides a consistent workflow that we recommend when attempting structure determination. The suggested order in which you undertake an initial analysis of the data follows i) mass spectrum, ii) IR spectrum, iii) ^{1}H NMR spectrum, and then iv) ^{13}C NMR spectrum. After this initial assessment it is likely that you will need to revert back to undertake more in-depth analyses of these data sets as you progressively assemble your structure. This process will be exemplified in the worked examples in Chapter 8, which follow the sequence we suggest here.

This checklist highlights several 'facts' which can often be determined with a high degree of certainty early on. The text in italics indicates secondary information (or words of caution) which can be noted at this stage but for which further evidence may be needed before assignments can be confirmed. At the end of this chapter, you can find a simple flowchart (Figure 7.2) showing the relationships between these key pieces of information that may be helpful in your data interpretation.

7.1 The mass spectrum

From the mass spectrum, identify . . .

a. The molecular ion peak which provides the monoisotopic mass of the compound. It must be the highest m/z value in the spectrum (except for its additional isotope peaks) and an odd-electron species. It must lead to fragments with logical neutral losses.

b. Isotope peaks for the molecular ion: do they indicate the presence of elements with characteristic isotope patterns (i.e. 'A' elements with no additional isotopes (e.g. F, I, P) or A+2 elements (e.g. Cl, Br, S))?

   The number of carbon atoms in the molecule may be predictable from the relative abundance of the A+1 isotope peak.

   You may also be able to gain an insight into the pattern of the molecular ion cluster. This can often indicate the number of halogen atoms present. For example, a 9:6:1 pattern, separated by two m/z units, is indicative of the presence of two chlorine atoms.

c. Whether the molecular ion is an odd integer value, which indicates the presence of nitrogen.

    Note that two nitrogen atoms will give an even molecular ion peak, so do not exclude the presence of nitrogen just because the molecular ion is even.

d. If any further information is available from the fragment masses, *e.g., are there clues from the chemical composition of ions and neutral losses? What do peak intensities tell us about the stability of fragments?*

7.2 The IR spectrum

From the IR spectrum, identify . . .

a. The presence, or absence, of absorption due to a carbonyl stretch.

    The range of carbonyl absorptions is typically 1650–1850cm^{-1}. Experience may help you further narrow down possible functional groups (e.g. amide or ester), but you should not guess too early on.

b. The presence or absence of absorptions due to an N–H, O–H, CON–H, or COO–H group, typically in the region 2700–3500 cm^{-1}.

    Be aware that absorptions due to CH stretching frequencies are ubiquitous in almost all organic molecules, and these appear around 3000 cm^{-1}.

c. Other stretches above 2000 cm^{-1} that are distinctive for certain functional groups, e. g. alkynes or nitriles.

7.3 The ^{1}H NMR spectrum

From the ^{1}H NMR spectrum identify . . .

a. The number of discrete resonances and their relative integrals.

    Make note of peaks that are clearly not part of the structure, such as reference compounds and solvent signals.

    Remember that resonances may be overlapping, so a 2 H resonance may be 2 × 1H resonances or a 1 × 2H resonance. Always examine expansions of the spectrum if available.

b. The chemical shifts of the resonances.

    These can be noted approximately—one decimal place is suggested at this stage. For multiplets and overlapping resonances over a broad chemical shift range, noting the range of the resonance(s) is useful, e.g. 1.2–2.1 ppm.

c. The splitting patterns of the resonances.

    If in doubt, it is better to assign the splitting pattern as 'not determined' or 'not clear' than mis-assign it at an early stage. If assigning a quartet, for example, don't forget to check that the peak heights match the 1:3:3:1 ratio indicated by Pascal's triangle. It may be that the resonance is in fact two distorted (roofed) doublets with similar chemical shifts.

    Note groups of peaks that may be indicative of specific chemical groups such as ethyl or isopropyl, but consider these assignments as tentative at the initial stages of analysis. Also look for very distinctive splittings, e.g. the rather large ^{1}H–^{1}H couplings associated with trans alkenes.

7.4 The ^{13}C NMR spectrum

From the ^{13}C NMR spectrum identify ...

a. The number of resonances.

> *Look for peaks that belong to the solvent—these will appear as multiplets due to their coupling with deuterium. Consider it possible that peaks of interest may be hidden within the solvent signal due to accidental overlap.*

> *It may be useful to note any particularly small or large peaks, especially in the aromatic region. This may give a clue as to the degree of substitution of the carbon atom in question, i.e. non-protonated carbons often display lower intensities.*

b. The chemical shifts of the resonances.

> *Make a note of whether the resonances are in the saturated or unsaturated part of the spectrum, or if they are suggestive of a carbonyl group.*

7.5 The degree of unsaturation and double bond equivalents

Identify the degree of unsaturation and double bond equivalents.

Data from the mass and NMR spectra can lead to the identification of a molecular formula for the molecule in question. This can, by simple calculation using the equation below, lead to what is known as 'degree of unsaturation' (DU) or 'double bond equivalent' (DBE), and represents the number of hydrogen molecules needed to reduce every unsaturated bond to a saturated one, and all rings to acyclic structures.

$$\text{Degree of unsaturation}\left(DU\right)=\frac{\left(2C+2\right)+N-X-H}{2}$$

in which C is the number of carbon atoms, N the number of nitrogen atoms, X the number of halogen atoms, and H the number of hydrogen atoms. Oxygen atoms are **not** included.

Thus, one double bond equivalent (DBE) would indicate that the molecule has one ring (e.g. cyclohexane) **or** one π bond (e.g. hexene). Two DBEs could be two rings, two double bonds or one triple bond, or one ring with a double bond (e.g. cyclohexene). Benzene thus requires four DBEs: three for the 'double bonds' and one for the ring (Figure 7.1).

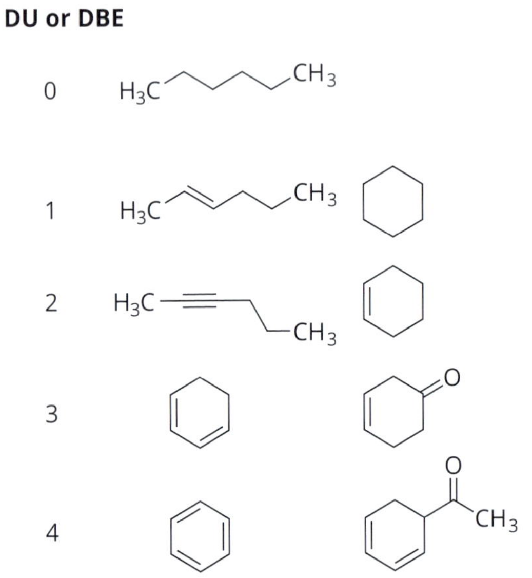

> The degree of unsaturation is a quick way of determining how many π bonds and/or rings are present in a molecule.

Whilst the number of π bonds and rings does not provide an immediate answer to a spectroscopic problem, this information can help point you in a particular direction, interpret data from the NMR spectrum, and ultimately to cross check with a possible structural solution.

DU or DBE

FIGURE 7.1 Examples of degrees of unsaturation (DU) or double bond equivalents (DBE).

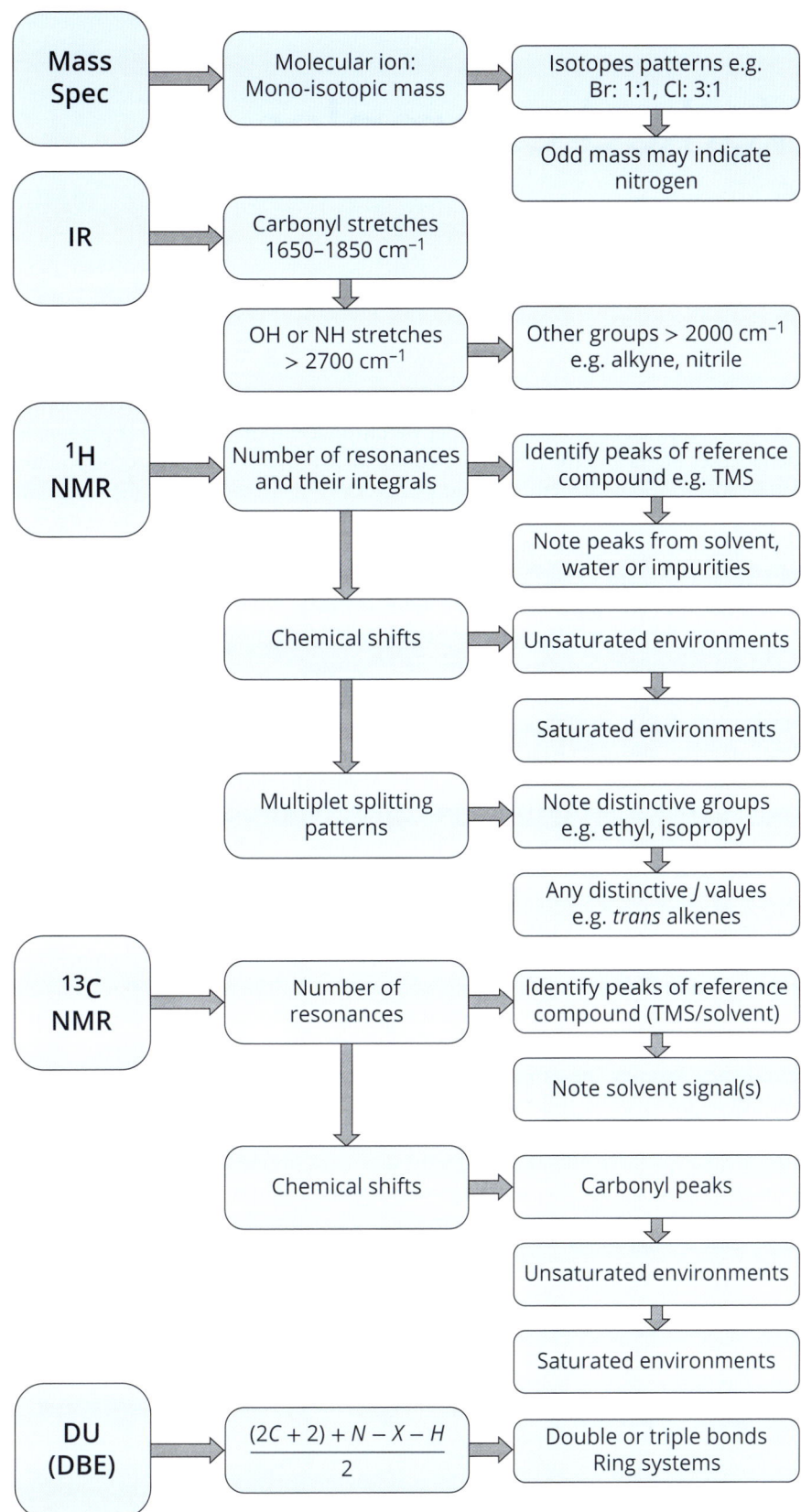

FIGURE 7.2 Analysis checklist flowchart. This overview provides a systematic and consistent approach to spectrum analysis with which to begin structure determination problems.

CHAPTER 8

Worked examples

The following worked examples illustrate the systematic approach for the analysis of spectroscopic and spectrometric data that we introduce in this book. This pragmatic approach focuses on ^{1}H and ^{13}C NMR spectra supported by data from the corresponding IR and mass spectra. Some structural assignments can be made with NMR data alone whilst others require data from multiple techniques. A few examples are more challenging, and we will find that a conclusive structure assignment is not possible using only the techniques we have covered so far. In Chapters 10 and 12, we will introduce additional NMR and mass spectrometry techniques that can help in such cases.

The spectra for the problems in this book have all been recorded from authentic samples and are presented in a consistent format. Occasionally this has led to 'shoulders' on the side of peaks, due to pixellation of the traces, perhaps inferring a splitting. In all cases, expansions are used to show true splitting patterns observed, which means that any signal without an expansion can be taken as either a singlet or undefined multiplet in the NMR data. In some cases, coupling constants are given with the multiplet expansions with the J values listed to the nearest 0.5 Hz. Integrals in proton spectra are shown as red traces with corresponding relative areas shown numerically under each multiplet. Carbon chemical shifts are tagged at the top of each peak.

For all examples, steps in the analysis process are presented sequentially as i) an *initial analysis*, followed by ii) a more ***detailed analysis*** and then finally iii) a ***conclusion*** in which the final structure is discussed. Some additional information is then provided as ***further rationalization*** to explain particular features in the spectra. Whilst interpretation of these finer points may not be essential in defining the structure (so that the further rationalization may be considered optional in the first instance), we provide these explanations to help you develop a fuller understanding of the spectroscopic data presented.

Whilst we provide this full analysis of each problem data set to guide you through the spectrum interpretation and structure determination, you may find that, as your skill develops, you may prefer to attempt to solve the problems yourself without referring to these explanations. You can then consult the guidance as needed to confirm that your own analysis is correct.

8.1 Example 1

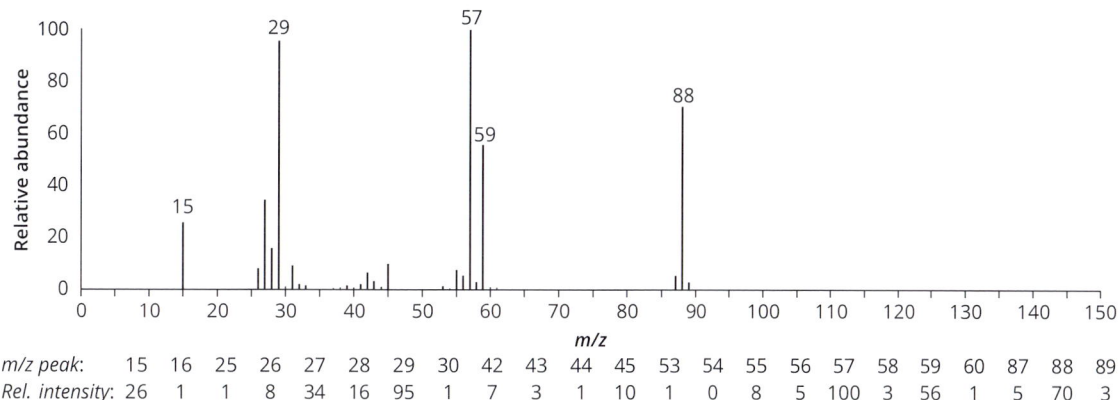

m/z peak:	15	16	25	26	27	28	29	30	42	43	44	45	53	54	55	56	57	58	59	60	87	88	89
Rel. intensity:	26	1	1	8	34	16	95	1	7	3	1	10	1	0	8	5	100	3	56	1	5	70	3

Mass spectrum

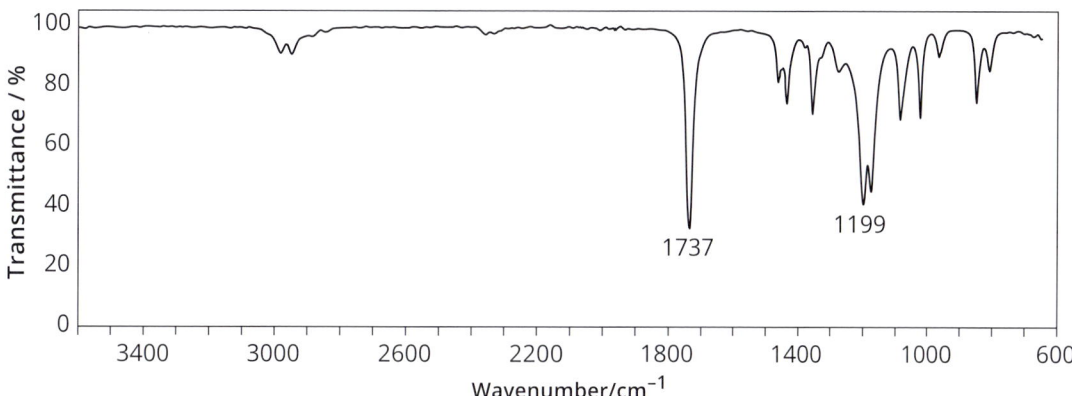

Infrared spectrum

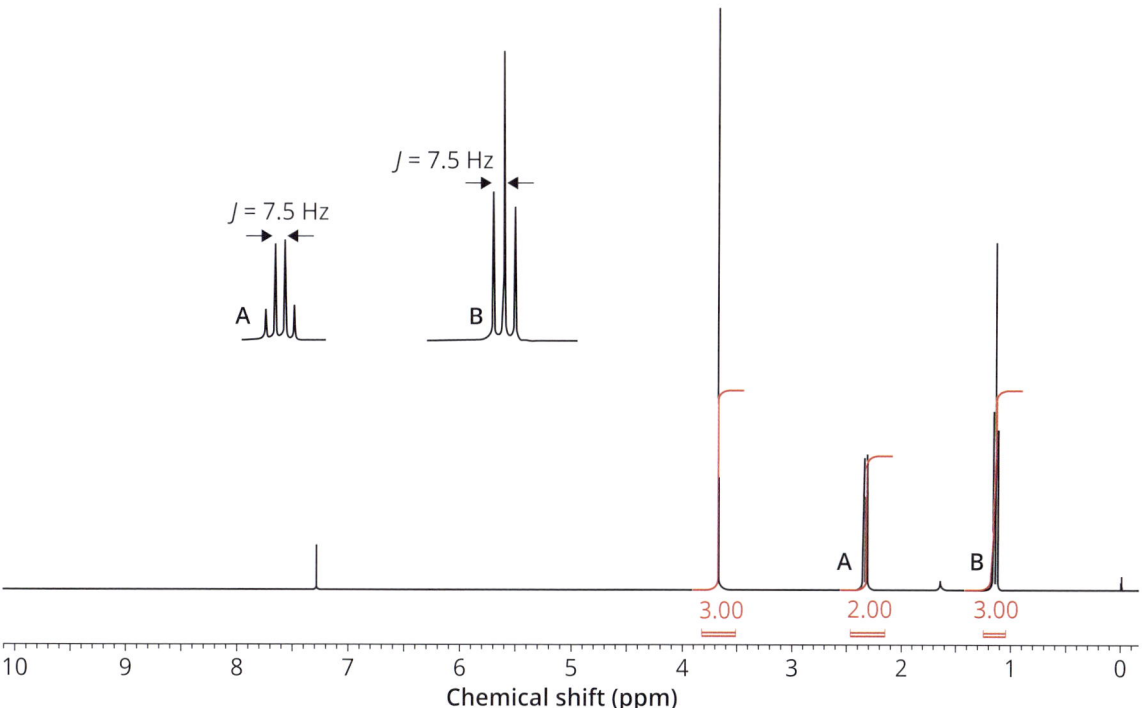

^{1}H NMR spectrum (400 MHz, CDCl$_3$)

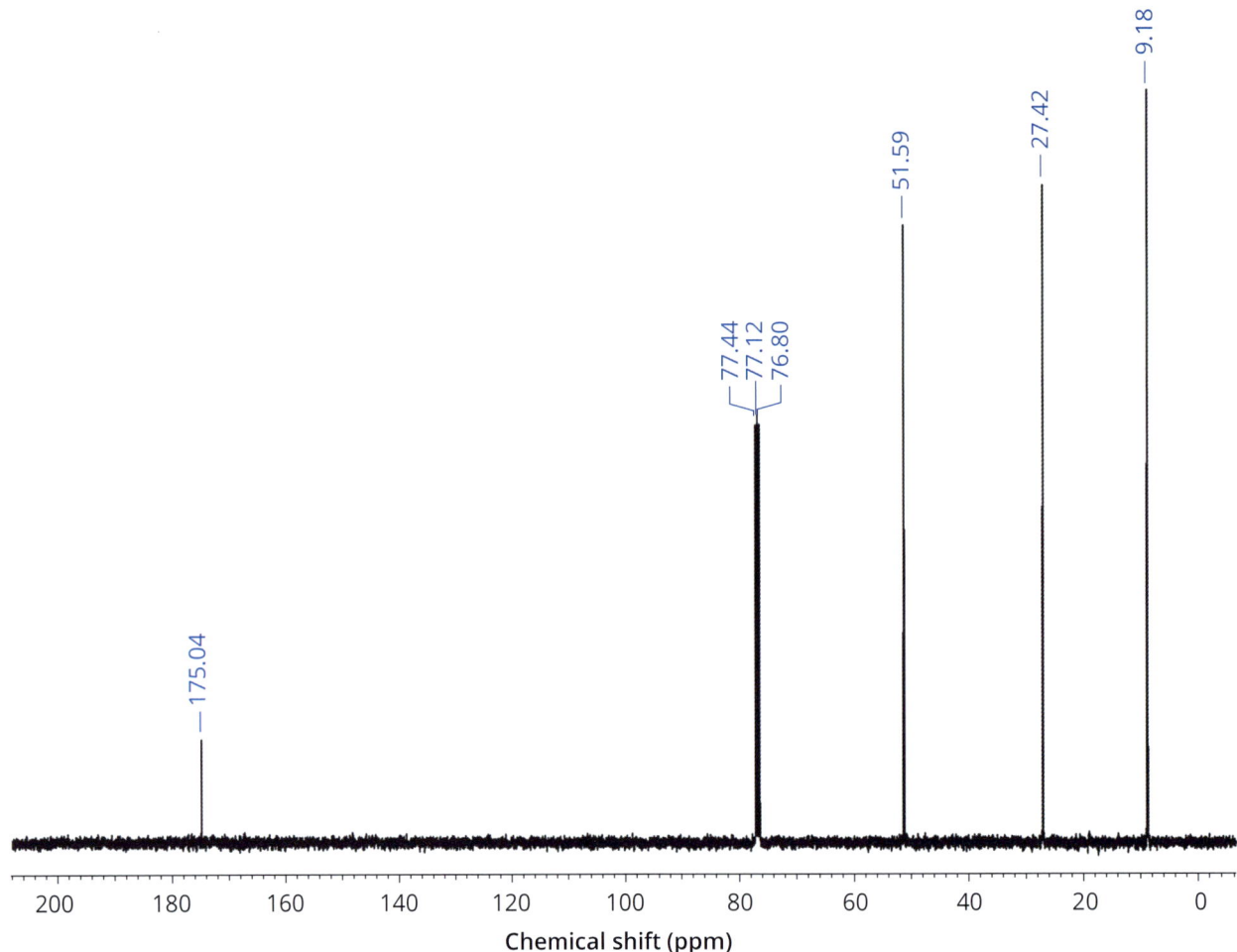

¹³C NMR spectrum (100 MHz, CDCl₃)

8.1.1 Initial analysis of Example 1

(a) The peak at *m/z* 88 in the mass spectrum likely corresponds to the molecular ion (M⁺•).

(b) The proposed molecular ion is an even integer value which, according to the nitrogen rule, indicates that either no nitrogen atoms or an even number of them (e.g. 2,4 etc) are present.

(c) The isotope pattern for the M⁺• indicates that there are no A+2 elements present (e.g. no chlorine or bromine).

(d) The number of carbon atoms in the molecular ion is predicted to be ~4 based on the relative abundances of the molecular ion and M+1 isotope peaks: e.g. (3/70)/0.011 = ~4.

(e) The infrared spectrum indicates the presence of a carbonyl group stretching frequency at ca. 1730 cm⁻¹. The type of carbonyl group cannot be determined for certain at this stage, but cross-checking with the ¹³C NMR for a signal between 165 and 220 ppm (typical of a carbonyl group) is quickly done and confirms the general assignment.

(f) The IR shows no clear-cut absorptions in the range 2500–3600 cm⁻¹ other than the ubiquitous CH stretch at ~3000 cm⁻¹.

(g) The IR shows a strong absorption around 1200 cm⁻¹ which may be due to a C–O stretch. However, this is typically the most crowded region of the spectrum, and such early assignments must therefore always be considered with great caution.

(h) The ^{1}H NMR shows three aliphatic signals, ignoring the trace $CHCl_3$ at 7.26 ppm and TMS at 0.00 ppm.

(i) The ratio of signal integrals is 3:2:3, which indicates the presence of eight hydrogen atoms (or a multiple).

(j) The ^{13}C NMR shows four signals—remember not to include the 1:1:1 triplet at 77 ppm due to the solvent $CDCl_3$—and the chemical shift ranges indicate three saturated and one unsaturated carbon environments. This matches the four carbon atoms predicted from the mass spectrum.

(k) The composition accounted for at this stage is C_4H_8O, which would provide an m/z value of 72 in the mass spectrum.

(l) From the conclusions in (a) the 'missing' mass is 16. This is likely to be due to an additional oxygen atom, and so the predicted molecular formula is $C_4H_8O_2$.

(m) From (l) DU = 1, thus there is one double bond present in the molecule. The C=O group already identified in (e) would account for the degree of unsaturation calculated.

8.1.2 Detailed analysis of Example 1

(n) ^{1}H NMR: There is a 3H singlet at 3.7 ppm. The 3H singlet is classically identified with an isolated methyl group. The chemical shift is indicative of a CH_3 group in an electron-withdrawing environment, and δ values around 4 ppm point towards it being an OCH_3.

(o) ^{1}H NMR: The splitting patterns of the 2H quartet at 2.3 ppm and the 3H triplet at 1.1 ppm, when taken together, indicate that an ethyl group is present, displaying a typical alkyl vicinal coupling constant of 7.5 Hz. The chemical shift of the quartet hints at the ethyl group being attached to a mildly electron-withdrawing group.

(p) ^{13}C NMR: There are three aliphatic signals at 9.2, 27.4, and 51.6 ppm. The chemical shifts of these signals mirror those observed in the ^{1}H NMR spectrum at 1.1, 2.3, and 3.7 ppm.

(q) ^{13}C NMR: A signal from a deshielded nucleus is seen at 175.0 ppm. This is consistent with a carbonyl group and confirms the deduction from the IR spectrum made above.

The analysis of the data indicates that the following structural elements are present, and these account for all of the observed proton and carbon signals in the NMR spectra.

1   **2**   **3**

8.1.3 Conclusion and final structure of Example 1

The molecular components identified match to the predicted molecular formula $C_4H_8O_2$ indicated by the m/z value of 88. Molecular components **2** and **3**, each with one unfilled valency, can be seen as 'edge pieces' in the molecular jigsaw, and the core of the molecule must therefore be made from fragment **1**. The molecular structure can be pieced together to give methyl propionate (**EG1**).

EG1

8.1.4 Further rationalization of Example 1

The fragment ion peaks at *m/z* 57 and *m/z* 59 in the mass spectrum correspond to the even-electron cation fragments produced via loss of CH_3O (**2**) and CH_3CH_2 (**3**) radicals respectively. This results from α-cleavage on either side of the carbonyl group. The *m/z* 59 peak results from loss of the alkyl group which is common for methyl esters. The prominent peak at *m/z* 29 results from inductive cleavage around the carbonyl group.

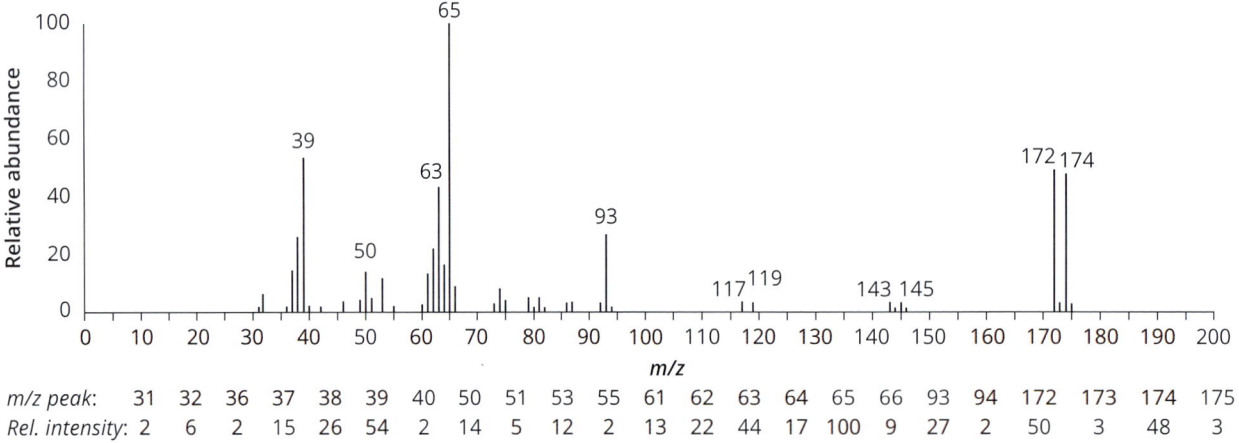

m/z 88   m/z 57   m/z 88   m/z 59

Note that, in the NMR spectrum, we can recognize the carbonyl chemical shift at 175 ppm as being indicative of the ester group in particular as this shift is too low for either a ketone or aldehyde group, both of which would be expected to occur above 190 ppm. We can also note that the proton shift of the ethyl CH_2 group at 2.3 ppm is consistent with it being adjacent to the carbonyl functionality that is responsible for the mild electron-withdrawing (deshielding) effect.

The proposed structure is also consistent with the intense and very obvious IR stretch at 1199 cm^{-1} which can be attributed to the single-bond C–O stretch of the ester group. This presents an example where a distinctive stretch that appears within the fingerprint region may be correlated with a functional group.

8.2 Example 2

m/z peak:	31	32	36	37	38	39	40	50	51	53	55	61	62	63	64	65	66	93	94	172	173	174	175
Rel. intensity:	2	6	2	15	26	54	2	14	5	12	2	13	22	44	17	100	9	27	2	50	3	48	3

Mass spectrum

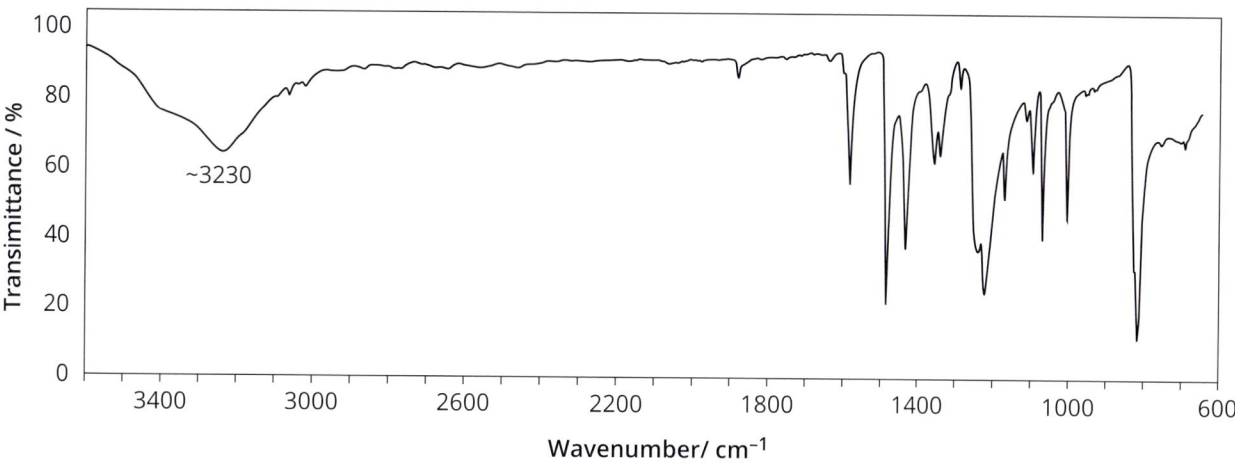

Infrared spectrum

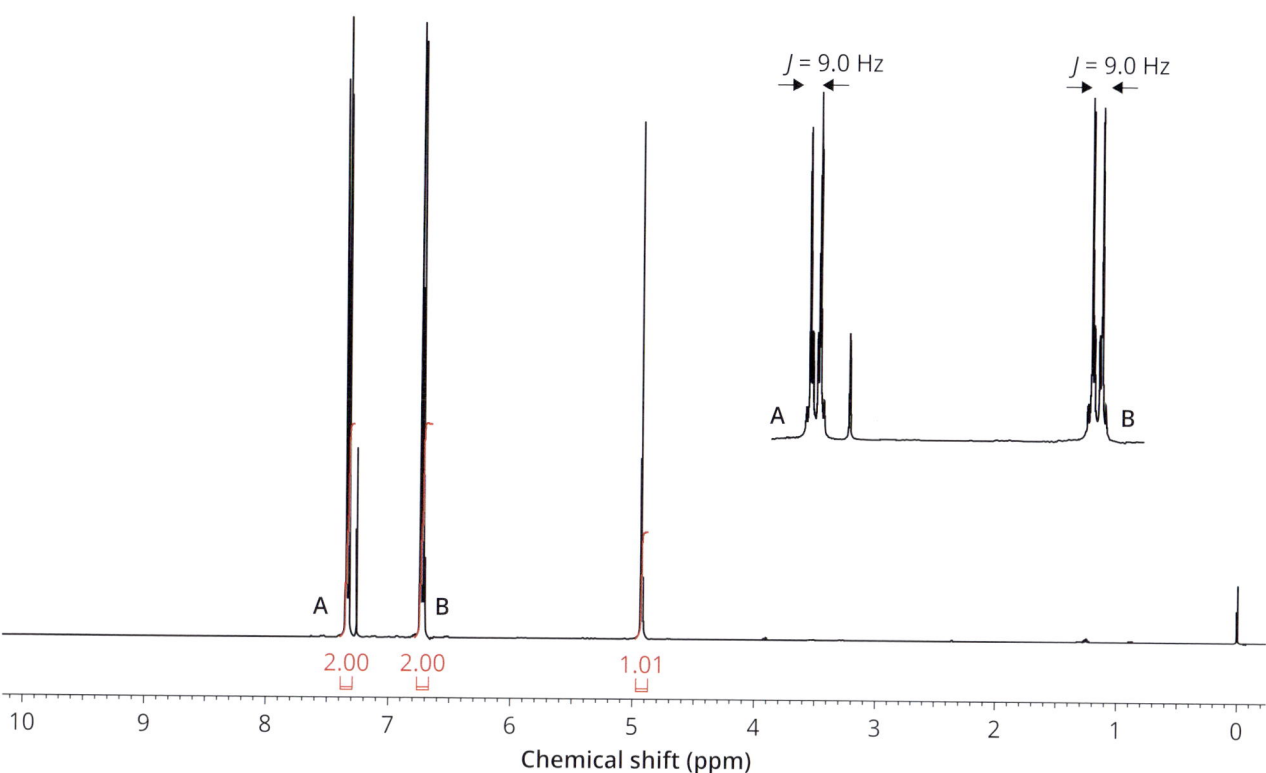

¹H NMR spectrum (400 MHz, CDCl₃)

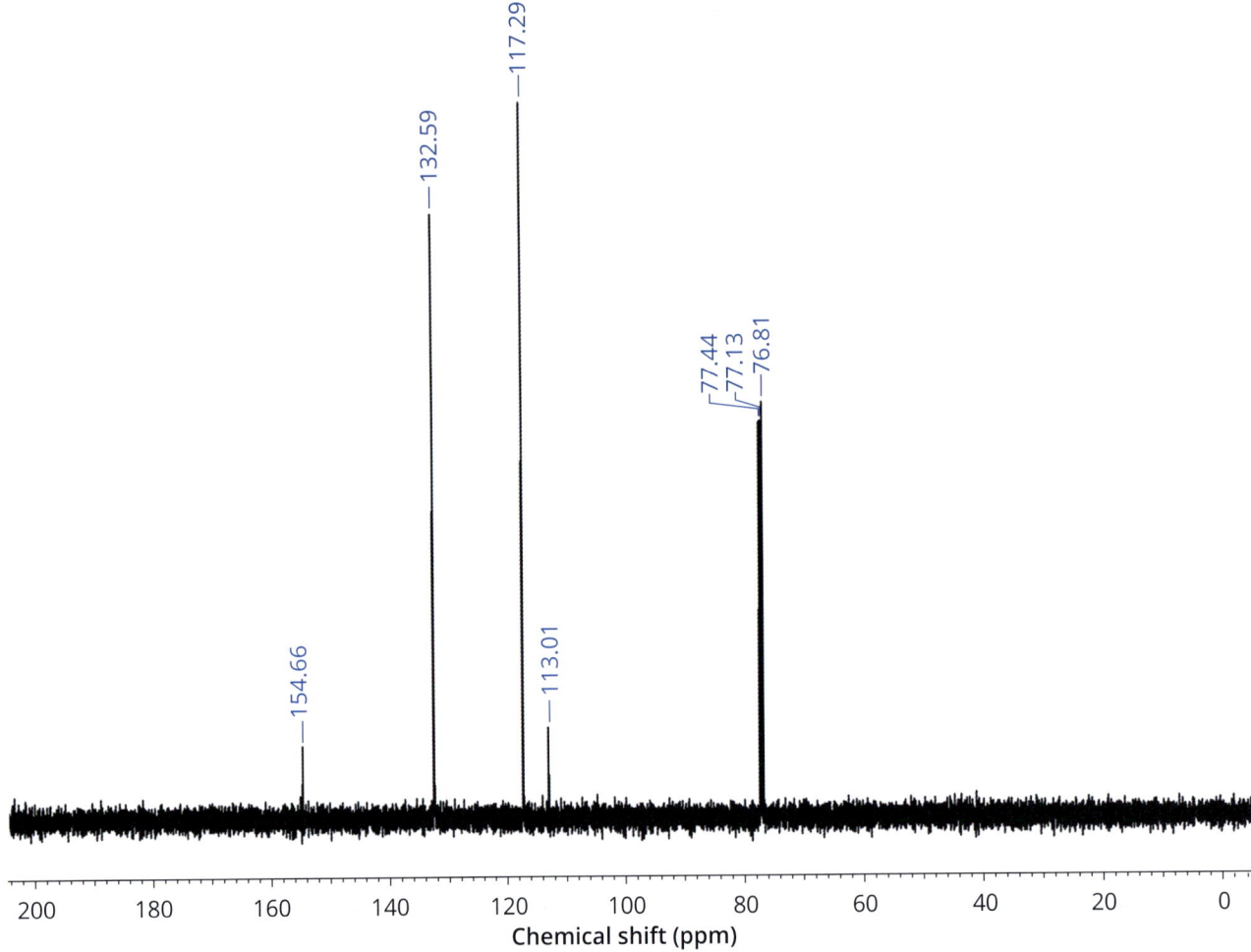

^{13}C NMR spectrum (100 MHz, CDCl$_3$)

8.2.1 Initial analysis of Example 2

(a) The two peaks at *m/z* 172 and *m/z* 174 in the mass spectrum may correspond to isotope peaks of the molecular ion (M$^{+\cdot}$).

(b) The 1:1 ratio of these peaks, spaced two mass units apart, indicates the presence of one bromine atom in the molecule.

(c) The molecular ion peaks are both even integer values, which suggests that either no nitrogen atoms, or an even number, are present.

(d) The number of carbon atoms in the molecular ion is predicted to be approximately 6 based on the relative abundances of the molecular ion isotope peaks: e.g. (3/48)/0.011 = 6 (rounded to nearest integer).

(e) The infrared spectrum shows a broad absorption around 3100–3400 cm^{-1}, but there is no strong absorption in the carbonyl region 1650–1850 cm^{-1}. This absorption indicates the presence of an OH group, but there is no carbonyl absorption present, which means that a carboxylic acid is not likely (this can be definitively ruled out by looking at the ^{13}C NMR spectrum).

(f) The ^{1}H NMR shows three signals, ignoring the singlets for trace CHCl$_3$ at 7.26 ppm and TMS at 0.00 ppm.

(g) The ratio of signal integrals in the ^{1}H NMR spectrum is 2:2:1, which indicates the presence of five hydrogen atoms (or a multiple).

(h) The ^{13}C NMR shows four signals and, based on the chemical shifts, all of these are in unsaturated carbon environments. Again, remember not to count the 1:1:1 triplet at 77 ppm due to the solvent CDCl$_3$.

(i) The composition accounted for at this stage is C$_4$H$_5$OBr, which adds to 148 for ^{79}Br (or 150 for ^{81}Br).

(j) From conclusion (a) and (i), the 'missing' mass is 24. Given that the number of ^{13}C signals is indicative of carbon environments and not carbon atoms in total, this could be due to two extra carbon atoms. The predicted molecular formula is therefore C$_6$H$_5$OBr.

(k) From (j) DU = 4 (don't forget to include the halogen in the calculation). From the ^{1}H NMR this is a good indicator of the presence of a benzene ring.

8.2.2 Detailed analysis of Example 2

(l) ^{1}H NMR: There are two 2H signals at 6.7 and 7.4 ppm. The chemical shift of these signals is characteristic of unsaturated CH signals from aromatic hydrogen environments. The symmetrical nature of these two 2H signals, as seen clearly in the expansion, is typical of the roofing pattern seen for a 1,4-disubstituted aromatic ring with a slightly larger $^3J_{HH}$ coupling of 9 Hz.

(m) ^{1}H NMR: There is a sharp 1H singlet at 4.9 ppm. The chemical shift of this uncoupled signal reveals a proton in a very electron-deficient environment, although not within the region of unsaturated CH signals typical of alkenes or aromatics. Based on the information from the infrared spectrum, and the fact that this is the last ^{1}H NMR signal to be assigned, it could be due to an OH signal. It is important to note that the broadness of exchangeable OH or NH signals in NMR may vary a lot depending on their inter- or intramolecular hydrogen bonding, which in turn may depend on the concentration of the sample and the solvent used. Thus, OH signals do not necessarily appear as broadened peaks.

(n) ^{13}C NMR: There are four signals between 113.0 and 154.7 ppm. The chemical shifts of these signals are typical of aromatic carbon atoms, and the symmetry of a 1,4-disubstituted aryl ring [see (l)] would account for all of these peaks. Whilst signals cannot be integrated in a standard ^{13}C spectrum, quaternary carbon atoms typically give lower intensity signals and thus the signals at 113.0 and 154.7 ppm may be assigned as belonging to the quaternary carbons.

The analysis of the data indicates that the following structural elements are present, and these account for all of the observed proton and carbon signals in the NMR spectra.

1   **2**   **3**

8.2.3 Conclusion and final structure of Example 2

The structural elements identified match the predicted molecular formula of C$_6$H$_5$OBr and give isotopic masses of 172 and 174. Structural elements **1** and **3**, each with one unfilled valency, can be seen as 'edge pieces' in the molecular jigsaw, and the core of the molecule must therefore be made from fragment **2**. The molecular structure can be pieced together to give 4-bromophenol (**EG2**).

EG2

8.2.4 Further rationalization of Example 2

The peak at *m/z* 93 corresponds to loss of a bromine radical via inductive cleavage. The remaining mass of the molecule, and relatively high abundance molecular ion, suggest that a single aromatic ring is present in the structure. The absence of a peak at *m/z* 91, corresponding to a tropylium ion, suggests that there is no alkyl substitution. The lower abundance peaks at *m/z* 143 and *m/z* 145 provide a clue, corresponding to a loss of 29. This represents a loss of COH from the molecular ion, which is characteristic of aromatic phenols. Taken together, the fragments in the mass spectrum provide supporting evidence for a bromine-substituted aromatic phenol.

From the NMR data we can understand the differences in the carbon chemical shifts for the aromatic ring by considering the influence of the attached heteroatoms. The highest shift at 154.7 ppm is caused by the strong electron-withdrawing (deshielding) effect of the directly attached oxygen, causing this carbon to stand away from the others. Conversely, the adjacent carbon is significantly shielded and so appears at only 117.4 ppm due to the resonance delocalization of the oxygen lone pair into the aromatic ring, as below. The same resonance shielding explains the lower shift of the corresponding proton resonance at 6.7 ppm. The carbon bearing the bromine is shielded directly by its large electron cloud, and so also appears at a lower chemical shift of 113.0 ppm (this is sometimes referred to as the 'heavy atom' effect).

8.3 **Example 3**

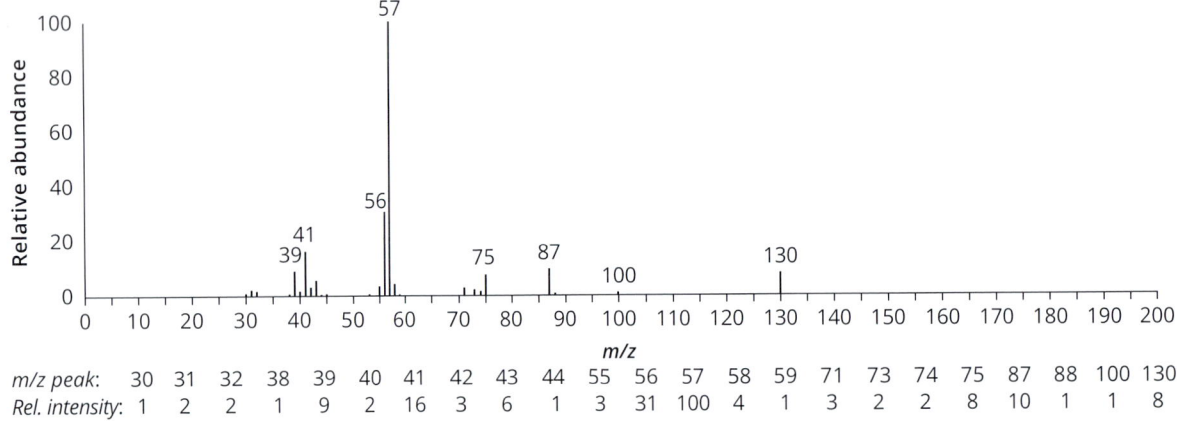

Mass spectrum

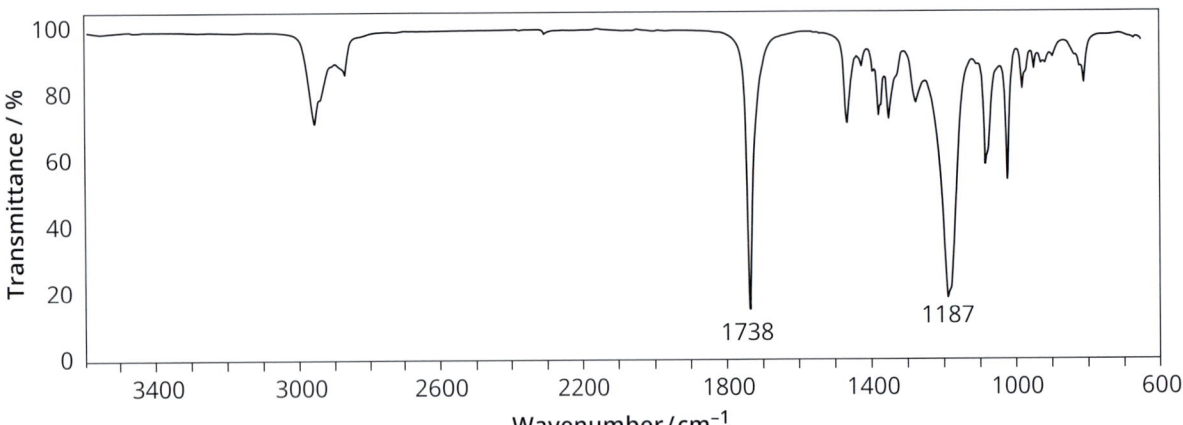

Infrared spectrum

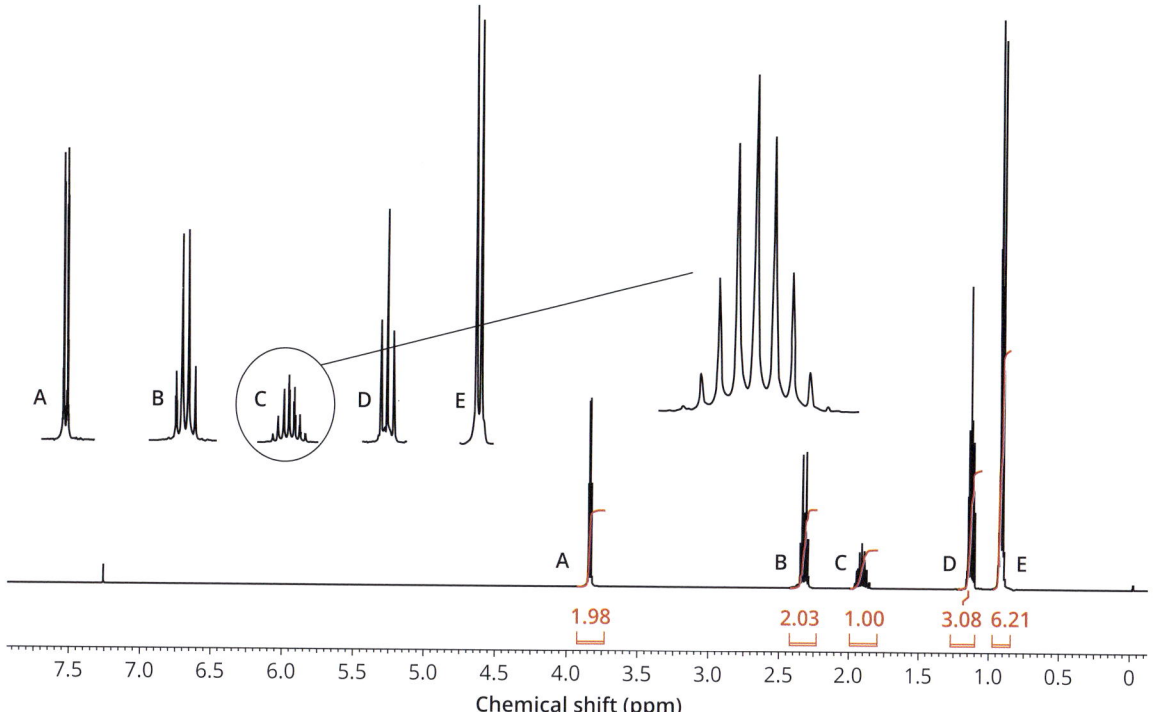

^{1}H NMR spectrum (400 MHz, CDCl$_3$)

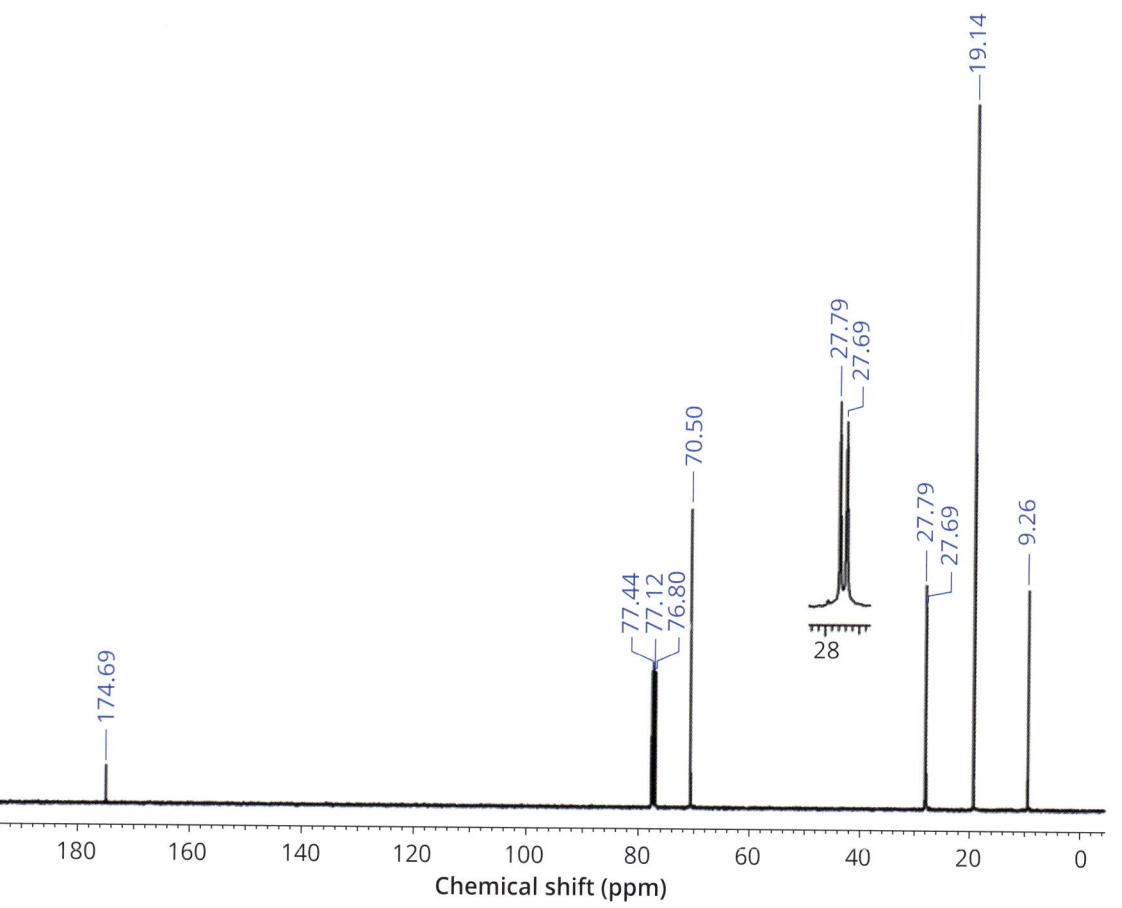

^{13}C NMR spectrum (100 MHz, CDCl$_3$)

8.3.1 Initial analysis of Example 3

(a) The peak at m/z 130 in the mass spectrum is likely to correspond to the molecular ion $(M^{+\bullet})$.

(b) An even $M^{+\bullet}$ indicates that either no nitrogen atoms or an even number of them (e.g. 2,4 etc) are present in the compound.

(c) A lack of M+2 isotope peaks for $M^{+\bullet}$ indicates that neither chlorine nor bromine is present.

(d) The relatively low abundance of the molecular ion suggests that it is unlikely to be aromatic.

(e) The infrared spectrum indicates the presence of a carbonyl group stretching frequency at ca. 1730 cm^{-1}. The type of carbonyl group cannot be determined for certain at this stage, but cross-checking with the ^{13}C NMR for a signal between 165 and 220 ppm (typical of a carbonyl group) is quickly done and confirms the general assignment.

(f) The IR shows a strong absorption around 1200 cm^{-1} which may be due to a C–O stretch. However, this is typically the most crowded region of the spectrum, and such early assignments must therefore always be considered with great caution.

(g) The IR shows no clear-cut absorptions in the range 2500–3600 cm^{-1} other than the ubiquitous CH stretch at ~3000 cm^{-1}.

(h) The ^{1}H NMR shows five signals, if we ignore the trace CHCl$_3$ at 7.26 ppm and TMS at 0.00 ppm.

(i) The ratio of signal integrals is 2:2:1:3:6, which indicates the presence of fourteen hydrogen atoms (or a multiple).

(j) The ^{13}C NMR shows six signals—remember not to include the 1:1:1 triplet at 77 ppm due to the solvent CDCl$_3$. The chemical shift ranges indicate one unsaturated (carbonyl) and five saturated carbon environments.

(k) The composition accounted for at this stage is $C_6H_{14}O$, which adds to 102 g/mol.

(l) From conclusions (a) and (k), we can derive that the difference in mass is 28 (130–102). This is likely to be due to an additional C+O or 2×N. If it is a C+O, then the lack of an eighth signal in the ^{13}C NMR spectrum means that two atoms are equivalent. Thus, the predicted molecular formula is either $C_7H_{14}O_2$ or $C_6H_{14}N_2O$.

(m) From (l), either formula gives DU = 1, thus there is one double bond present in the molecule [the C=O identified in (e)].

8.3.2 Detailed analysis of Example 3

(n) ^{1}H NMR: The spectrum reveals five aliphatic signals: a 2H doublet at 3.7 ppm, a 2H quartet at 2.3 ppm, a 1H nonet at 1.9 ppm, a 3H triplet at 1.1 ppm, and a 6H doublet at 0.9 ppm. It is important to note that signals with many coupling nuclei can be easily mis-assigned unless a detailed expansion is examined. In this case, the nonet shows a 1:8:28:56:70:56:28:8:1 pattern in which the outer lines are hardly discernible.

(o) ^{13}C NMR: The signal at 174.7 ppm is typical of a carbonyl carbon, leaving five aliphatic signals. It is therefore likely that these five carbons bear the five types of aliphatic hydrogen atom already seen in the ^{1}H NMR spectrum. The deshielded signal at 70.5 is indicative of a carbon atom attached to an electron-withdrawing group, likely an oxygen atom, and thus could be associated with the doublet at 3.7 ppm. The molecular formula containing two oxygen atoms is therefore now favoured $(C_7H_{14}O_2)$,

(p) ^{1}H NMR: In cases where there are several signals, it is often easiest to start analysis at the very right-hand side of the spectrum. These shielded signals are most likely to represent the end of an aliphatic group or chain and are thus a sensible place to start.

(q) ^{1}H NMR: The 6H doublet at 0.9 ppm is characteristic of the two equivalent methyl groups in an isopropyl group [–CH(CH$_3$)$_2$]. The corresponding 1H signal could therefore be the nonet at 1.9 ppm, which in turn will need to be coupled to another two hydrogen atoms for the correct splitting pattern.

(r) ^{1}H NMR: The 3H triplet is characteristic of a methyl group next to two identical protons, i.e. –CH$_2$CH$_3$. The corresponding signal with compatible integration and multiplicity is the quartet seen at 2.3 ppm.

(s) ^{1}H NMR: Four signals appear to have been joined as two pairs, leaving the 2H signal at 3.7 ppm. This signal is a doublet implying a CH$_2$ with a single, neighbouring coupling nucleus. This CH$_2$ must clearly be connected to the CH giving rise to the 1H nonet at 1.9 ppm. As noted above, the chemical shift is characteristic of a CH$_2$ attached to a strongly electron-withdrawing group and is suggestive of a –OCH$_2$– group.

The analysis of the data indicates that the following structural elements are present, and these account for all of the observed proton and carbon signals in the NMR spectra.

1   **2**   **3**

8.3.3 Conclusion and final structure of Example 3

The structural elements identified match the predicted molecular formula and weight of C$_7$H$_{14}$O$_2$ and 130 g/mol respectively. The carbonyl group is revealed as an ester. The molecular structure can be pieced together to give isobutyl propionate (**EG3**).

EG3

8.3.4 Further rationalization of Example 3

Note that we can now recognize the carbonyl chemical shift at 174.7 ppm as being indicative of the ester group in particular, as this shift is too low for either a ketone or aldehyde group, both of which would be expected to occur above 190 ppm. The carbon at 70.5 ppm together with the proton at 3.7 ppm arise from the significant deshielding associated with neighbouring oxygen atoms, and such shift combinations are often indicative of oxygen substitution (although bear in mind that precise values will vary!).

The low abundance of the molecular ion in the mass spectrum, and its even m/z value, support the assumption that the compound is a non-aromatic structure that does not contain nitrogen. The fragmentation mass spectrum appears to be relatively simple with a base peak at m/z 57 which characterizes loss of [C$_2$H$_5$CO] via α-cleavage around the carbonyl of the ester leading to the larger of the two possible radical losses. This provides evidence in support of the EG3 structural assignment. The peak at m/z 87 represents α-cleavage initiated via the oxygen of the ester linkage.

The proposed structure is also consistent with the intense and very obvious IR stretch at 1187 cm^{-1} which can be attributed to the single-bond C–O stretch of the ester group. This presents another example where a distinctive stretch appearing within the fingerprint region may be correlated with a functional group.

3.7 ppm and 70.5 ppm

m/z 130 m/z 57

m/z 130 m/z 87

8.4 **Example 4**

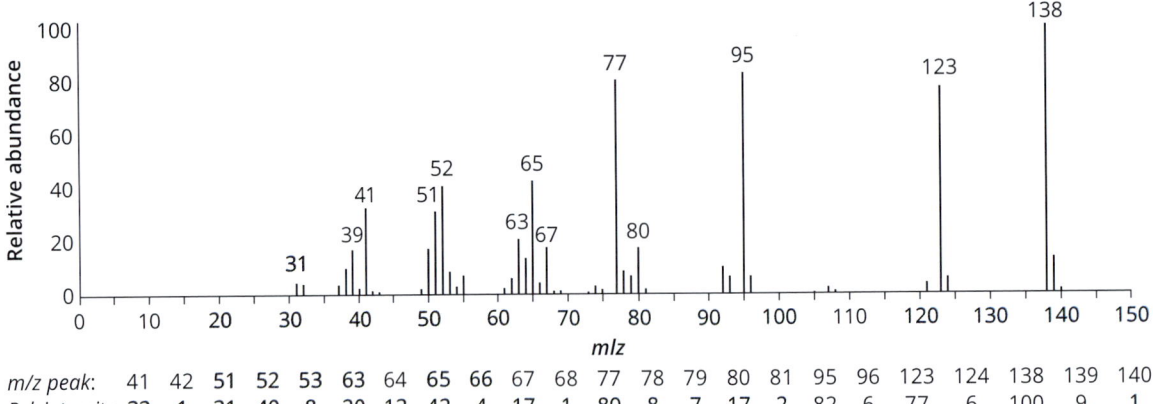

m/z peak:	41	42	51	52	53	63	64	65	66	67	68	77	78	79	80	81	95	96	123	124	138	139	140
Rel. intensity:	32	1	31	40	8	20	13	42	4	17	1	80	8	7	17	2	82	6	77	6	100	9	1

Mass spectrum

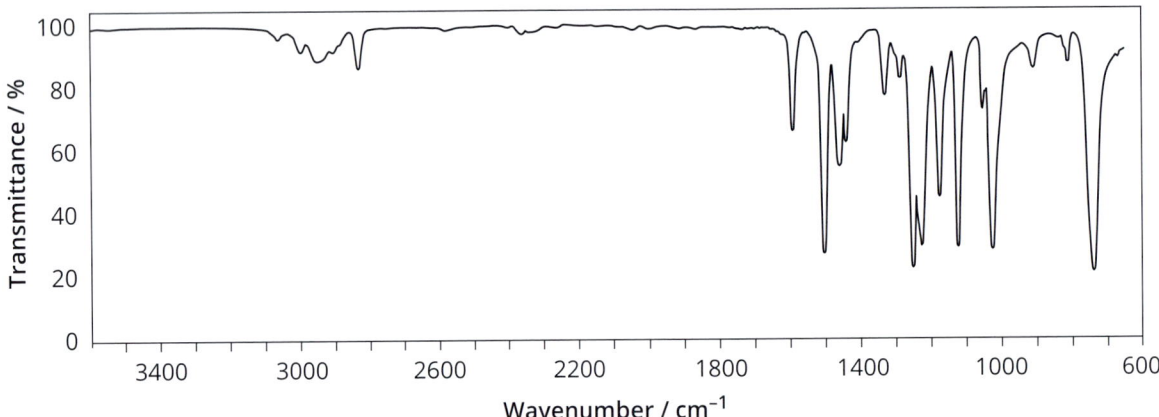

Infrared spectrum

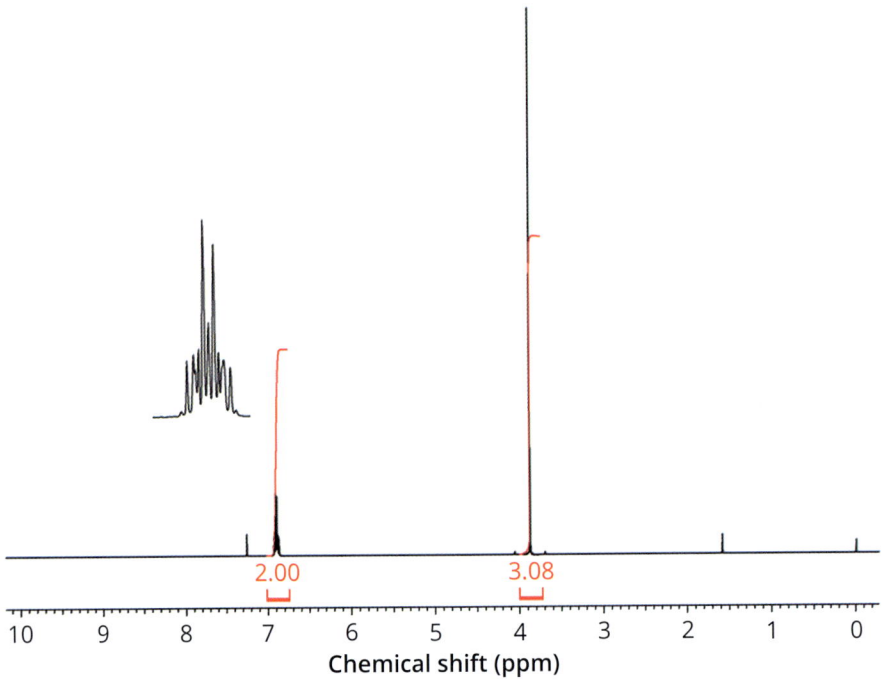

¹H NMR spectrum (400 MHz, CDCl₃)

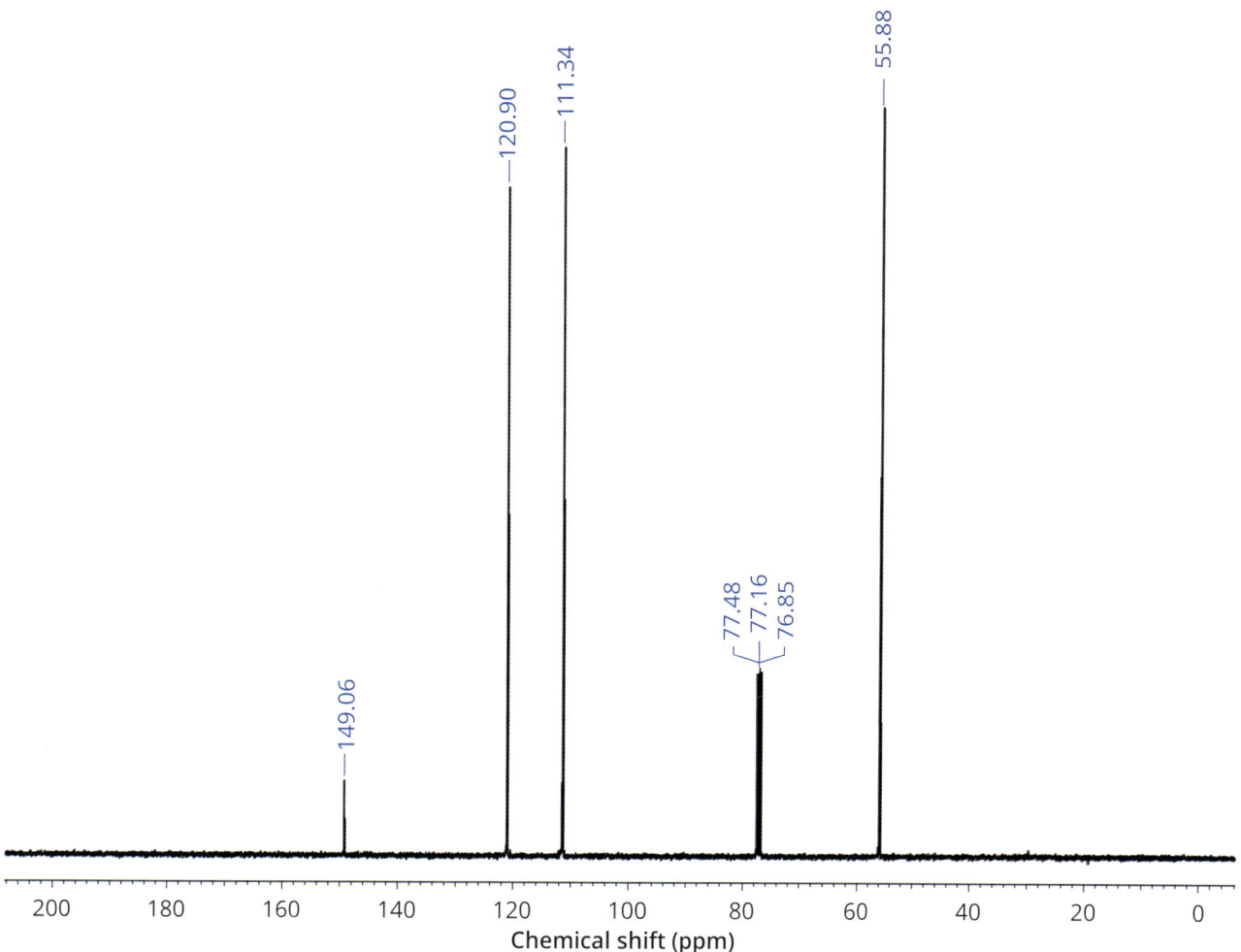

¹³C NMR spectrum (100 MHz, CDCl₃)

8.4.1 Initial analysis of Example 4

(a) The abundant peak at *m/z* 138 in the mass spectrum is likely to represent the molecular ion (M⁺˙).

(b) An even M⁺˙ indicates that either no nitrogen atoms or an even number of them (e.g. 2,4 etc) are present in the compound.

(c) The isotope pattern for M⁺˙ indicates that neither chlorine nor bromine is present.

(d) The high abundance of the proposed molecular ion suggests that the compound may be aromatic.

(e) Based on the relative abundance of M and M+1 peaks in the mass spectrum, the estimated number of carbon atoms in the molecular ion is 8.

(f) In the infrared spectrum there is no strong absorption in the carbonyl region 1650–1850 cm⁻¹; the molecule therefore does not contain a carbonyl group.

(g) The IR shows no clear-cut absorptions in the range 2500–3600 cm⁻¹ other than the ubiquitous CH stretch at 3000 cm⁻¹.

(h) The ¹H NMR shows only two signals which are a multiplet and a singlet—if we ignore the trace CHCl₃ at 7.26 ppm, a small water impurity at 1.6 ppm, and TMS at 0.00 ppm.

(i) The ratio of signal integrals is 2:3, which indicates the presence of five hydrogen atoms (or a multiple).

(j) The ^{13}C NMR shows four signals—remember not to include the 1:1:1 triplet at 77 ppm due to the solvent $CDCl_3$. The chemical shift ranges indicate three unsaturated and one saturated carbon environments.

(k) The constitution accounted for at this stage is only C_4H_5, which adds to just 53 g/mol. Clearly there is a lot of 'missing' mass as this is less than half of the observed mass of the molecular ion. This indicates a molecule with a large degree of symmetry and/or additional atoms not seen in the NMR spectra.

8.4.2 Detailed analysis of Example 4

(l) ^1H NMR: There is a singlet at 3.9 ppm. The uncoupled signal integrates to 3H, but could well be 6H—see (k) above—indicative of an isolated methyl group or groups. The chemical shift is indicative of a CH_3 group in an electron-withdrawing environment and a δ value around 4 ppm points towards it being an OCH_3. There may be one or two identical methoxy groups in this molecule.

(m) ^1H NMR: There is a 2H multiplet around 6.9 ppm. This chemical shift is in the region of unsaturated CH signals, but as the signal could be worth 4H a disubstituted benzene ring is a possibility.

(n) ^{13}C NMR: There is one aliphatic signal at 55.9 ppm. The chemical shift of this signal is supportive of the assignment of a methoxy group in the ^1H NMR spectrum.

(o) ^{13}C NMR: Three signals are seen at 149.1, 120.9, and 111.3 ppm that are consistent with the assignment of unsaturated CH signals in the ^1H NMR spectrum.

(p) At this stage the simplicity of the apparent NMR spectra with an m/z for the molecular ion of 138 strongly indicates a molecule with a degree of symmetry. The first estimate of the molecular constitution, C_4H_5, may therefore be doubled to C_8H_{10}, or 106. The missing mass is 32 (138–106), which is likely to be two oxygen atoms in two identical methoxy groups [see (l)] and fits conclusions from the mass spectra in (a) and (e).

(q) For $C_8H_{10}O_2$, the DU = 4. This is further indication of the presence of a benzene ring, as derived from the ^1H NMR and suggested by the high abundance of the molecular ion in the mass spectrum.

(r) If the assumption that the molecule contains a disubstituted benzene ring is correct, then the ^{13}C NMR spectra rules out a 1,4 or 1,3-substitution pattern. In neither of these cases would only three aromatic carbon signals/environments be observed. This points towards a 1,2-disubstituted benzene ring; the proton shifts are too similar in this case to allow interpretation of the overlapped multiplet patterns.

The analysis of the data indicates that the following structural elements are present.

1   **2**   **3**

8.4.3 Conclusion and final structure of Example 4

The two identical methoxy groups and the symmetry of the aromatic ring account for all of the observed proton and carbon signals in the NMR spectra. They also give an additive molecular formula of $C_8H_{10}O_2$ (138 g/mol) which matches the molecular ion in the mass spectrum. Structural elements **1** and **3**, each with one unfilled valency, can be seen as 'edge pieces' in the molecular jigsaw, and the core of the molecule must therefore be made from fragment **2**. The molecular structure can be pieced together to give 1,2-dimethoxybenzene (EG4).

EG4

8.4.4 Further rationalization of Example 4

It is possible to rationalize the different aromatic carbon chemical shifts by considering the influence of the attached oxygen atoms of the OMe groups. The strong electron-withdrawing (deshielding) effect of oxygen moves the attached carbon to a large chemical shift at 149.4 ppm. However, the resonance donation of the oxygen lone pair into the aromatic ring will lead to shielding of the adjacent carbon, causing this to appear at a much smaller shift of 111.3 ppm.

The proton multiplet has a rather complex appearance due to overlap of all the proton resonances. The fact that these couple to one another causes additional distortions of the multiplet structures (along the lines of those discussed in section 2.3.3), making their direct interpretation essentially impossible.

The high intensity molecular ion peak in the mass spectrum suggests an aromatic compound, and this is supported by the fragment peak at *m/z* 77, which is indicative of a phenyl cation derived from α-cleavage of a substituted aromatic compound. The fragmentation spectrum is relatively complex at lower masses (e.g. including *m/z* 65, a diagnostic peak for a substituted benzene ring). The presence of the fragment peak at *m/z* 123 corresponds to loss of a methyl radical. Given the high abundance of this fragment, it is likely to be bound to an electronegative atom that facilitates fragmentation via α-cleavage. An oxygen–carbon bond is likely as there is no evidence of A+2 elements or nitrogen present. This suggests the presence of a methoxy group. The fragment at *m/z* 95 represents loss of 28 (CO) from *m/z* 123. Aromatic ethers, such as a benzene ring with a substituted methoxy group, commonly lose a methyl radical followed by CO. Here the loss of a second methoxy group is indicated in the mass spectrum by peaks at *m/z* 80 (–methyl radical) and *m/z* 52 (–CO). Note that *m/z* 80 is an even integer value suggesting an odd-electron fragment ion. The latter in particular would be difficult to interpret unambiguously from the mass spectrum alone but is useful confirmation alongside the NMR data, which is more challenging to interpret in this example.

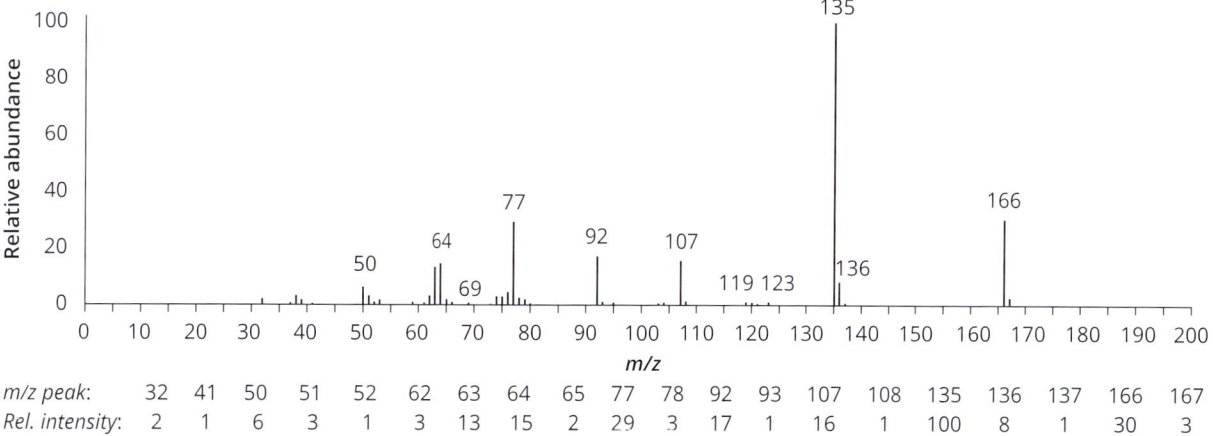

m/z 138   *m/z* 123   *m/z* 95

8.5 **Example 5**

Mass spectrum

m/z peak:	32	41	50	51	52	62	63	64	65	77	78	92	93	107	108	135	136	137	166	167
Rel. intensity:	2	1	6	3	1	3	13	15	2	29	3	17	1	16	1	100	8	1	30	3

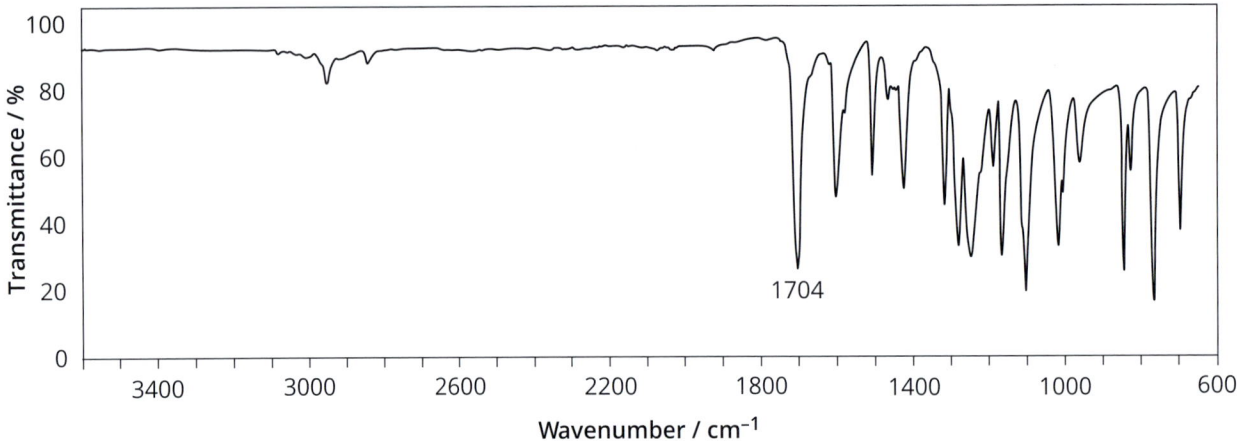

Infrared spectrum

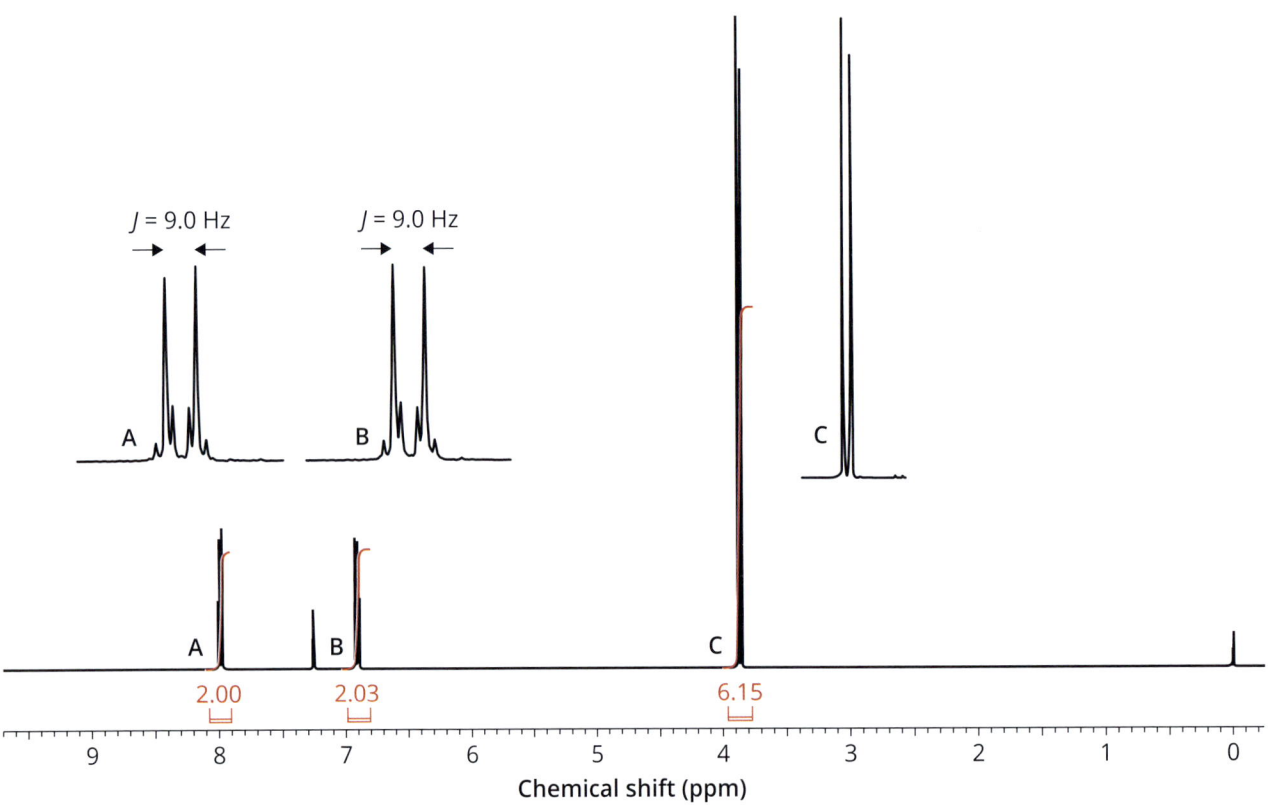

¹H NMR spectrum (400 MHz, CDCl₃)

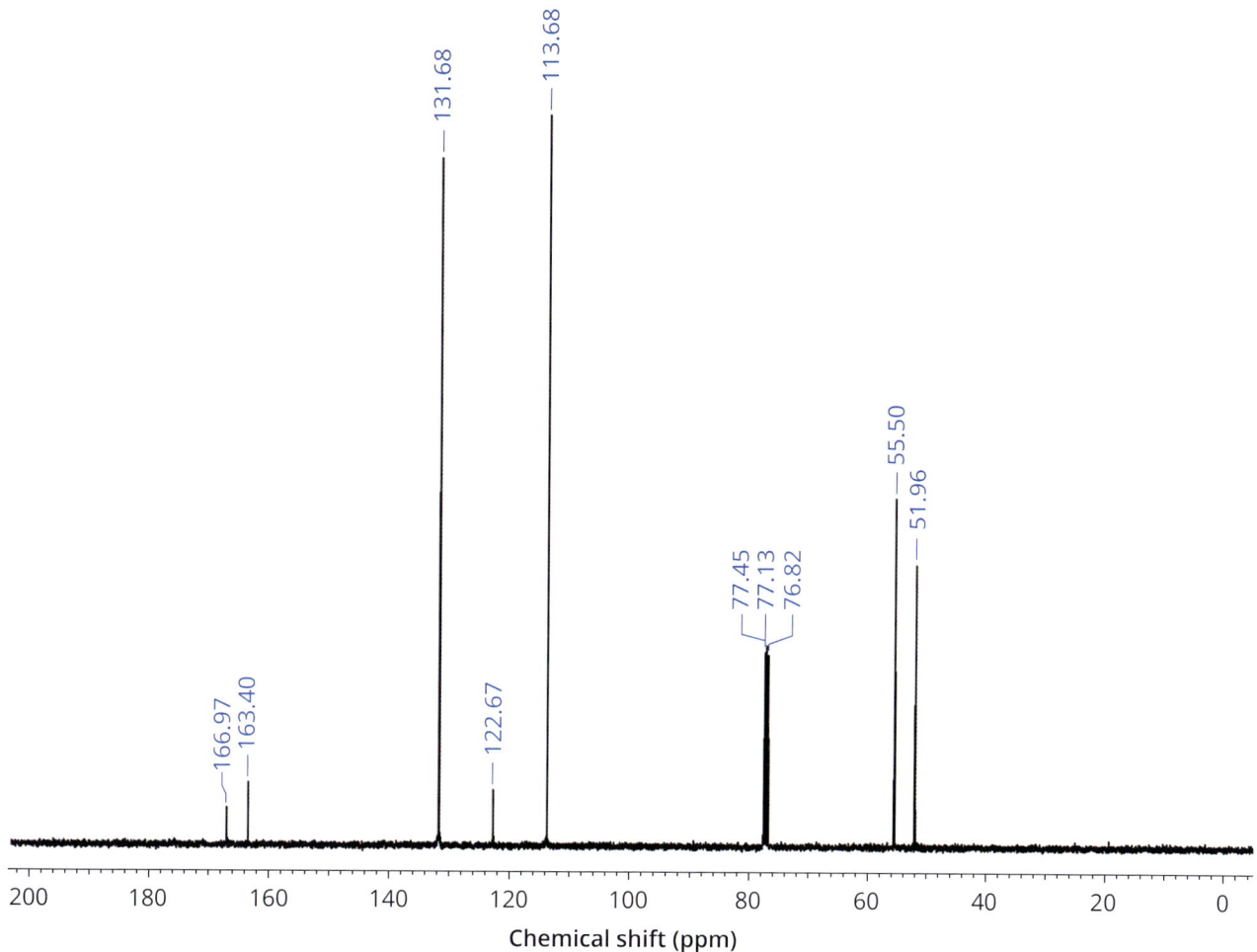

^{13}C NMR spectrum (100 MHz, CDCl$_3$)

8.5.1 Initial analysis of Example 5

(a) The peak at *m/z* 166 in the mass spectrum is likely to represent the molecular ion (M$^{+•}$).

   An even M$^{+•}$ indicates that either no nitrogen atoms or an even number of them (e.g. 2,4 etc) are present in the compound.

(b) The isotope pattern for M$^{+•}$ indicates that neither chlorine nor bromine is present.

(c) Based on the relative abundance of its M and M+1 peaks in the mass spectrum, the estimated number of carbon atoms in the molecular ion is ~9.

(d) In the infrared spectrum, there is a strong absorption in the carbonyl region at ~1700 cm^{-1}.

(e) The IR shows no clear-cut absorptions in the range 2500–3600 cm^{-1} other than the ubiquitous CH stretch at 3000 cm^{-1}.

(f) The ^{1}H NMR shows three signals (if we ignore the trace CHCl$_3$ at 7.26 ppm and TMS at 0.00 ppm). Two of these are doublet-like signals (8.0 and 6.9 ppm) and the third one appears to be a doublet (3.9 ppm).

(g) The ratio of signal integrals is 2:2:6, which indicates the presence of ten hydrogen atoms (or a multiple).

(h) The ^{13}C NMR shows seven signals—remember not to include the 1:1:1 triplet at 77 ppm due to the solvent $CDCl_3$. The chemical shift ranges indicate five unsaturated and two saturated carbon environments. The infrared spectrum already indicates that one of the signals will be a carbonyl group; the small peak at 167.0 is the most likely candidate for this.

(i) The composition accounted for at this stage is only $C_7H_{10}O$, which adds to 110 g/mol. This indicates a molecule with some degree of symmetry and additional atoms not identifiable from the NMR spectra.

8.5.2 Detailed analysis of Example 5

(j) ^1H NMR: There are two signals at 8.0 and 6.9 ppm, each integrating to 2H. The signals are in the aromatic region of the spectrum, and the slight upwards distortion of the 'inner half' of each signal reveals the classic, centrosymmetric 'roof top' pattern characteristic of a 1,4-disubstituted aromatic ring with $^3J_{HH} = 9$ Hz. You might mistake these for a pair of doublets at first glance; the chemical shifts are quite far apart and so the roofing is shallow. However, close examination of the expansion does not show the 'clean' coupling expected of a pair of doublets, but rather the additional characteristic fine structure of an AA'BB' pattern.

(k) ^{13}C NMR: We can now re-evaluate the ^{13}C spectrum and molecular composition. Such a 1,4-disubstituted system would be expected to provide four carbon environments (163.4, 131.7, 122.7, 113.7) for a 6-carbon aryl ring. Thus, our molecular composition can now be revised to $C_9H_{10}O$ (which would provide a m/z in the mass spectrum at 134).

(l) From $C_9H_{10}O$ we can work out that DU = 5. The aromatic ring accounts for 4 DBE and the last one is for the carbonyl group [see (e)].

(m) ^1H NMR: The remaining unassigned signal at 3.8 ppm integrates to 6H. As these lines are so close together it may be mistaken for a doublet, but this can be ruled out straight away when one considers that there is no other unassigned signal which could bear the coupling H atom (see **EG3** for an example containing the isopropyl group $CHMe_2$). This example shows how important it is to realize that integration of the spectrum will most likely have been done automatically by the NMR software. Signals will be integrated together simply because they have similar chemical shifts and not because they arise from a single chemical environment.

(n) ^1H NMR: The two singlets at ~3.8 ppm are each worth 3H and this is indicative of two methyl groups. The chemical shifts indicate that both are deshielded and attached to an electron-withdrawing group, which is likely to be oxygen in both cases.

(o) ^{13}C NMR: There are two aliphatic signals at 51.9 and 55.5 ppm. The chemical shift of these signals is supportive of the assignment of two different OCH_3 groups in the ^1H NMR spectrum above. The molecular formula can be determined as $C_9H_{10}O_3$ (166 g/mol).

The analysis of the data indicates that the following structural elements are present, and these account for all of the observed proton and carbon signals in the NMR spectra.

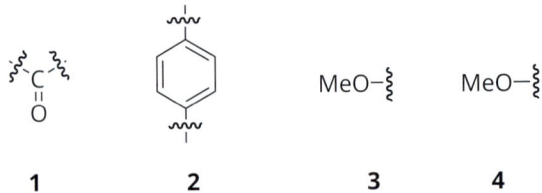

     1   **2**   **3**   **4**

8.5.3 Conclusion and final structure of Example 5

The structural elements identified match the predicted number of carbon atoms, full molecular formula, and *m/z* predicted from the mass spectrum and molecular ion at *m/z* 166. Structural elements **3** and **4**, each with one unfilled valency, can be seen as 'edge pieces' in the molecular jigsaw, and the core of the molecule must therefore be made from structural elements **1** and **2**. The molecular structure can be pieced together to give methyl 4-methoxybenzoate (**EG5**).

8.5.4 Further rationalization of Example 5

It is possible to rationalize the different aromatic carbon chemical shifts by considering the influence of the attached oxygen atoms of the OMe group and the carbonyl group. The strong electron-withdrawing (deshielding) effect of oxygen moves the attached carbon to a large chemical shift at 163.4 ppm. However, the resonance donation of the oxygen lone pair into the aromatic ring will lead to shielding of the adjacent carbon, causing this to appear at a much smaller shift of 113.7 ppm. This resonance effect also explains the smaller shift of the attached proton at 6.9 ppm.

The substantially larger chemical shift of the other aromatic protons at 8.0 ppm is caused by another resonance delocalization effect—this time due to the attached carbonyl group. This allows electron density to move out of the aromatic ring onto the carbonyl oxygen, which causes deshielding of the adjacent position and moves the proton to a larger shift (and likewise for its attached carbon at 131.7 ppm).

The mass spectrum in this case has similarities to the one for Example 4. The presence of a fragment peak at *m/z* 77 is indicative of a phenyl cation derived from α-cleavage of a substituted aromatic compound. However, here we do not see loss of the methyl group but rather the whole methoxy group via α-cleavage, indicated by the base peak at *m/z* 135. We also see loss of the methoxy group followed by CO through inductive cleavage to yield the *m/z* 107 fragment.

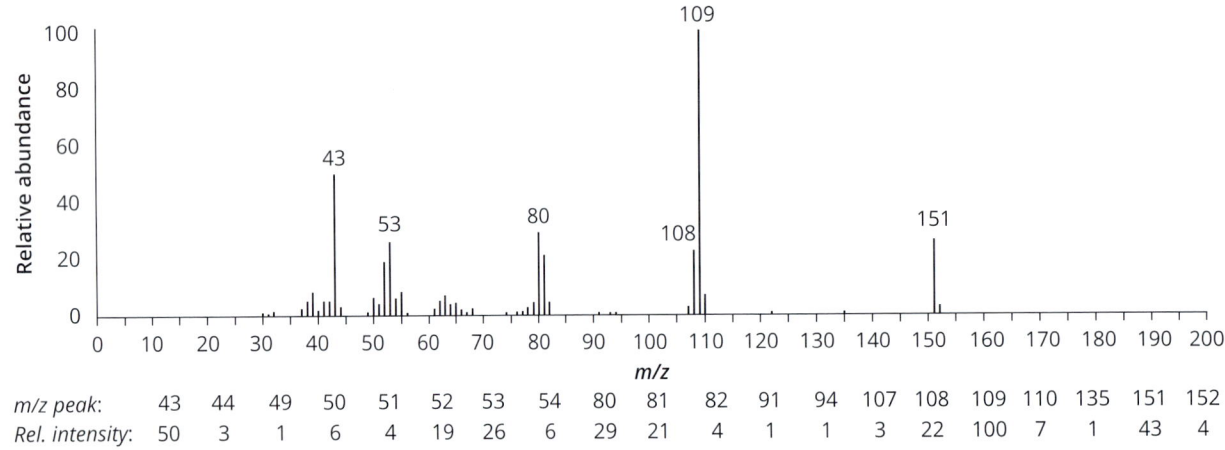

m/z 107   *m/z* 166   *m/z* 135

Methyl 4-methoxybenzoate has 3 different oxygen atoms where non-bonding electrons can be removed to form a radical cation. In practice the carbonyl oxygen is often favoured due to its ability to resonance-stabilize the charge, but it is likely that there is a heterogenous mixture of isobaric radical cations present based on the different oxygen atoms available, leading to multiple fragmentation mechanisms represented in the mass spectrum.

8.6 **Example 6**

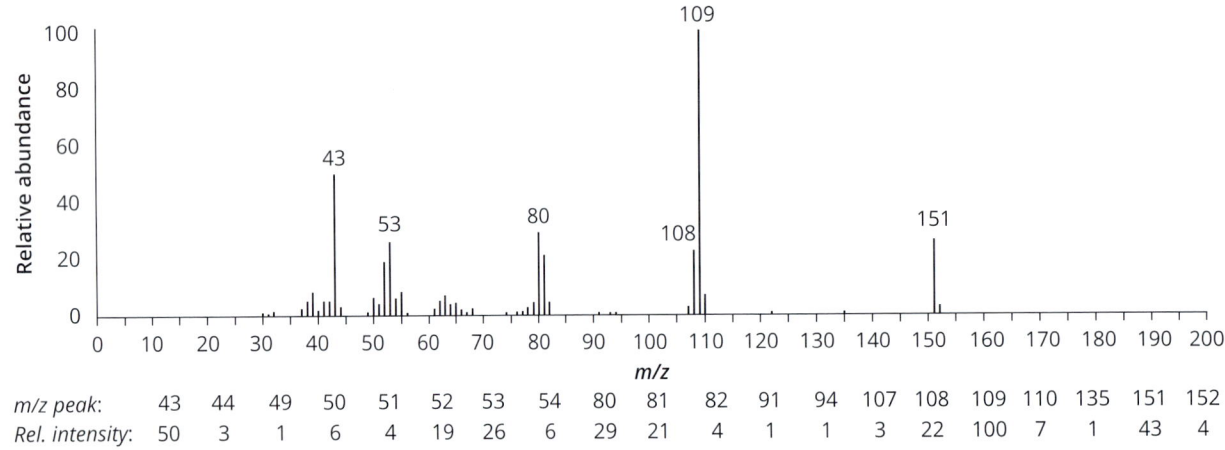

m/z peak:	43	44	49	50	51	52	53	54	80	81	82	91	94	107	108	109	110	135	151	152
Rel. intensity:	50	3	1	6	4	19	26	6	29	21	4	1	1	3	22	100	7	1	43	4

Mass spectrum

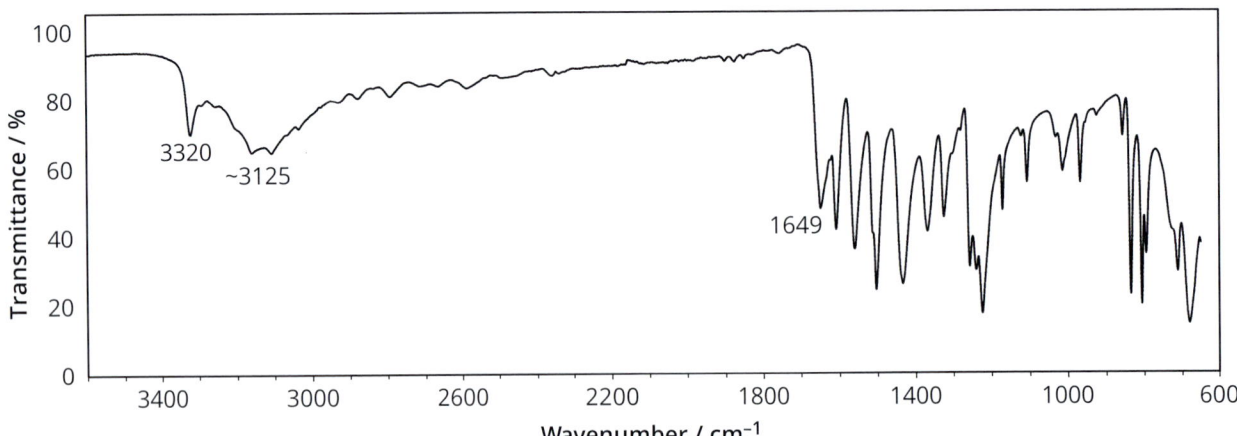

Infrared spectrum

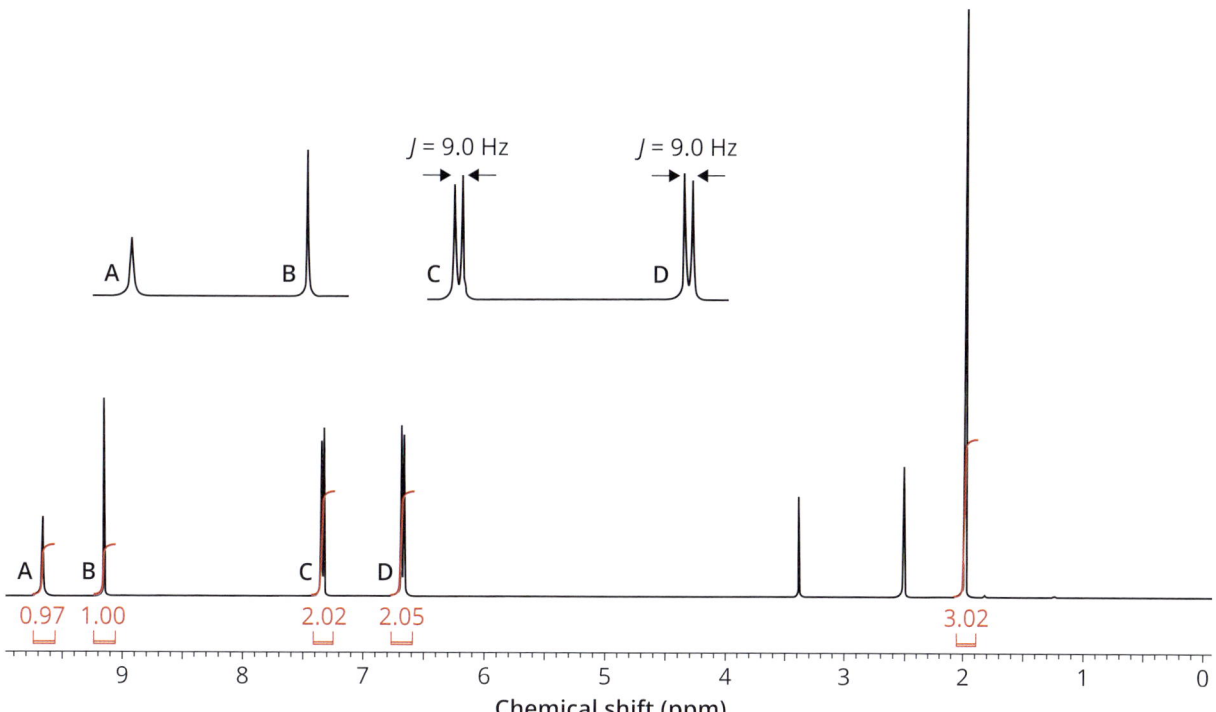

¹H NMR spectrum (400 MHz, *d*₆-DMSO)

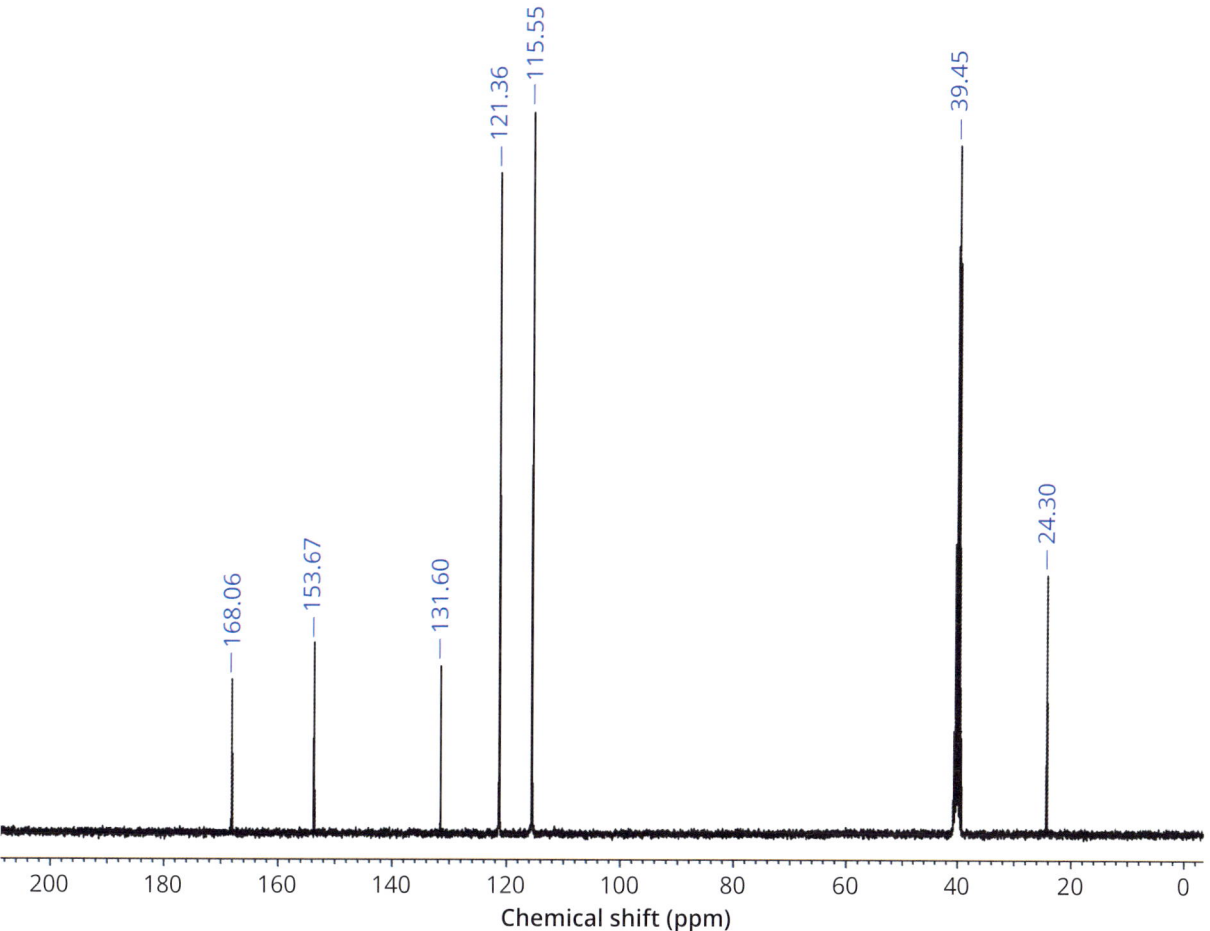

¹³C NMR spectrum (100 MHz, *d*₆-DMSO)

8.6.1 Initial analysis of Example 6

(a) The peak at *m/z* 151 in the mass spectrum is likely to represent the molecular ion ($M^{+\bullet}$).

(b) An odd $M^{+\bullet}$ indicates that at least one nitrogen atom is present (or an odd multiple, e.g. 3, 5, etc).

(c) The isotope pattern $M^{+\bullet}$ suggests that neither chlorine nor bromine is present.

(d) The abundance of the proposed molecular ion suggests that the compound may be aromatic.

(e) The number of carbon atoms in the molecular ion is estimated to be ~8 based on the relative abundance of its M and M+1 peaks in the mass spectrum.

(f) In the infrared spectrum there is an absorption at 1650 cm^{-1} in the carbonyl region; the molecule therefore contains a carbonyl group.

(g) The broad absorption in the range 3000–3400 cm^{-1} in the infrared spectrum indicates the presence of a hydrogen-bonded group, and is typical of an OH. An overlapping sharp peak at 3300 cm^{-1} suggests the possibility that another type of Het-H group may be present (where Het is a heteroatom such as N, S, etc).

(h) The ^{1}H NMR shows five signals. The spectrum was run as a solution in d_6-DMSO, and so shows in addition the characteristic residual protonated solvent peak at 2.5 ppm (i.e. CD_2HSOCD_3) and a water peak at 3.3 ppm. DMSO is very hygroscopic, which means that, unless a new bottle of deuterated solvent is used, the water peak can often be very prominent.

(i) The ratio of signal integrals is 1:1:2:2:3, which indicates the presence of nine hydrogen atoms (or a multiple).

(j) The ^{13}C NMR shows six signals—remember not to include the peak at 39.5 ppm due to the solvent d_6-DMSO. The chemical shift ranges indicate five unsaturated and one saturated carbon environments.

(k) The molecular composition accounted for at this stage is $C_6H_9NO_2$, which would provide a molecular weight of 127 g/mol, which is 24 mass units short of the observed molecular ion (*m/z* 151). Indeed, from point (e) we can derive that the missing mass is due to two additional carbon atoms that have not yet been accounted for, and so the predicted molecular formula is $C_8H_9NO_2$.

(l) From (k), we can conclude that DU = 5. The unsaturated signals in the ^{1}H and ^{13}C spectra, and the relatively high intensity of the molecular ion, indicate that an aromatic ring is likely to be present. Together with the double bond from the C=O identified in (f), this accounts for the degree of unsaturation.

8.6.2 Detailed analysis of Example 6

(m) ^{1}H NMR: There are two signals at 7.3 and 6.7 ppm, each of which integrates to 2H. The signals are in the aromatic region of the spectrum, and the slight upwards distortion of the 'inner half' of each signal reveals the classic, centrosymmetric roofing pattern characteristic of a 1,4-disubstituted aromatic ring with $^3J_{HH} = 9$ Hz. You might mistake these for a pair of doublets at first glance; again the chemical shifts are quite far apart and so the roofing is shallow. However, close examination of the expansion does not show the 'clean' coupling expected of a pair of doublets, but rather the broadened peaks at the baseline that are indicative of additional smaller splittings seen in an AA'BB' pattern.

(n) ^{1}H NMR: Two singlets are seen at 9.2 and 9.7 ppm, though the latter is quite broad compared to the former. Broadened signals can be indicative of H atoms attached directly to electronegative atoms, such as OH, NH, etc, but the chemical shift range is also typical of an aldehyde hydrogen atom (H–C=O). The IR spectrum gives evidence of both C=O and OH/NH signals. So, at this stage, it is not totally clear whether both types of groups are represented by these signals.

(o) ^{1}H NMR: A sharp, 3H singlet is seen at 2.0 ppm. This is typical of a methyl singlet, and the chemical shift reveals that it is neither attached directly to an electronegative atom (e.g. O, N, etc), nor to an aliphatic sp^3 carbon centre. Thus, it appears that the CH$_3$ is attached to an unsaturated group (e.g. HC–C=C, HC–Ar, or HC–C=O).

(p) ^{13}C NMR: We can re-evaluate the ^{13}C spectrum based on information from the ^{1}H NMR and IR spectra. The signal at 168.1 ppm is the likely contender for the carbonyl carbon (identified in the IR), and the only aliphatic peak (24.3 ppm) that of the methyl carbon.

(q) ^{13}C NMR: A 1,4-disubstituted aromatic system will provide four carbon environments, and four such signals can be seen at 153.7, 131.6, 121.4, and 115.6 ppm. The two smaller and two larger peaks hint that the ring is disubstituted (see section 3.4).

The analysis of the data for **EG6** indicates that the following structural elements are present, and these account for all of the observed proton and carbon signals in the NMR spectra, and match the molecular formula C$_8$H$_9$NO$_2$.

8.6.3 Conclusion and final structure of Example 6

Putting these structural elements together and deducing a structure for **EG6** is easier once the nature of the carbonyl functional group is resolved. The chemical shift of the ^{13}C signal (168.1 ppm) is at the lower end of the range typical for this group. If an aldehyde group was present, its ^{13}C C=O signal would be >190 ppm. Whilst it is not wise to 'learn' chemical shift tables, it is worth recalling that the ketone and aldehyde carbonyl groups have ^{13}C signals consistently above 190 ppm, whereas the higher oxidation acids, esters, amides, etc typically have signals in the 165–180 ppm range. This analysis indicates that the two 1H singlets in the ^{1}H spectrum above 9 ppm are due to an NH **8** and OH **9**, as deduced from the IR, and that the carbonyl group is not an aldehyde.

The stretching frequency of the C=O in the IR spectrum is at 1649 cm^{-1}. This low value is typical of an amide, and thus structural elements **1** and **8** can be combined (recall

the more electrophilic carbonyl groups, e.g. acid chlorides or anhydrides, have stretching frequencies at the higher end of the range, around 1750–1800 cm^{-1}). This combination now leaves the amide fragment and aromatic ring fragment as 'centre pieces' of the jigsaw, which can be linked in one of two ways as shown below. The methyl **3** and hydroxyl **9** are 'edge pieces' of the jigsaw.

The piece of evidence that points to one of these possibilities is the chemical shift of the NH. Anilides have consistently higher NH chemical shifts (typically > 8.5 ppm) than benzamides (< 8.5 ppm), and so even if the NH cannot be assigned specifically to one of the peaks above 9 ppm, the benzamide is ruled out.

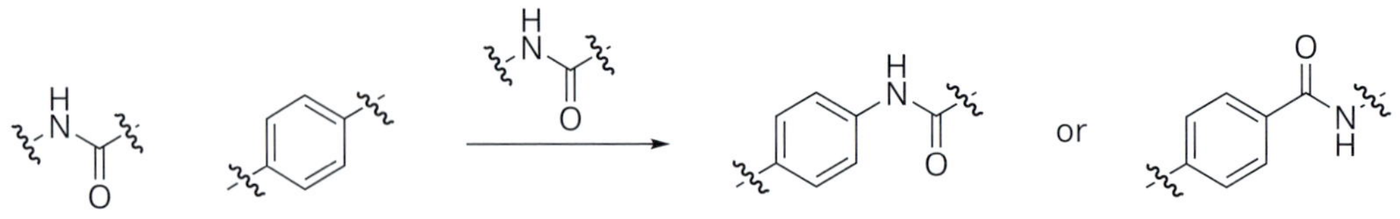

The methyl fragment **3** has a chemical shift which means it could either be linked to the carbonyl group **1**, or connected to the aromatic ring **2**. However, in reality, the hydroxyl group can only be linked to the ring. If it was attached to nitrogen, a carbamic acid would be the result. Carbamic acids are generally unstable, and these readily lose carbon dioxide, as shown below. The molecular structure can be pieced together to give *N*-(4-hydroxyphenyl) acetamide (paracetamol or acetaminophen, **EG6**).

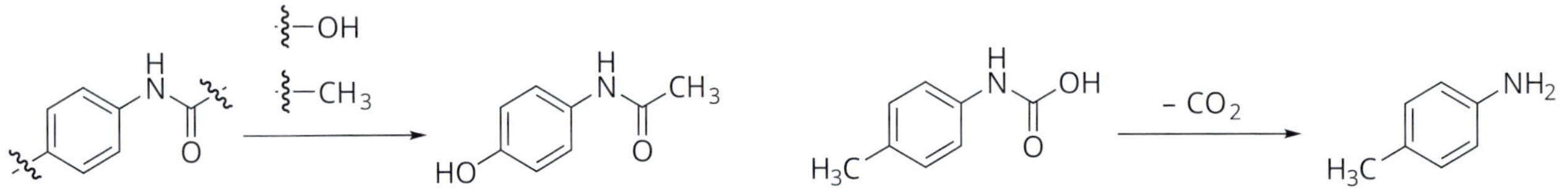

8.6.4 Further rationalization of Example 6

From the NMR data, it is possible to rationalize the different aromatic carbon and proton chemical shifts by considering the influence of the attached heteroatoms. The strong electron-withdrawing (deshielding) effect of oxygen moves the attached carbon to a large chemical shift at 153.7 ppm, whereas the less strongly withdrawing nitrogen has a less significant effect, leaving its attached carbon at 131.6 ppm. However, the resonance donation of the oxygen lone pair into the aromatic ring leads to shielding of the adjacent carbon, causing this to appear at a much smaller shift of 115.6 ppm. This resonance effect also explains the smaller shift of the attached proton at 6.7 ppm; note the similar behaviour seen in Example 5. These values rule out having the methyl group attached directly to the aromatic ring but instead confirm the location of the hydroxyl group.

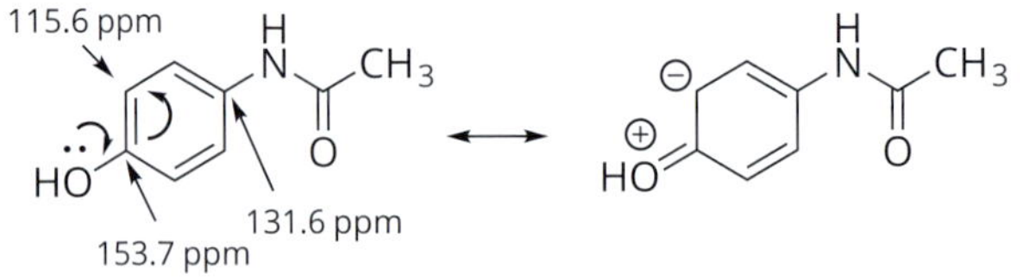

Note also that the possible benzamide structure can also be ruled out on the basis of the aromatic proton chemical shifts since a carbonyl group attached to the aromatic ring would cause the adjacent protons to resonate at chemical shifts above 7.3 ppm and closer to 8 ppm, as seen in Example 5.

Although it is seemingly a relatively simple fragmentation mass spectrum with a small number of prominent peaks, the range of heteroatoms across multiple functional groups make this example challenging to interpret. The base peak in the mass spectrum is at *m/z* 109 and results from a hydrogen transfer (rearrangement) from the methyl group on the acetyl moiety followed by α-cleavage initiated by the charge on the amide nitrogen. The resulting distonic radical cation may further fragment, leading to a peak at *m/z* 80. The prominent low mass peak at *m/z* 43 likely results from α-cleavage around the carbonyl oxygen radical cation, as illustrated.

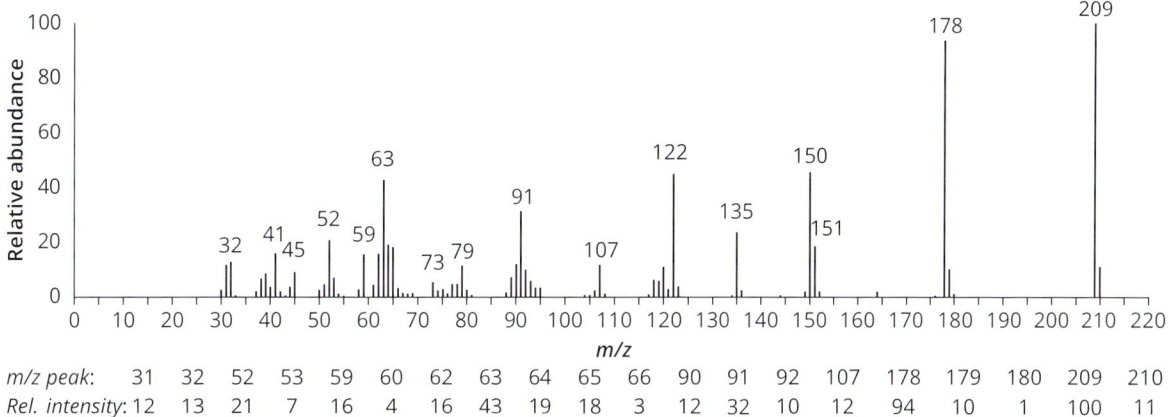

m/z 151   *m/z* 109   *m/z* 80

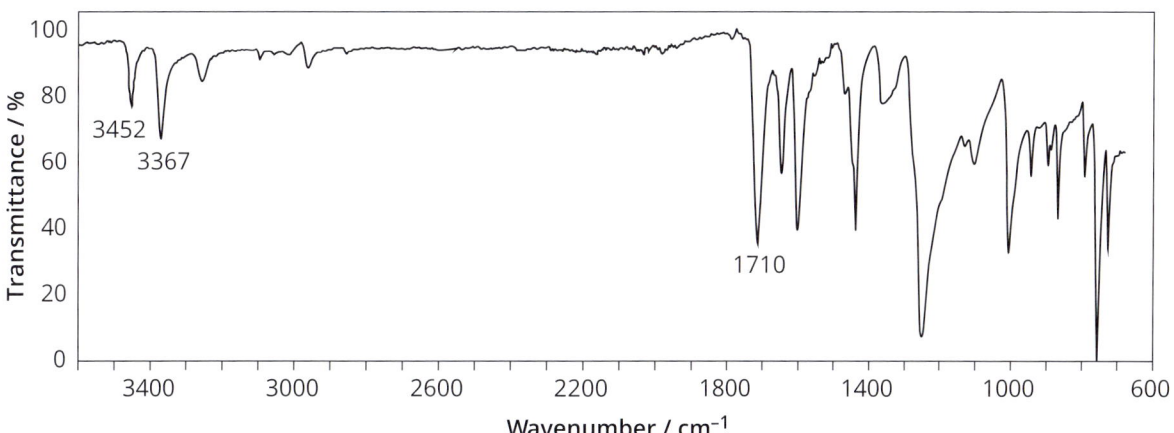

m/z 151   *m/z* 43

8.7 **Example 7**

Mass spectrum

m/z peak:	31	32	52	53	59	60	62	63	64	65	66	90	91	92	107	178	179	180	209	210
Rel. intensity:	12	13	21	7	16	4	16	43	19	18	3	12	32	10	12	94	10	1	100	11

Infrared spectrum

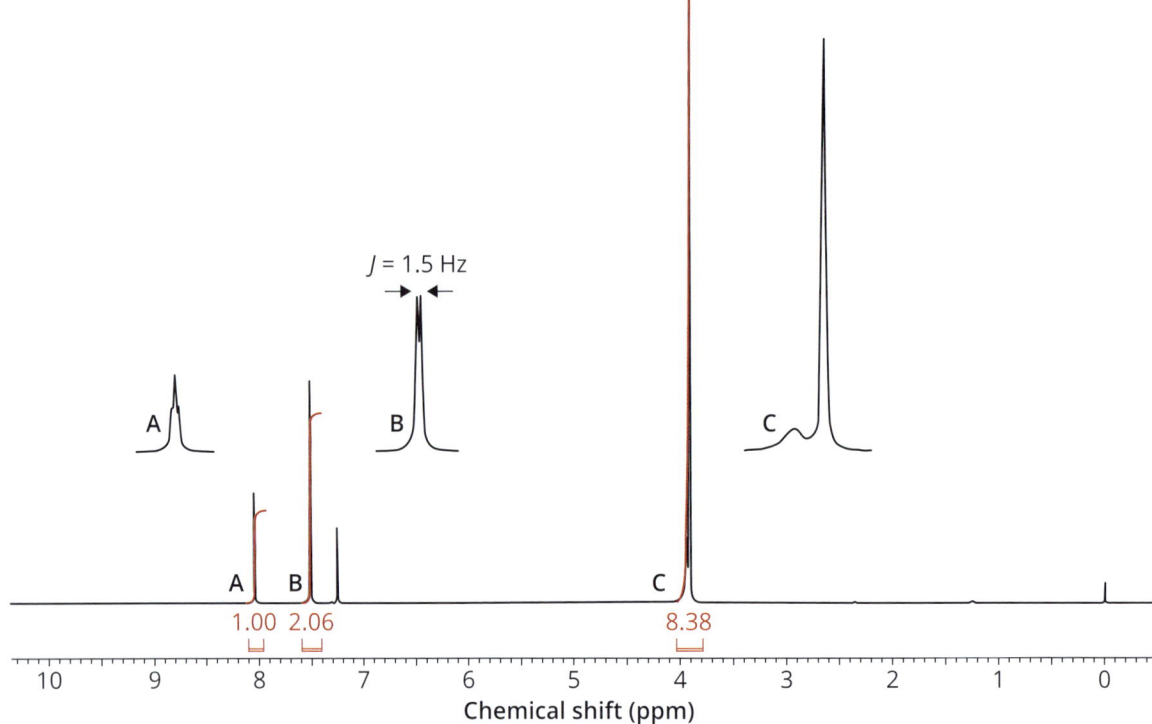

^{1}H NMR spectrum (400 MHz, CDCl$_3$)

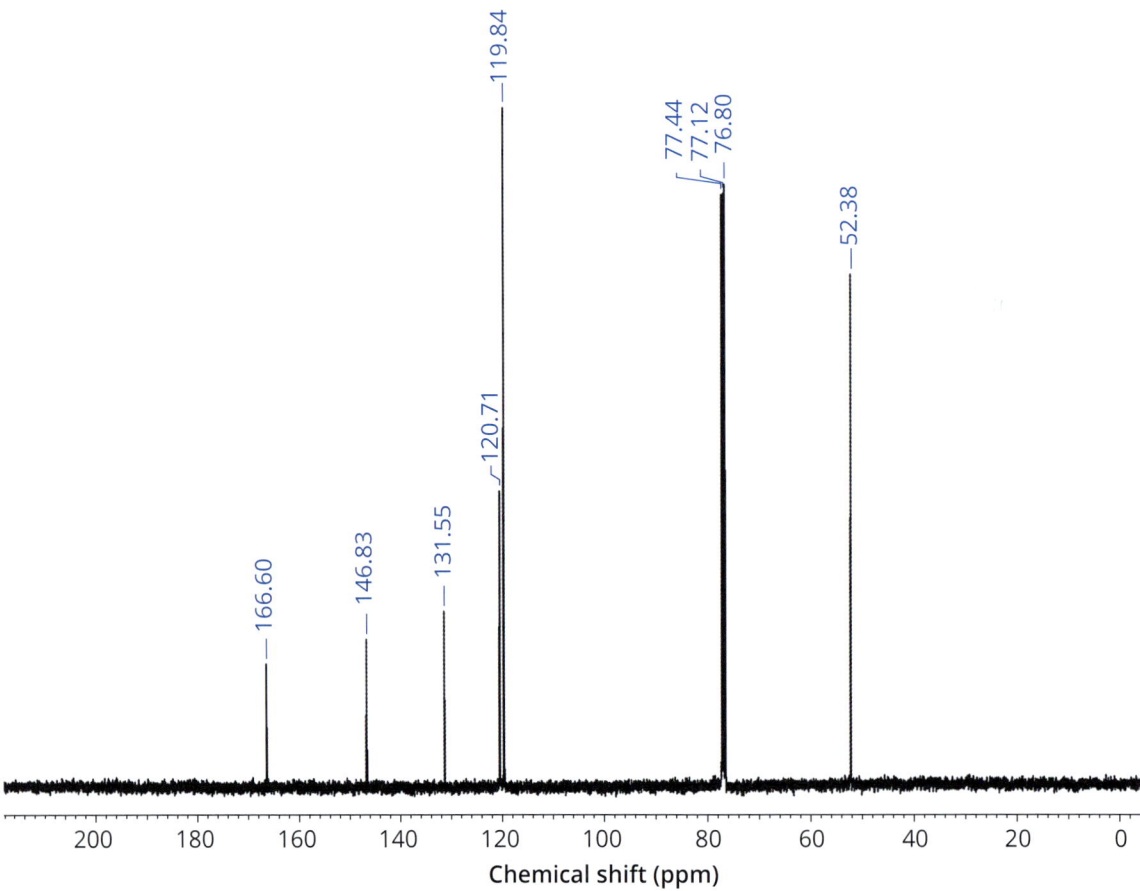

^{13}C NMR spectrum (100 MHz, CDCl$_3$)

8.7.1 Initial analysis of Example 7

(a) The peak at m/z 209 in the mass spectrum is likely to represent the molecular ion ($M^{+\bullet}$).

(b) An odd integer $M^{+\bullet}$ value indicates that at least one nitrogen atom is present in the compound, or there could be any odd number (e.g. 3, 5, etc).

(c) The abundance of the proposed molecular ion suggests that the compound may be aromatic.

(d) The number of carbon atoms in the molecular ion is ~10 based on the relative abundance of its M and M+1 peaks in the mass spectrum.

(e) The isotope pattern for $M^{+\bullet}$ indicates that neither chlorine nor bromine are present.

(f) In the infrared spectrum there is a strong absorption at 1710 cm^{-1} in the carbonyl region; the molecule therefore contains a carbonyl group.

(g) There are some sharp absorptions between 3350 and 3450 cm^{-1}, typical of the frequency range for Het-H bonds (where Het is an electronegative atom such as O or N). However, O–H stretches are typically broad for alcohols and phenols (cf. **EG2** and **EG6**), or very broad for carboxylic acids (cf. **EG10** later). These sharper signals are indicative of N–H stretching absorptions, and the fact that there are more than one is typical of a primary amine group. These give rise to two peaks due to symmetric and asymmetric NH stretching modes, and this spectrum thus indicates the presence of a primary amino group.

(h) The ^{1}H NMR shows four signals, including the broad peak at around 4.0 ppm, ignoring the trace CHCl$_3$ at 7.26 ppm and TMS at 0.00 ppm.

(i) The ratio of signal integrals is 1:2:8, though looking closely it can be noted that a broad peak at 4.0 is co-integrated with a sharp singlet at a slightly smaller chemical shift. The presence of eleven hydrogen atoms (or a multiple) is nevertheless indicated.

(j) The ^{13}C NMR shows six signals—remember not to include the 1:1:1 triplet at 77 ppm due to the solvent CDCl$_3$. The chemical shift ranges indicate five unsaturated and one saturated carbon environments.

(k) The molecular constitution at this stage is only $C_6H_{11}NO$, which adds up to 113 g/mol—or, in other words, 96 mass units short of the observed molecular ion (209 g/mol). This indicates a molecule with some degree of symmetry and thus additional atoms that are not immediately identifiable from the NMR spectra.

8.7.2 Detailed analysis of Example 7

(l) ^{1}H NMR: The signals at 8.1 ppm (integrating to 1H) and 7.5 ppm (integrating to 2H) are in the unsaturated region of the spectrum. Both show evidence of couplings, a triplet for the former and a doublet for the latter signal. However, the magnitude of the couplings is small, around 1.5 Hz, thus pointing neither towards larger *trans* or smaller *cis* couplings found in alkenes, nor the very common $^3J_{HH}$ (*ortho*) couplings seen in many aromatic systems. Instead, couplings of this magnitude (< 2 Hz) are indicative of $^4J_{HH}$ (*meta*) couplings in aromatic systems (section 2.3.2).

(m) The number of signals in the aromatic region totals 3H, which is indicative of a trisubstituted ring.

(n) Upon inspection of the expansion, the signals around 4.0 ppm reveal a broad singlet and a sharp singlet. Compared to the aromatic signals, these two integrate to 8H. A broad singlet is indicative of an H-bonded group such as the NH$_2$ group revealed by the IR spectrum, leaving the large, sharper singlet worth 6H. In real situations, the integral could be carefully split to get these ratios, but as the spectrum trace does not return to the baseline it would always be an estimate.

(o) For such a simple signal, the large 6H integration for the singlet at 3.9 ppm is an indicator of symmetry within the molecule [see (k)], possibly arising from two identical CH_3 groups. The chemical shift of this signal is indicative of H atoms attached to or very near an electronegative atom such as oxygen; thus, this signal could be due to two identical methoxy groups.

(p) ^{13}C NMR: There are five unsaturated carbon environments, with the signal at 166.6 ppm typical of the carbonyl group indicated from the IR spectra.

(q) The four signals at 146.8, 131.6, 120.7, and 119.8 ppm are characteristic of aromatic CH signals. As there are fewer than six signals, this is indicative of a ring with one plane of symmetry.

(r) The single peak in the saturated carbon region of the spectrum is at 52.4 ppm. This chemical shift is consistent with the carbon being attached to oxygen, and thus being partner to the CH_3O singlet at 3.9 ppm in the ^1H NMR spectrum.

(s) As noted, the integral of the methyl singlet is worth 6H. So, with two methoxy groups the molecular constitution can now be revised to $C_9H_{11}NO_3$, which is still 28 mass units short of the observed molecular ion (*m/z* 209). The 28 mass units could either be from two nitrogen atoms or one carbon plus one oxygen. Given that conclusion (d) indicated 10 carbon atoms in the structure, the likeliest scenario is that an extra carbonyl group is present. Since only a single carbonyl *environment* is indicated by the carbon-13 NMR, this suggests that the two OCH_3 groups must belong to identical methyl esters.

(t) The molecular formula at this stage is now $C_{10}H_{11}NO_4$, which is commensurate with the molecular ion at *m/z* 209 in the mass spectrum and the predicted number of carbon atoms in conclusion (d). A calculation gives DU = 6. The aromatic ring would account for 4 DBE with the remainder from the two carbonyl groups.

The analysis of the NMR data for **EG7** indicates that the following structural elements are present, and these account for all of the observed proton and carbon signals in the NMR spectra, and match the molecular formula $C_{10}H_{11}NO_4$.

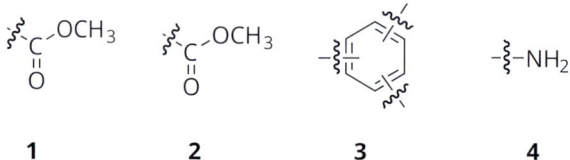

| 1 | 2 | 3 | 4 |

8.7.3 Conclusion and final structure of Example 7

The evidence so far indicates that the molecule **EG7** contains two methyl ester groups, an amino group, and a trisubstituted ring. The presence of only four signals in the aromatic region of the ^{13}C spectrum shows that the ring must have one plane of symmetry, and the magnitude of coupling indicates that there is only $^4J_{HH}$ (meta) coupling. The 2H doublet and 1H triplet are characteristic of a 1,3,5-trisubstituted aromatic system, and thus **EG7** can be identified as dimethyl 5-aminoisophthalate.

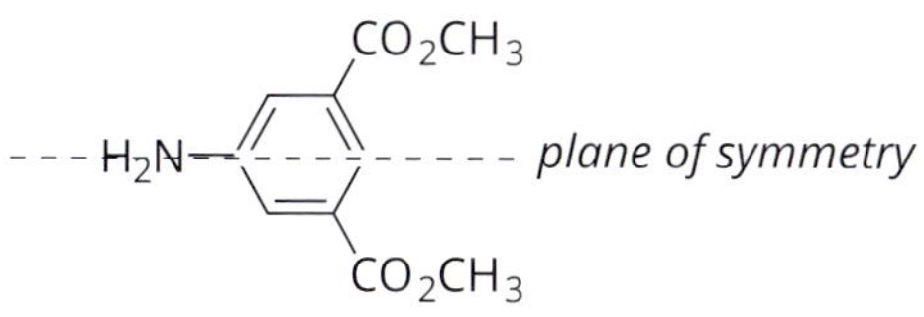

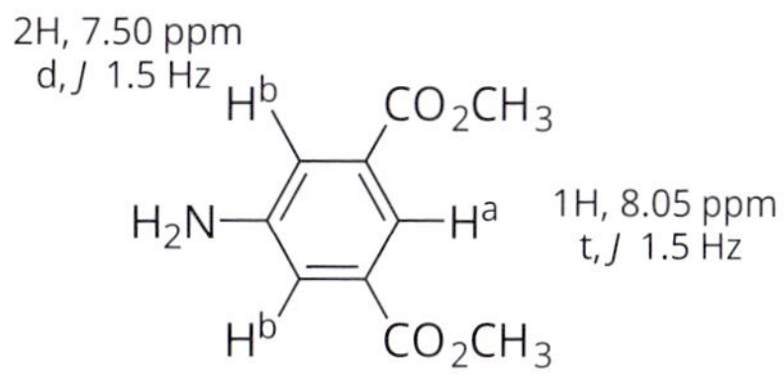

8.7.4 Further rationalization of Example 7

The high chemical shift of 8.05 ppm observed for H^a may be rationalized by the resonance delocalization of electron density from the aromatic ring onto the neighbouring ester groups, leading to a substantial deshielding at this position:

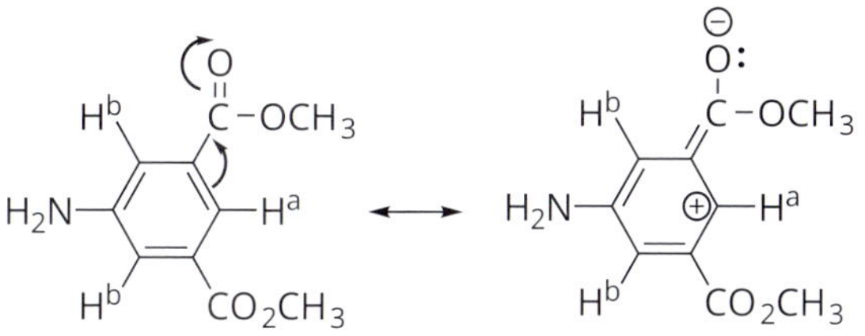

Whilst a similar mechanism would also lead to deshielding of the H^b protons, this is balanced somewhat by the electron donation from the nitrogen group, which provides a competing shielding effect and therefore causes these protons to appear at 7.5 ppm:

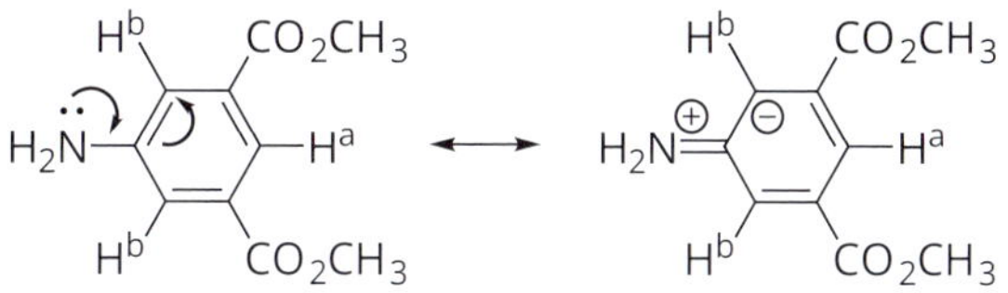

The presence of the mildly electron-withdrawing nitrogen also causes the attached carbon to appear at 146 ppm. The two H^b protons are equivalent due to the symmetry plane and so show no coupling with each other, hence appearing as a doublet from $^4J_{HH}$ coupling only to H^a. The NH_2 resonance at 4 ppm is very broad due to exchange of these protons with dissolved water in the solvent; this process tends to be accelerated by trace levels of acid that occur commonly in chloroform.

This is a relatively complex mass spectrum representing many fragment peaks. The abundant molecular ion peak indicates the presence of an aromatic ring, and the high intensity fragment ion peaks at higher m/z values suggest ring-stabilized fragments. The main fragment peaks in the mass spectrum are accounted for by α-cleavage and inductive cleavage when the charge and unpaired electron of the molecular ion are located on the carbonyl oxygen. They provide supporting evidence for the presence of an $-OCH_3$ group via α-cleavage and CO_2CH_3 (i.e. for **1** above). Note that the odd m/z peaks (m/z 135 and 107) indicate loss of nitrogen and/or rearrangements. m/z 91 indicates a tropylium ion, which provides supporting evidence for a substituted aromatic structure. Fragments at lower m/z values with lower abundances suggest that a ring-opening with rearrangements is likely.

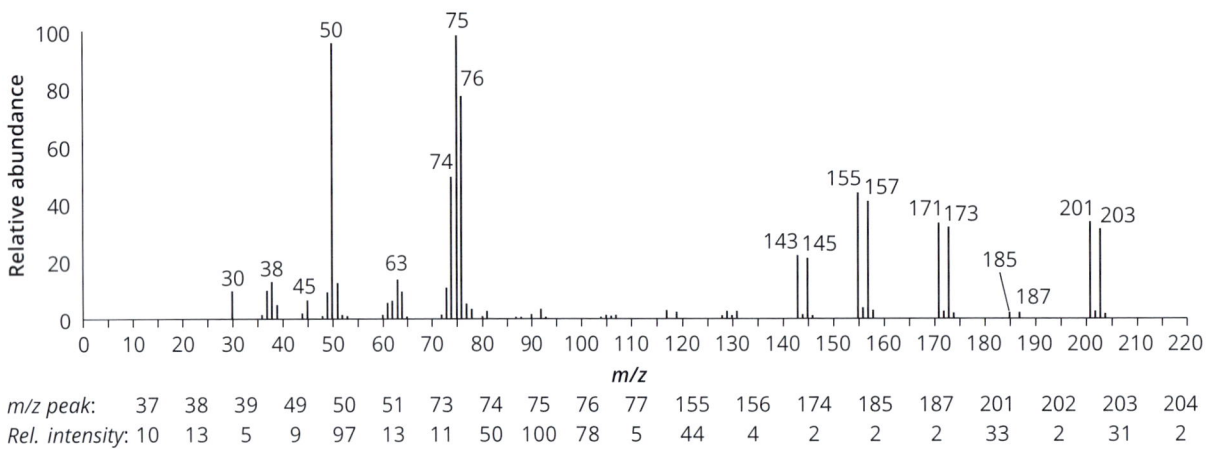

m/z 178   m/z 209   m/z 59

m/z 150

8.8 **Example 8**

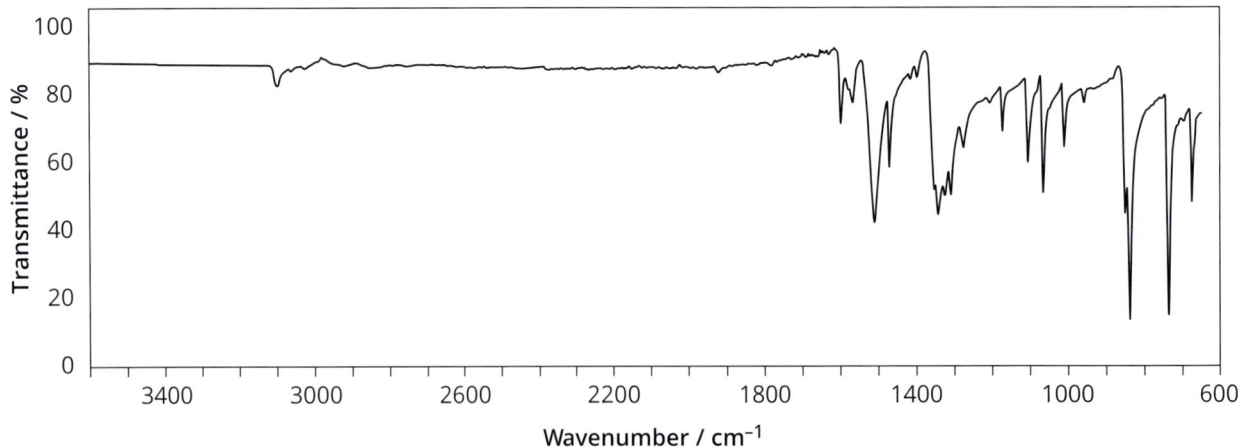

m/z peak:	37	38	39	49	50	51	73	74	75	76	77	155	156	174	185	187	201	202	203	204
Rel. intensity:	10	13	5	9	97	13	11	50	100	78	5	44	4	2	2	2	33	2	31	2

Mass spectrum

Infrared spectrum

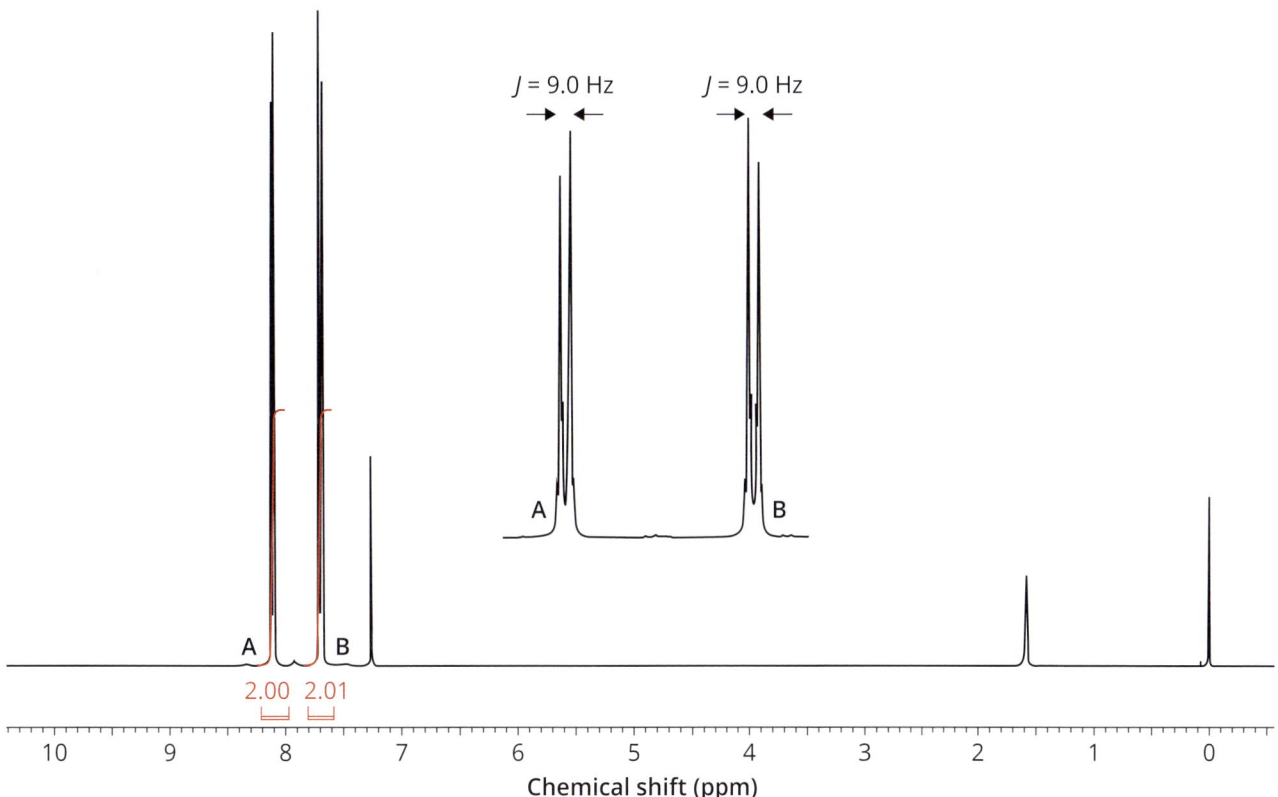

¹H NMR spectrum (400 MHz, CDCl₃)

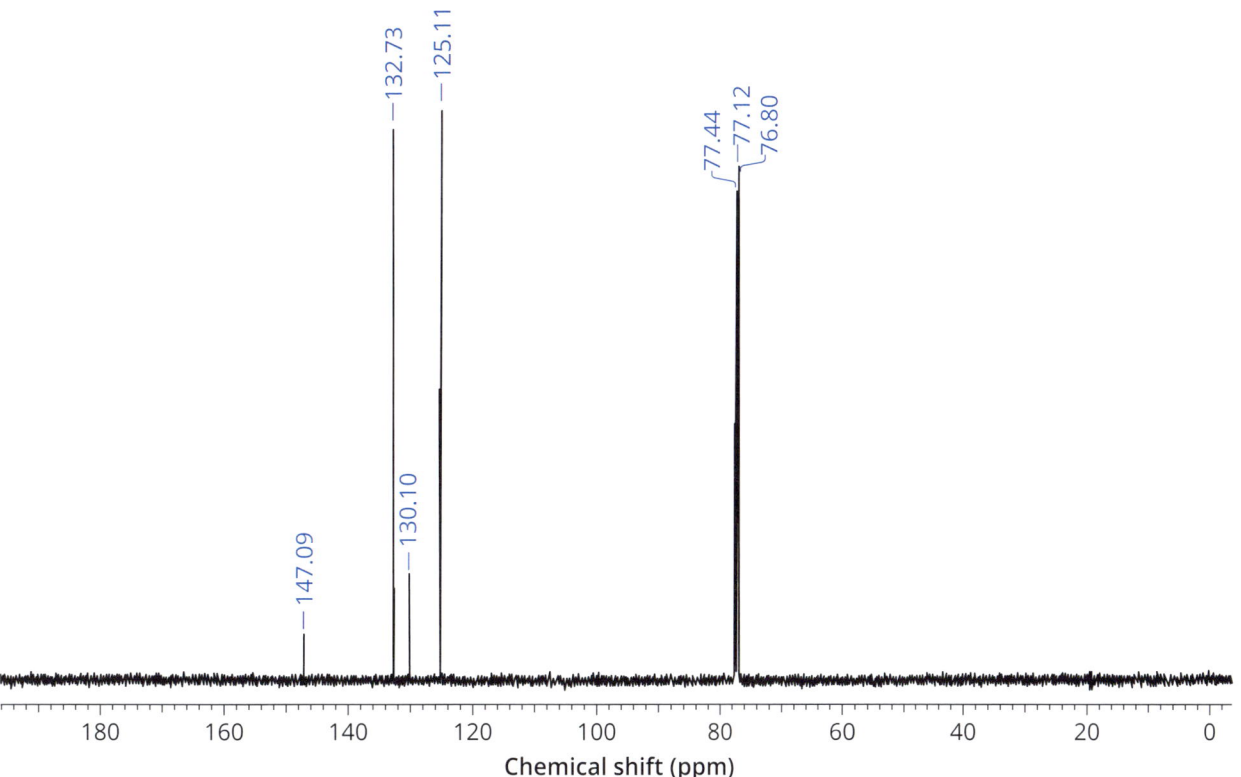

¹³C NMR spectrum (100 MHz, CDCl₃)

8.8.1 Initial analysis of Example 8

(a) The 1:1 ratio of the isotope peaks at *m/z* 201 and 203 in the mass spectrum corresponds to molecular ions ($M^{+\bullet}$) containing both isotopic forms of a single bromine atom.

(b) An odd $M^{+\bullet}$ indicates that at least one nitrogen atom is present, or there could be any odd multiple of these (e.g. 3, 5, etc).

(c) The relatively high abundance of the molecular ion peaks suggests that the compound may be aromatic.

(d) The number of carbon atoms in the molecular ion is ~6 based on the relative abundance of its M and M+1 peaks in the mass spectrum.

(e) The infrared spectrum is relatively featureless above 1600 cm^{-1} apart from CH stretches around 3100 cm^{-1}. There is no strong absorption in the carbonyl region 1650–1850 cm^{-1} and no absorption indicating the presence of an OH or NH$_{(2)}$ group.

(f) The ^{1}H NMR shows two signals, if we ignore the singlets for trace CHCl$_3$ at 7.26 ppm, water at 1.60 ppm, and TMS at 0.00 ppm.

(g) The ratio of signal integrals in the ^{1}H NMR spectrum is 2:2, which indicates the presence of four hydrogen atoms (or a multiple).

(h) The ^{13}C NMR shows four signals, and based on the chemical shifts all of them are in unsaturated carbon environments. Again, remember not to include the 1:1:1 triplet at 77 ppm due to the solvent CDCl$_3$.

(i) The composition accounted for at this stage is C$_4$H$_4$NBr, which adds to 145 for ^{79}Br (or 147 for ^{81}Br). From conclusion (a) the 'missing' mass is therefore 56.

8.8.2 Detailed analysis of Example 8

(j) ^{1}H NMR: Two 2H signals can be seen at 7.7 and 8.1 ppm. The chemical shift of the signals is characteristic of unsaturated CH signals from aromatic hydrogen environments. The symmetrical nature of these two 2H signals, as seen clearly in the expansion, is typical of the roofing pattern seen with a 1,4-disubstituted aromatic ring with $^3J_{HH} = 9$ Hz.

(k) ^{13}C NMR: There are four signals between 125.1 and 147.1 ppm. The chemical shifts of these signals are typical of aromatic carbon atoms, and the symmetry of a 1,4-disubstituted aryl ring [see (j)] would account for all of these peaks. Whilst signals cannot be integrated in a standard ^{13}C spectrum, quaternary carbon atoms typically give lower intensity signals and thus the signals at 130.1 and 147.1 ppm may be assigned to one of these.

(l) A revised molecular composition is C$_6$H$_4$NBr, which adds to 169 for ^{79}Br (or 171 for ^{81}Br) and is 32 mass units short of the molecular ion.

(m) The missing 32 mass units are unlikely to come from carbon atoms (as 6 are predicted in conclusion (d)). However, these mass units could easily come from two oxygen atoms, yielding the molecular formula C$_6$H$_4$NO$_2$Br. From this formula, we can derive that the DU = 5 (don't forget to include the halogen!). Four DBE can be assigned to the aromatic ring, leaving one remaining DBE to be assigned.

(n) As there are no carbon atoms unaccounted for, the last DBE cannot be from an alkene or carbonyl π bond. This leaves the N=O group as the only option, and this is likely part of a nitro group to 'use up' the last oxygen.

The analysis of the data indicates that the following structural elements are present, and these account for all of the observed proton and carbon signals in the NMR spectra.

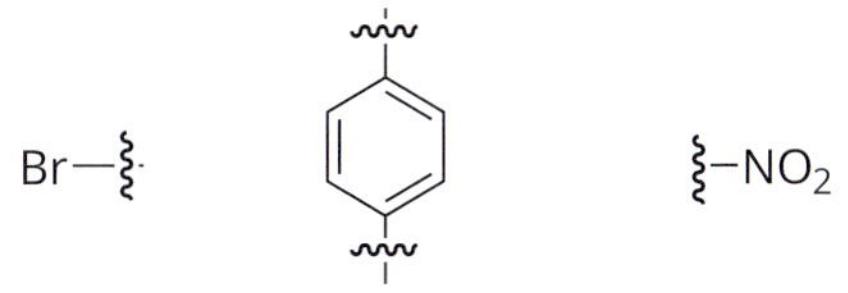

1   **2**   **3**

These findings are commensurate with the evidence from NMR.

8.8.3 Conclusion and final structure of Example 8

The structural elements identified match the predicted molecular formula and weight of $C_6H_4NO_2Br$ and 201 or 203 g/mol respectively. Structural elements **1** and **3**, each with one unfilled valency, can be seen as 'edge pieces' in the molecular jigsaw, and the core of the molecule must therefore be made from structural element **2**. The molecular structure can be pieced together to give 1-bromo-4-nitrobenzene (**EG8**).

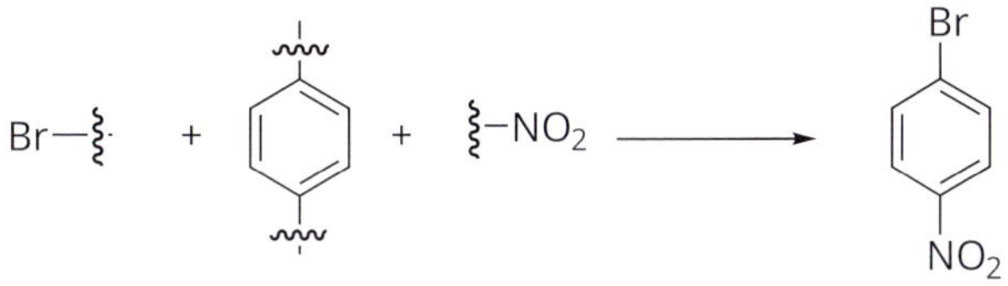

EG8

8.8.4 Further rationalization of Example 8

In relation to the NMR spectrum, the aromatic carbon with the largest chemical shift (147.1 ppm) may be attributed to that carrying the electronegative $-NO_2$ group which is responsible for the deshielding effect on this centre.

The mass spectrum shows a number of fragment peaks that retain the bromine atom (e.g. m/z 171/173 with neutral loss of 30 and m/z 155/157 with neutral loss of 46) as indicated by bromine's characteristic isotope peaks. The loss of 30 corresponds to NO, which is corroborated by two further pieces of information: i) the fragments are odd integer values, suggesting that they don't contain nitrogen (lost in the neutral fragment); ii) the loss of 46, indicated by the peak at m/z 155, is indicative of the loss of NO_2. Loss of NO and NO_2, and a high abundance molecular ion peak, suggest an aromatic nitro compound. As the Br atom is retained in the cation fragments, it is likely that the Br is substituted elsewhere on the aromatic ring. Evidence to support this is provided by the fragment peak at m/z 76, which corresponds to loss of a bromine radical from m/z 155/157 (i.e. loss of NO_2 radical followed by a Br radical). The peaks at m/z 143/145 correspond to loss of CNO_2 and ring opening.

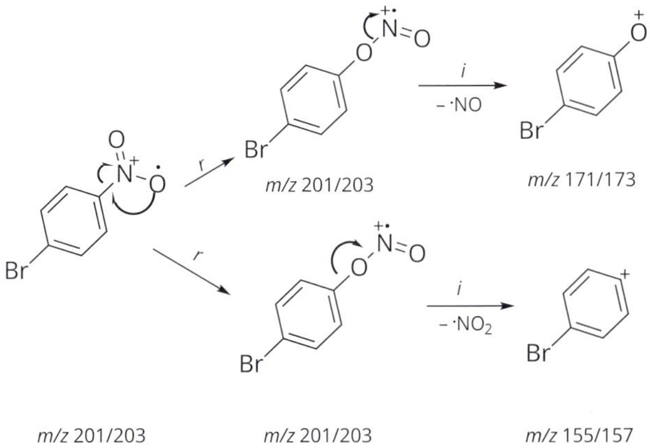

8.9 **Example 9**

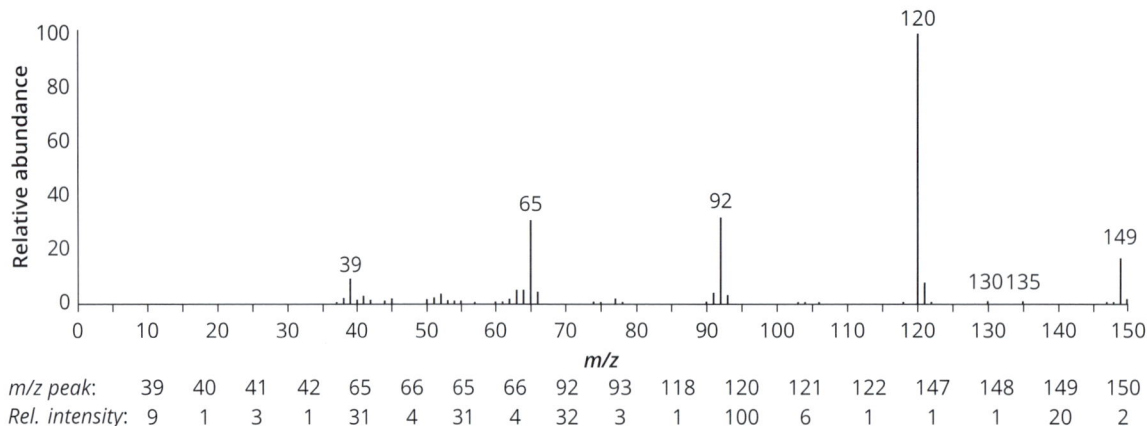

Mass spectrum

m/z peak:	39	40	41	42	65	66	65	66	92	93	118	120	121	122	147	148	149	150
Rel. intensity:	9	1	3	1	31	4	31	4	32	3	1	100	6	1	1	1	20	2

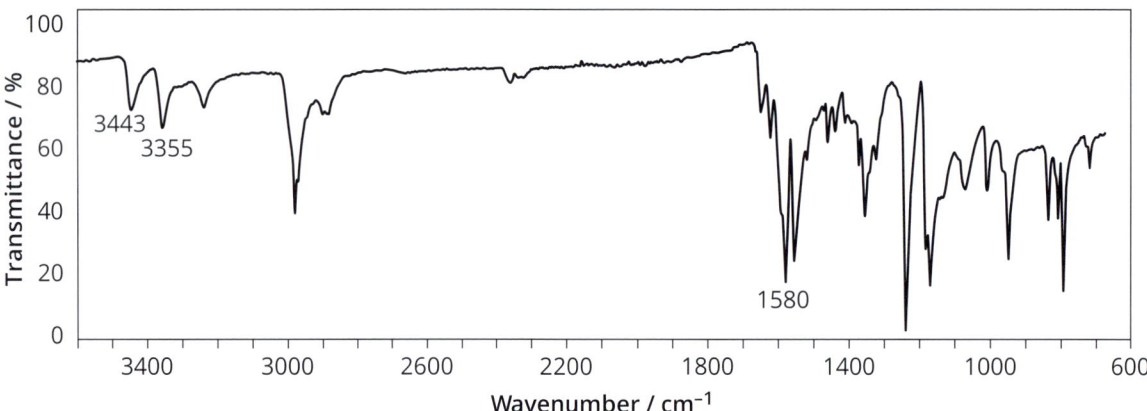

Infrared spectrum

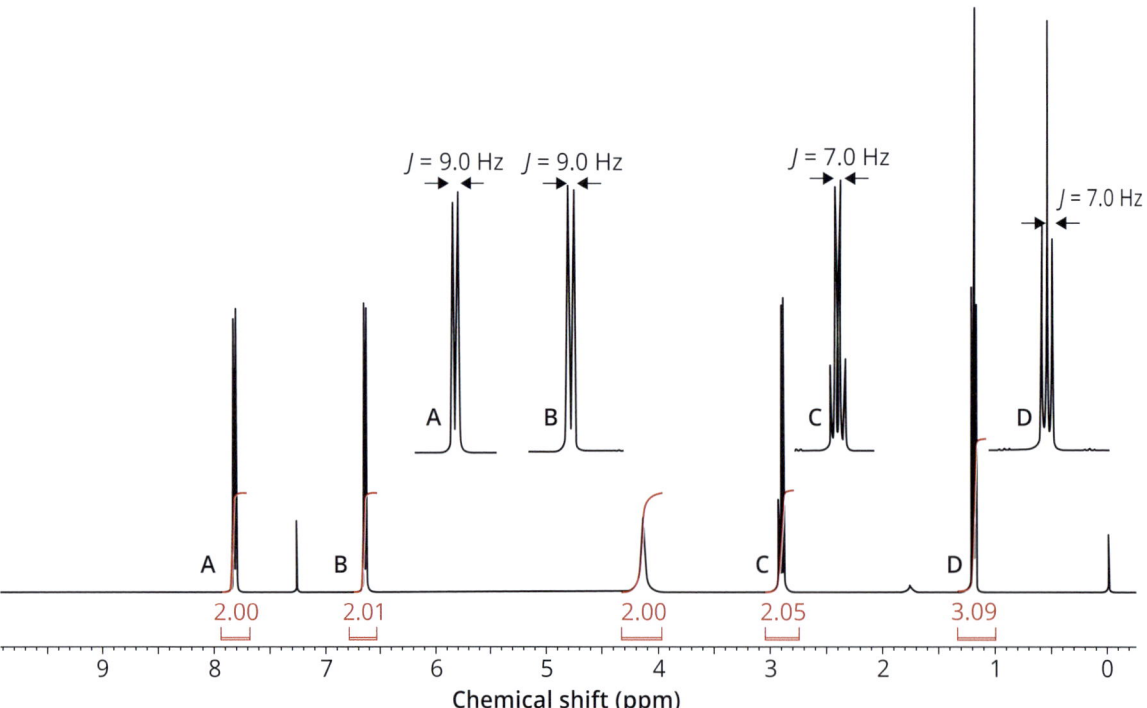

¹H NMR spectrum (400 MHz, CDCl₃)

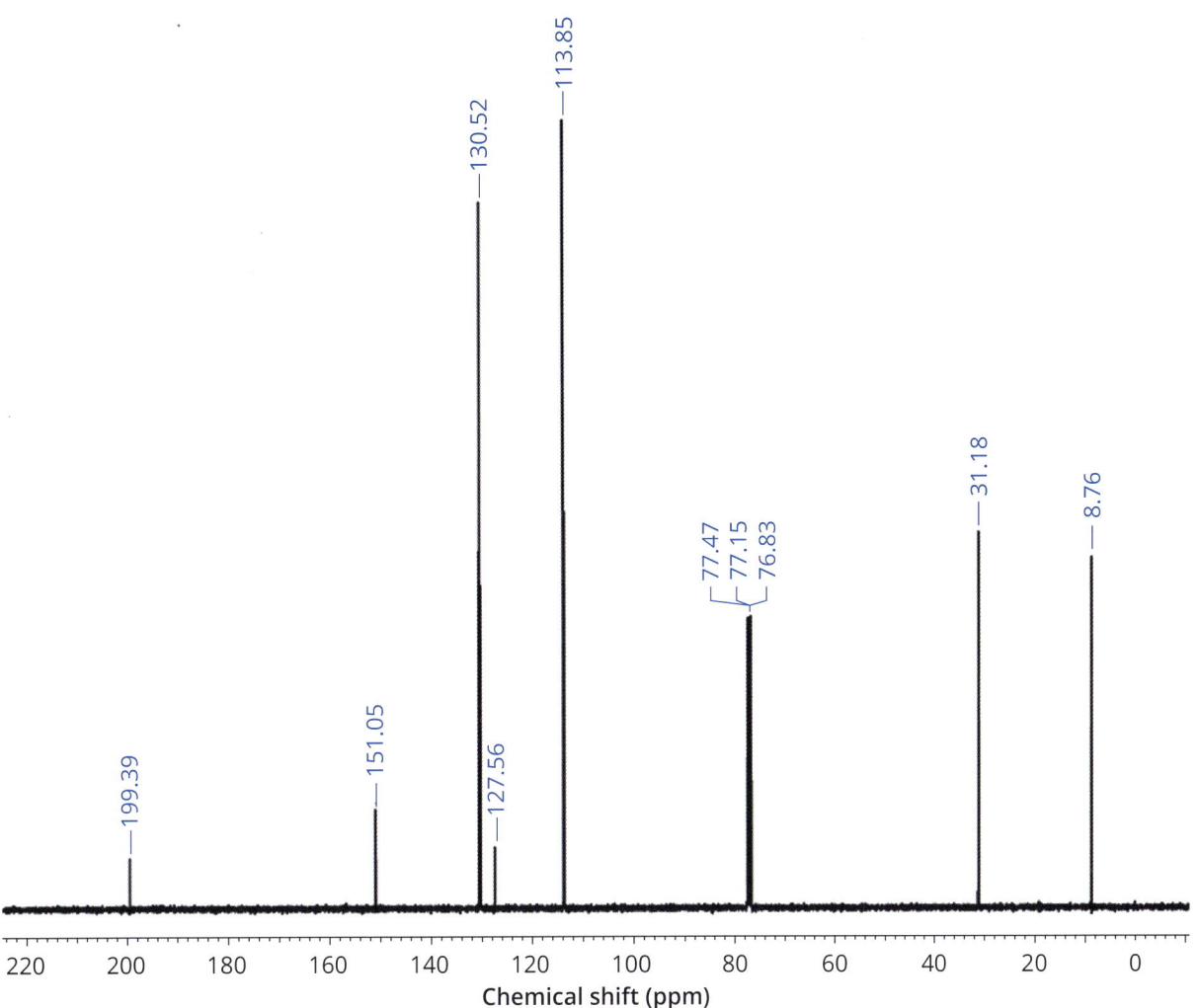

¹³C NMR spectrum (100 MHz, CDCl₃)

8.9.1 Initial analysis of Example 9

(a) The peak at *m/z* 149 in the mass spectrum corresponds to the molecular ion (M⁺·).

(b) The isotopic abundances suggest that neither chlorine nor bromine are present.

(c) An odd M⁺· indicates that at least one nitrogen atom is present in the compound, or there could be any odd multiple of these (e.g. 3, 5, etc).

(d) The presence of *m/z* 65, and a fairly abundant molecular ion, suggests that the compound may be aromatic.

(e) The number of carbon atoms in the molecular ion is ~9 based on the relative abundance of its M and M+1 peaks in the mass spectrum.

(f) The infrared spectrum doesn't show a strong absorption in the range 1650–1850 cm⁻¹, but a cross-check with the ¹³C NMR reveals a signal at 199.4 ppm. This latter observation is very typical of a ketone or aldehyde group, but the lack of a 'matching' absorption in the IR range is puzzling at first sight. At this stage, it is best to carefully note this apparent discrepancy and move on before making a firm conclusion.

(g) The region around 2950–3000 cm⁻¹ in the IR spectrum shows absorptions typical of the CH stretching frequency, and there are also sharp, but not very intense, absorptions around 3400 cm⁻¹. These are typical of the frequency range for Het-H bonds (where Het is an electronegative atom such as O or N). In example **EG7** it was noted that O–H stretches are typically broad for alcohols and phenols, or very broad for carboxylic acids. These sharper signals are therefore indicative of N–H stretching absorptions. As there is more than one absorption, this further points towards a primary amine group being present.

(h) The ^{1}H NMR shows five signals, if we ignore the trace $CHCl_3$ at 7.26 ppm and TMS at 0.00 ppm.

(i) The ratio of signal integrals is 2:2:2:2:3, which indicates the presence of eleven hydrogen atoms (or a multiple).

(j) The ^{13}C NMR shows seven signals, and the chemical shift ranges indicate two saturated and five unsaturated carbon environments. Remember not to count the $CDCl_3$ 1:1:1 triplet at 77 ppm.

(k) Of the five unsaturated carbon signals, one is in the carbonyl region (199.4 ppm) and the other four are in the alkene/aromatic region.

(l) The composition accounted for so far is $C_7H_{11}NO$, which adds to a mass of 125.

(m) From conclusions (a) and (l), the difference in mass is 24. This is most probably due to two extra carbon atoms, as indicated by conclusion (e), which would give a molecular formula of $C_9H_{11}NO$.

(n) For $C_9H_{11}NO$, the DU = 5. From (k), and elsewhere, the presence of an aromatic ring is indicated, and there is further evidence of another DBE which could be the carbonyl group. These groups would account for the overall degree of unsaturation calculated.

8.9.2 Detailed analysis of Example 9

(o) ^{1}H NMR: There are two signals at 7.8 and 6.6 ppm, each of which integrates to 2H. The signals are in the aromatic region of the spectrum, and the slight roofing distortion of each signal reveals the classic, centrosymmetric pattern characteristic of a 1,4-disubstituted aromatic ring with $^3J_{HH}$ = 9 Hz (an AA'BB' pattern).

(p) ^{1}H NMR: A broad singlet can be seen at 4.1 ppm. Broadened signals are often indicative of H atoms attached directly to electronegative atoms, such as OH, NH, etc. With the integration being 2H, and the IR indicating the presence of an amino group [see (g)], this signal is most likely due to an NH_2 group.

(q) ^{1}H NMR: The signals at 2.9 and 1.2 ppm are due to a 2H quartet and a 3H triplet with J_{HH} = 7 Hz. In combination these are indicative of the presence of an ethyl (CH_2CH_3) group. The chemical shift of the quartet will be useful in assigning the connectivity in due course.

(r) ^{13}C NMR: We can re-evaluate the ^{13}C spectrum based on information from the ^{1}H NMR and other spectra. The signal at 199.4 ppm has been identified as a carbonyl carbon, and the two aliphatic peaks at 31.2 and 8.8 ppm must be paired with the ethyl group identified from the ^{1}H NMR.

(s) ^{13}C NMR: A 1,4-disubstituted aromatic system will provide four carbon environments, and four such signals can be seen at 151.1, 130.5, 127.6, and 113.9 ppm. The two smaller and two larger peaks add support to the conclusion that the ring is disubstituted (see section 3.2), as the smaller peaks at 151.1 and 127.6 ppm are due to the quaternary (substituted) carbons, and the larger peaks due to the unsubstituted carbons.

The analysis of the data for **EG9** indicates that the following structural elements are present, and these account for all of the observed proton and carbon signals in the NMR spectra, and match the molecular formula $C_9H_{11}NO$.

1　　　　**2**　　　　**3**　　　　**4**

8.9.3 Conclusion and final structure of Example 9

The structural elements identified match the predicted molecular formula and weight of $C_9H_{11}NO$ and 149 g/mol respectively. Structural elements **3** and **4**, each with one unfilled valency, can be seen as 'edge pieces' in the molecular jigsaw, and the core of the molecule must therefore be made from structural elements **1** and **2**. The amino and ethyl groups could be connected in two different arrangements, giving either 4-ethylbenzamide or a 4-aminophenyl ketone. The chemical shift of the CH_2 group is in the region of 2–3 ppm and could be explained by either of these structures. However, the chemical shift of the carbonyl group (199.4 ppm) is greater than 190 ppm and will therefore only fit the ketone and not the amide. Recall that the carbon atoms in more electrophilic carbonyl groups (ketone/aldehyde) have ^{13}C signals at consistently higher chemical shifts than those in the less electrophilic examples, such as an amide, as we also noted in the analysis of Example 6. The molecular structure can be pieced together to give 1-(4'-aminophenyl)propan-1-one (**EG9**).

4-ethylbenzamide

1-(4'-aminophenyl)propan-1-one
EG9

This structural assignment is supported by the fragmentation peaks observed in the mass spectrum. The fragment peak at m/z 120 (the base peak) represents α-cleavage from the aromatic ketone (a loss of 29 corresponding to CH_2CH_3). **Note**: the odd m/z 149 radical cation indicates the presence of an odd number of nitrogen atoms. Fragmentation (via single-bond cleavage) leads to even m/z fragments, indicating an odd number of nitrogen atoms and suggesting that the even m/z 120 and m/z 92 also contain an odd number of nitrogen atoms (e.g. have not been lost via the neutral fragments). The other significant fragment peak at m/z 65 is an odd m/z value, suggesting loss of an odd number of nitrogen atoms in the neutral fragment. (Note that an odd m/z value fragment could also be formed via rearrangement to give a distonic radical cation fragment such as in a McLafferty rearrangement—but this is not likely here). The fragmentation supports the **EG9** structural assignment.

m/z 149 m/z 120

8.9.4 Further rationalization of Example 9

The IR spectrum does not show a C=O stretching frequency in the range most commonly expected of these groups, yet the proposed product structure contains one. Why is this so? This example provides a good case for showing why trying to learn from data tables is the wrong approach, and trying to fit an answer around an early deduction (the IR does not show a typical absorption due to a carbonyl stretching frequency) can be very unhelpful. The data tables always provide *typical* ranges, but the more narrowly those ranges are defined, the more likely there will be examples that don't 'fit the rule'. Having broader ranges, on the other hand, means that the ranges overlap and become less useful. The reason behind the apparent lack of a C=O absorption in this example is that the stretching frequency of a carbonyl group is heavily dependent on conjugation, as this may greatly affect the localization of the π electrons in the C=O bond. As can be seen below, the amino group has a lone pair of electrons which are de-localized through the ring and carbonyl π bond. It is also for this reason that an amide C=O stretching frequency is much lower (ca. 1650 cm^{-1}) than that of, say, an aldehyde, ketone, or ester.

This delocalization through the structure also explains the observed NMR chemical shifts. The position adjacent to the amino group exhibits rather small shifts of 6.6 and 113 ppm for ^{1}H and ^{13}C respectively, due to shielding effects from the nitrogen lone pair. Conversely, the carbon bearing the nitrogen is deshielded at 151 ppm due to the electronegativity of the nitrogen, providing further evidence for the attachment of this heteroatom directly to the aromatic ring:

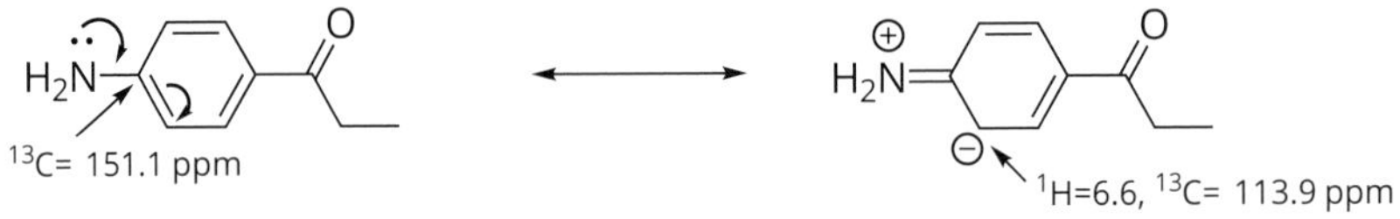

^{13}C= 151.1 ppm

^{1}H=6.6, ^{13}C= 113.9 ppm

The aromatic protons adjacent to the carbonyl group have a much larger chemical shift of 7.8 ppm due to the resonance relocation of electron density onto the ketone functional group:

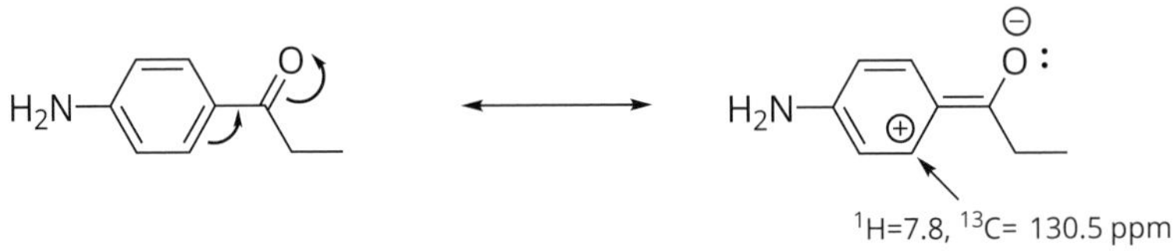

^{1}H=7.8, ^{13}C= 130.5 ppm

The fragment represented by the peak at *m/z* 92 in the mass spectrum has an even integer value indicating that its chemical composition contains an odd number of nitrogen atoms (assuming 1 nitrogen atom rather than 3 are present in the radical cation). This fragment corresponds to the loss of 57, via a mechanism involving inductive-cleavage and indicating a slightly less favoured route by its lower abundance compared to *m/z* 120, where α-cleavage on the other side of the carbonyl is more favourable.

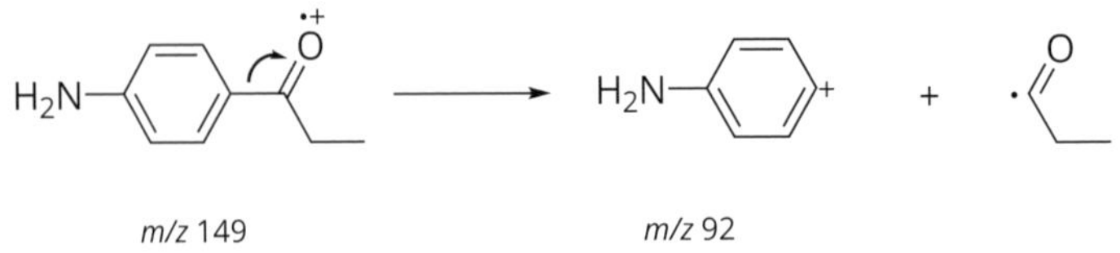

m/z 149

m/z 92

8.10 Example 10

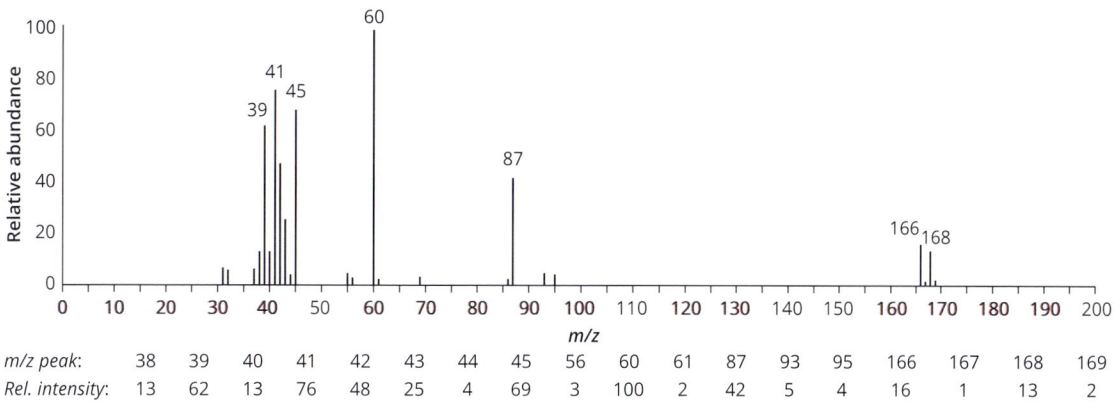

m/z peak:	38	39	40	41	42	43	44	45	56	60	61	87	93	95	166	167	168	169
Rel. intensity:	13	62	13	76	48	25	4	69	3	100	2	42	5	4	16	1	13	2

Mass spectrum

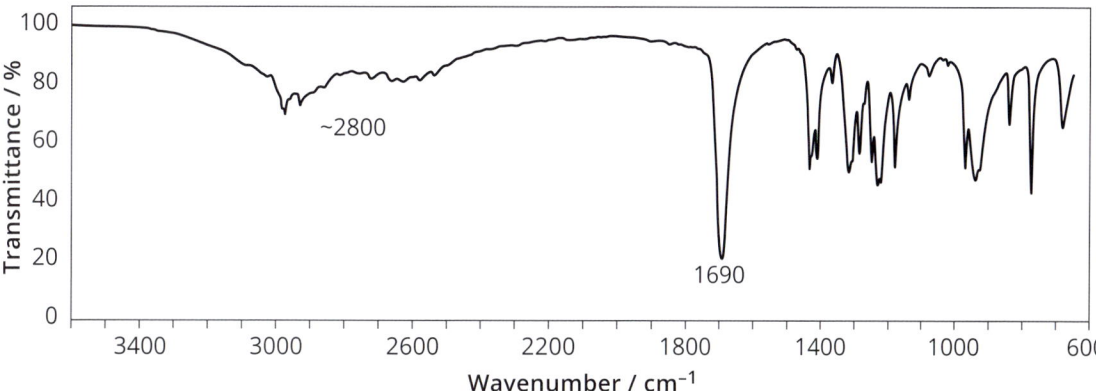

Infrared spectrum

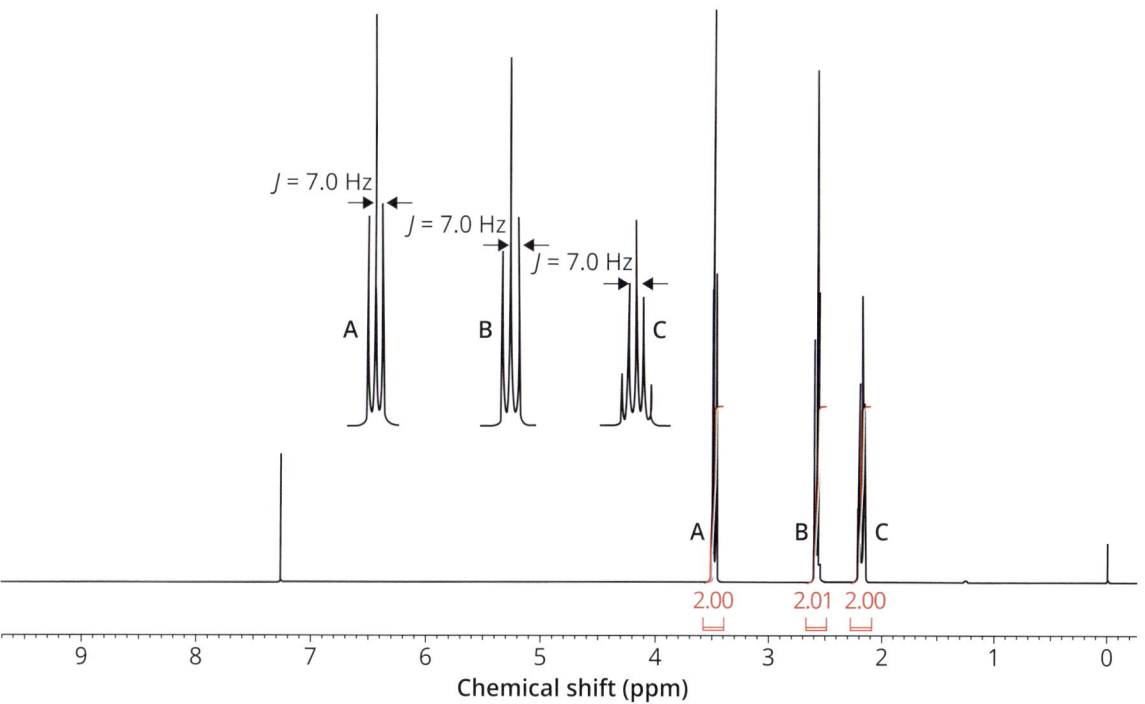

^{1}H NMR spectrum (400 MHz, CDCl$_3$)

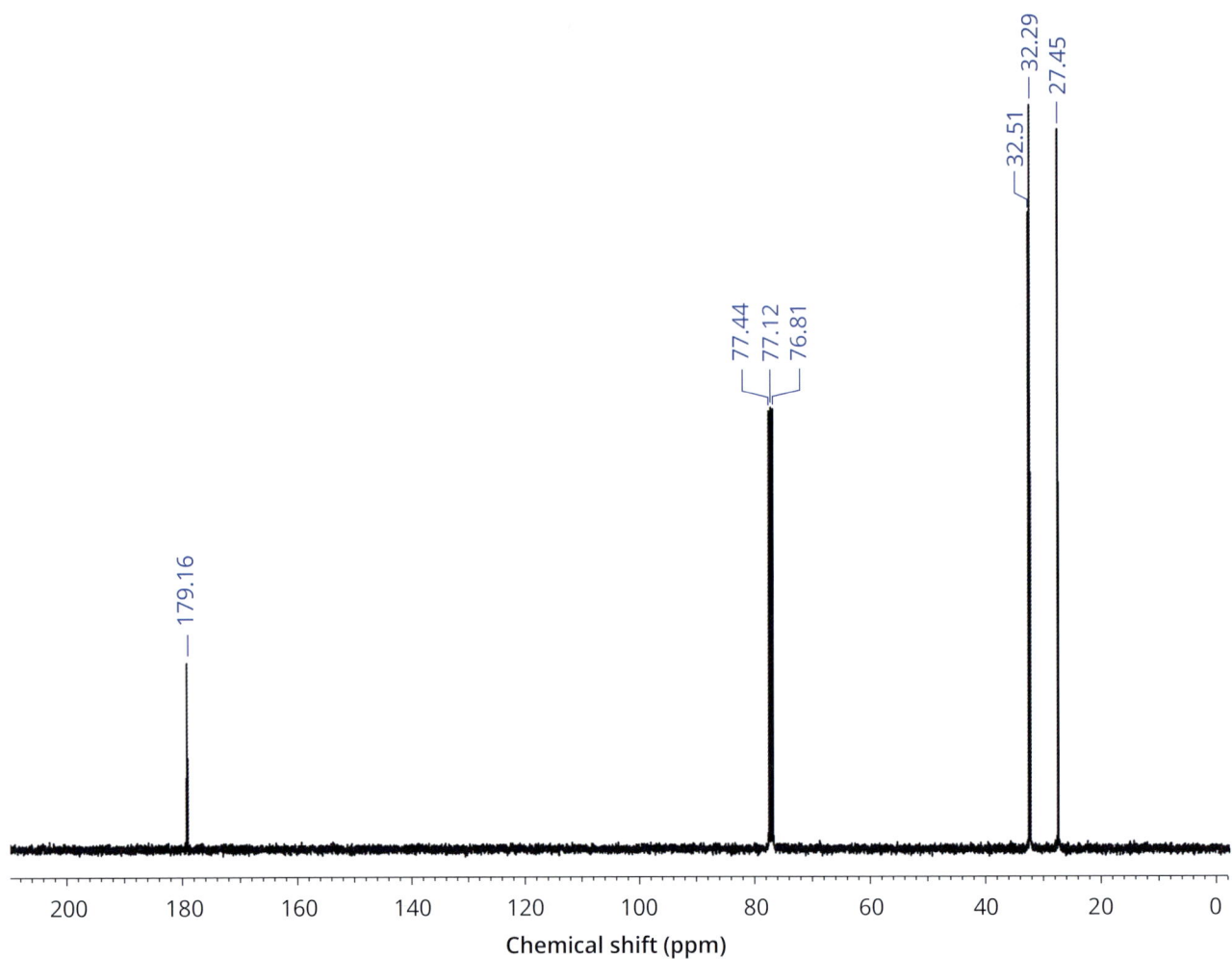

^{13}C NMR spectrum (100 MHz, CDCl$_3$)

8.10.1 Initial analysis of Example 10

(a) From the mass spectrum, the peaks at m/z 166 and m/z 168 correspond to isotopes of the molecular ion (M$^{+\cdot}$).

(b) From the mass spectrum, the isotope peaks at m/z 166 and m/z 168 have similar intensities and are separated by two mass units, indicating the presence of one bromine atom in the structure.

(c) An even M$^{+\cdot}$ indicates that either no nitrogen atoms or an even number of them are present in the compound.

(d) The molecular ion peaks are relatively low in abundance in the mass spectrum, suggesting that the compound may not contain an aromatic ring structure.

(e) The number of carbon atoms in the molecular ion is ~4 based on the relative abundance of its M and M+1 peaks in the mass spectrum.

(f) The IR shows a very broad, jagged absorption in the range 2500–3200 cm^{-1} which is characteristic of an OH stretch in a carboxylic acid; this observation is supported further by the presence of an absorption due to a C=O stretch at 1690 cm^{-1}.

(g) The ^{1}H NMR shows three signals, if we ignore the trace CHCl$_3$ at 7.26 ppm and TMS at 0.00 ppm.

(h) The ratio of signal integrals is 2:2:2, which indicates the presence of six hydrogen atoms (or a multiple).

(i) The ^{13}C NMR shows four signals—remember not to include the 1:1:1 triplet at 77 ppm due to the solvent CDCl$_3$. The chemical shift ranges indicate one unsaturated and three saturated carbon environments. The infrared spectrum already indicates that one of the signals will be a carbonyl group, and the small peak at 179.2 ppm is easily assigned in this case.

(j) The estimated molecular composition at this stage is C$_4$H$_6$, with the IR spectrum indicating that there is a carboxylic acid (+O$_2$H) and the mass spectrum revealing the presence of Br and 4 carbon atoms. This gives C$_4$H$_7$BrO$_2$ and, using ^{79}Br, this gives a molecular weight of 166 g/mol.

(k) For C$_4$H$_7$BrO$_2$, we can calculate that the DU = 1. Thus, only one double bond or ring needs accounting for, and a carbonyl group that satisfies this requirement has already been identified [see (f) and (i)].

8.10.2 Detailed analysis of Example 10

(l) ^{1}H NMR: Of the three signals, all in the aliphatic region, visible in this spectrum, the 2H pentet has the smallest δ value at 2.15 ppm. The chemical shift and splitting pattern are indicative of a CH$_2$ flanked by four neighbouring hydrogen atoms which are likely to be in the form of two more CH$_2$ groups, i.e. a propyl unit -CH$_2$CH$_2$CH$_2$-.

(m) ^{1}H NMR: The couplings of the two triplets at 3.5 and 2.6 ppm match what would be expected for the two end groups of the propyl identified in (l). The 7.0 Hz couplings between these protons are consistent with $^3J_{HH}$ values seen for aliphatic alkyl groups. The chemical shift of each is determined by what else is attached at each end, but as the difference is relatively large this could be of use in assigning the signals in the final structure.

(n) The IR absorption at 2500–3200 cm^{-1} indicates the presence of a carboxylic acid. This observation was supported by the presence of both a C=O stretch and a carbonyl carbon signal in the ^{13}C NMR spectrum, as we noted in (f) and (i). However, no corresponding OH signal was observed in the ^{1}H NMR spectrum. This example reminds us that very acidic hydrogen atoms can undergo exchange processes and, as a result, signals can be very broad, sometimes to the extent that they are not seen at all. Overall, the evidence firmly points to a carboxylic acid, even though the OH is not observed in the ^{1}H NMR spectrum.

(o) ^{13}C NMR: The signal at 179.2 ppm has been assigned already as the carbonyl carbon, leaving three aliphatic signals at 27.5, 32.3, and 32.5 ppm. The chemical shifts of these signals mirror the three signals observed in the ^{1}H NMR spectrum.

The analysis of the data indicates that the following structural elements are present, and these account for all of the observed proton and carbon signals in the NMR spectra.

$$-CH_2-CH_2-CH_2- \qquad\qquad C{\overset{OH}{\underset{O}{\big|}}} \qquad\qquad -Br$$

1	2	3

From the mass spectrum, the fragment ion at m/z 87 indicates the loss of a bromine radical (79/81 Da) from the molecular ion by charge site-driven cleavage. The fragment peak at m/z 45 corresponds to further loss of CO_2H, providing supporting evidence for the presence of the structural elements **2** and **3** determined from NMR. These fragments indicate that there is a terminal bromine atom and a separate carboxyl group associated with a non-aromatic structure.

8.10.3 Conclusion and final structure of Example 10

The structural elements identified match the predicted molecular formula and weight of $C_4H_7BrO_2$ and 166 or 168 g/mol respectively. Structural elements **2** and **3**, each with one unfilled valency, can be seen as 'edge pieces' in the molecular jigsaw, and so the core of the molecule must therefore be structural element **1**. The molecular structure can be pieced together to give 4-bromobutanoic acid (**EG10**).

EG10

8.10.4 Further rationalization of Example 10

The mass spectrum shows a base peak at m/z 60. At first, this peak is not easy to rationalize, but its even integer m/z value provides an important clue. The nitrogen rule states that an even integer radical cation contains no nitrogen atoms (or an even number of nitrogen atoms), and this will result in odd m/z fragments upon radical loss. The lack of nitrogen thus indicates a rearrangement (breaking of more than 1 bond), which is common for carboxylic acids and esters. The peak therefore likely results from a McLafferty rearrangement. The base peak at m/z 60 also clearly does not contain a bromine atom due to the lack of its characteristic M+2 isotope pattern. So, the bromine has clearly been lost in the neutral fragment created through the McLafferty rearrangement.

m/z 166/168   *m/z* 60

It is possible to assign specifically the two triplets from the ¹H NMR spectrum. The greater electronegativity of the bromine deshields the neighbouring methylene group to a greater extent than the carboxyl group, hence we can provide the following proton assignments:

The three aliphatic signals in the ¹³C NMR spectrum, however, are much closer and impossible to assign reliably. But a 2D proton–carbon correlation spectrum (known as HSQC) would allow the definitive assignment of the carbon signals, if that were required, by correlating them with their attached protons (whose shift assignments we already know). You can learn more about this experiment in Chapter 10 on 2D NMR methods.

8.11 **Example 11**

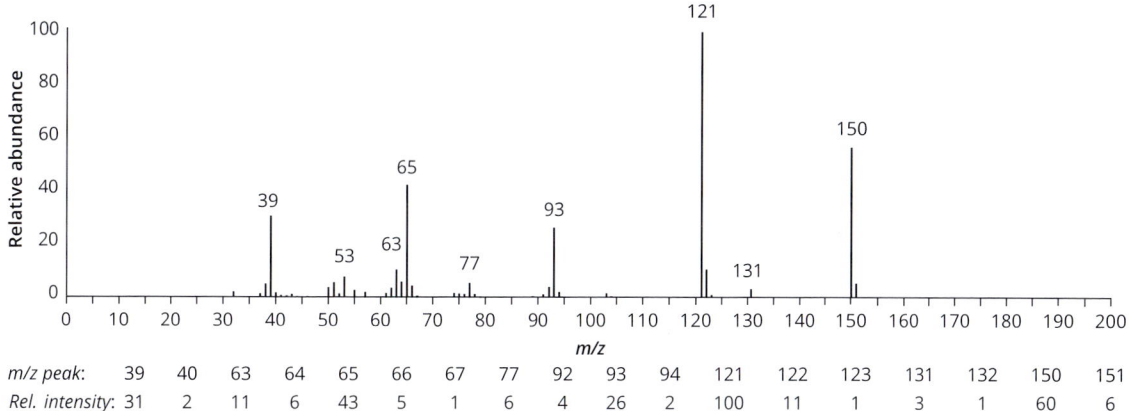

Mass spectrum

m/z peak:	39	40	63	64	65	66	67	77	92	93	94	121	122	123	131	132	150	151
Rel. intensity:	31	2	11	6	43	5	1	6	4	26	2	100	11	1	3	1	60	6

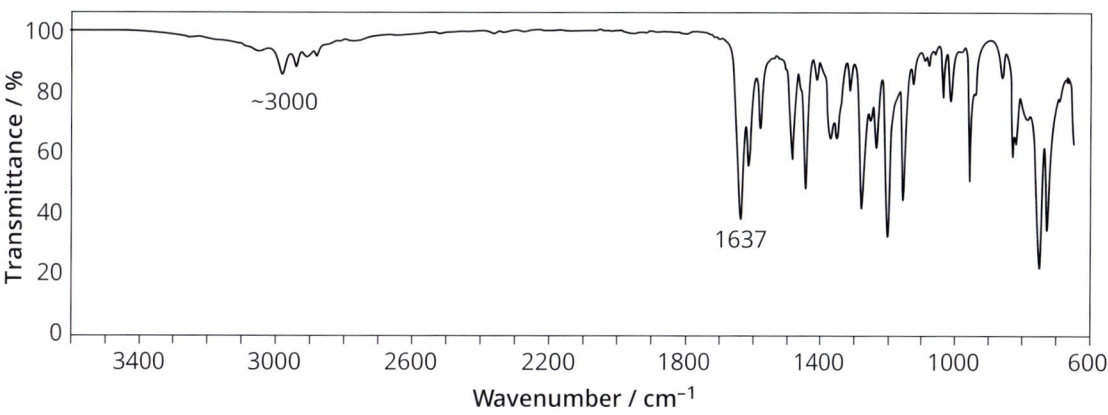

Infrared spectrum

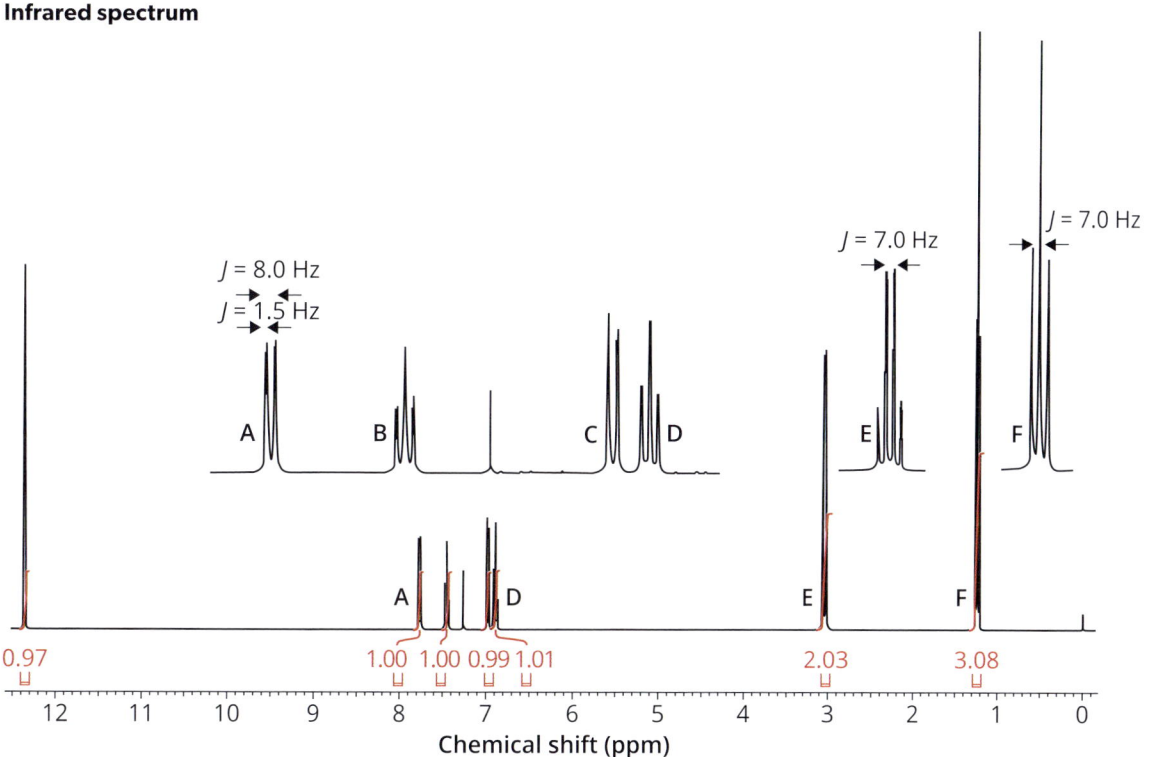

¹H NMR spectrum (400 MHz, CDCl₃)

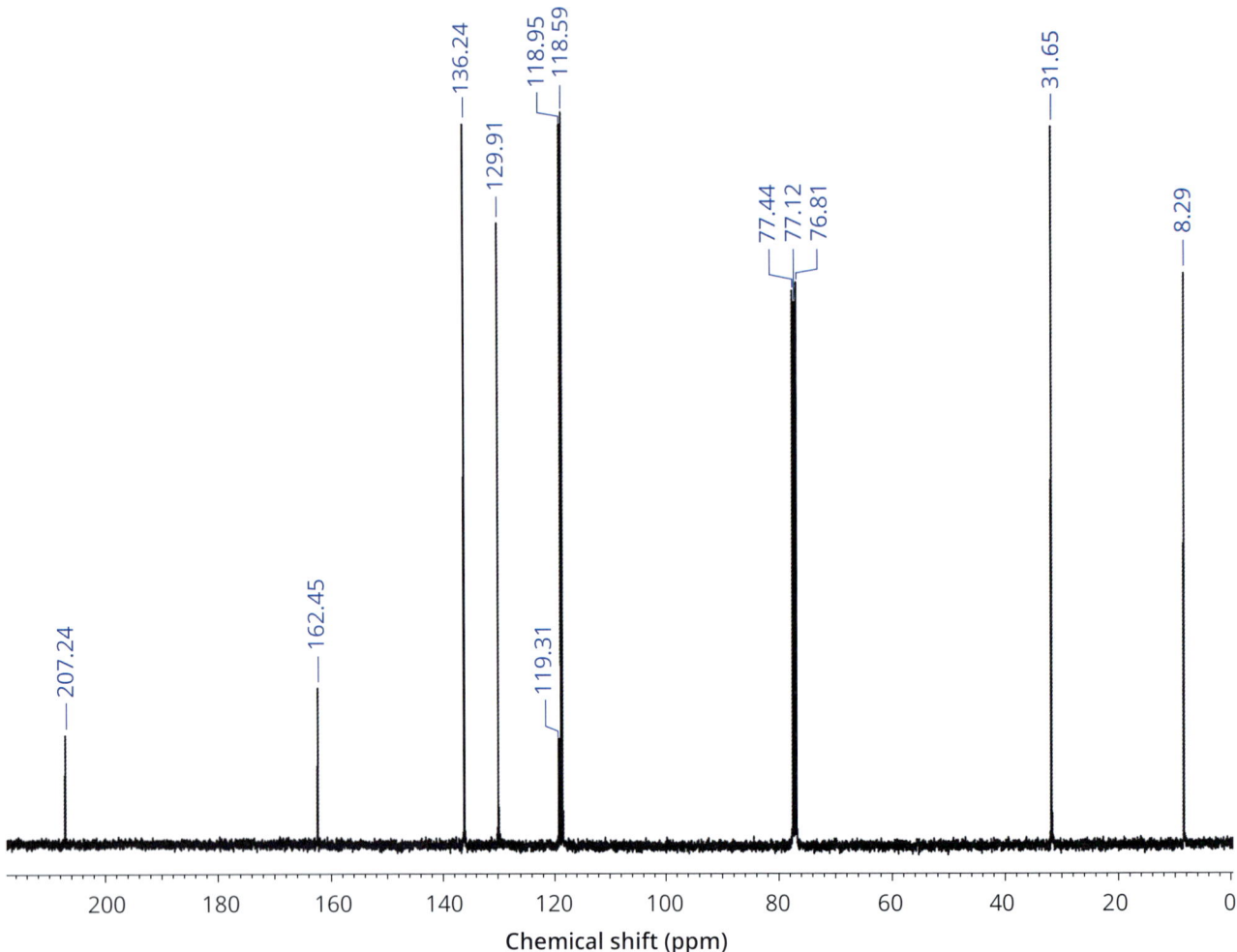

^{13}C NMR spectrum (100 MHz, CDCl$_3$)

8.11.1 Initial analysis of Example 11

(a) The peak at m/z 150 in the mass spectrum is likely to represent the molecular ion (M$^{+\bullet}$).

(b) The isotopic abundances suggest that neither chlorine nor bromine are present.

(c) An even M$^{+\bullet}$ indicates that either no nitrogen atoms or an even number of them (e.g. 2, 4, etc) are present in the compound.

(d) A prominent molecular ion peak suggests that the compound may be aromatic.

(e) The number of carbon atoms in the molecular ion is predicted to be ~9 based on the relative abundance of M and M+1 peaks in the mass spectrum.

(f) The infrared spectrum indicates the presence of a carbonyl group due to the strong absorption at ~1640 cm^{-1}. Whilst this absorption is quite low for a carbonyl group, which casts some doubt on this conclusion, a cross-check with the ^{13}C NMR reveals a signal at 207 ppm. This kind of signal is typical of a ketone or aldehyde group and therefore confirms the general assignment.

(g) The region around 3000 cm^{-1} in the IR spectrum shows quite a broad absorption, but it is neither very intense nor over a very great range.

(h) The ^{1}H NMR shows seven signals. We should ignore the trace CHCl$_3$ at 7.26 ppm and TMS at 0.00 ppm.

(i) The ratio of signal integrals is 1:1:1:1:1:2:3, which indicates the presence of ten hydrogen atoms (or a multiple).

(j) ^{13}C NMR shows nine signals, and the chemical shift ranges indicate two saturated and seven unsaturated carbon environments. Remember that we should not count the $CDCl_3$ 1:1:1 triplet at 77 ppm.

(k) Of the unsaturated carbon signals, six are in the aromatic region and one in the carbonyl region (207.2 ppm).

(l) The estimated molecular constitution so far is $C_9H_{10}O$, which adds up to a molecular weight of 134 and corresponds to the number of carbon atoms predicted by the isotope peak abundances in the mass spectrum.

(m) From conclusion (a) and (l), we can deduce that the difference in mass is 16 (150–134). This is most probably due to an extra oxygen atom, which would give a tentative molecular formula of $C_9H_{10}O_2$.

(n) For $C_9H_{10}O_2$, we can calculate that the DU = 5. From (d), (f), (k), and elsewhere, the presence of an aromatic ring and a carbonyl group are indicated, and these would account for the degree of unsaturation calculated.

8.11.2 Detailed analysis of Example 11

(o) ^1H NMR: The signal at 12.4 ppm is a singlet 1H. Given that the signal is above 10 ppm, it must arise from a deshielded and uncoupled nucleus—e.g. an –OH group. A carboxylic acid immediately comes to mind; however, this signal is sharp whereas acids typically give a very broadened signal. Also, a carboxylic acid would give a jagged IR absorption over a wide range (2500–3200 cm^{-1}, O–H stretch). This is not clearly evident in the IR spectrum where the absorption is not that intense.

(p) ^1H NMR: The signals at 7.8 ppm (doublet, 1H), 7.4 ppm (triplet, 1H), 6.9 ppm (doublet, 1H), and 6.8 ppm (triplet, 1H) are chemical shifts typical of unsaturated CHs. Their J_{HH} values of 8.0 and 1.5 Hz suggest the presence of both ^3J and fine ^4J coupling, supporting the idea that these are in an aromatic environment, as such longer-range coupling can often be observed in aromatic environments. The major pattern of two doublets and two triplets is consistent with a 1,2-disubstituted aromatic system.

(q) ^1H NMR: The signals at 3.0 ppm (quartet, 2H) and 1.2 ppm (triplet, 3H) sharing a 7.0 Hz coupling are consistent with an ethyl group (CH_3CH_2–). The chemical shift of the CH_2 indicates that it is attached to a moderately electron-withdrawing group, but it is not likely to be an oxygen atom as this would give a shift closer to 4 ppm.

(r) The ^{13}C NMR spectrum supports the idea of a disubstituted aromatic ring. There are four significantly more intense signals (unsubstituted ArC at 118.6, 119.0, 129.9 and 136.2 ppm) and two weaker ones (substituted ArC at 119.3 and 162.5 ppm).

(s) ^{13}C NMR: The two aliphatic signals (8.3 and 31.7 ppm) are consistent with the ethyl group identified in the ^1H NMR spectrum.

(t) The heavily deshielded signal at 207.2 ppm has already been assigned to a carbonyl group, and such a large shift (>190 ppm) is typical of a ketone.

The analysis of the data indicates that the following structural elements are present, and these account for all of the observed proton and carbon signals in the NMR spectra.

| 1 | 2 | 3 | 4 |

The mass spectrum provides a series of fragment peaks which help confirm the presence of a substituted aromatic ring; e.g. a relatively high abundance molecular ion peak, along with characteristic peaks at m/z 77 and m/z 65 corresponding to a phenyl cation and its breakdown product. The base peak is a fragment at m/z 121 corresponding to a loss of 29 which could be HCO or C_2H_5 but is most likely to correspond to loss of the ethyl group as shown by the NMR (**4**). The peak at m/z 93 corresponds to a phenolic carbocation produced from the loss of 57 from the molecular ion, suggesting that the aromatic ring has an OH substituent. A loss of 57 corresponds to the loss of a C_3H_5O radical, which provides supporting evidence that the loss of 29 was due to C_2H_5 and not CHO.

m/z 150   *m/z* 121

m/z 150   *m/z* 93

8.11.3 Conclusion and final structure of Example 11

The structural elements identified match the predicted molecular formula and mass of $C_9H_{10}O_2$ and 150 g/mol respectively. Adding these structural elements together is straightforward once it is determined that the OH group is not part of a carboxylic acid. Thus, the OH must be one of the substituents on the 1,2-disubstituted aromatic ring. The molecular structure can be pieced together to give 1-(2'-hydroxyphenyl)propan-1-one (**EG11**).

8.11.4 Further rationalization of Example 11

The OH group gave a sharp signal in the ^{1}H NMR spectrum, which is also very deshielded at 12.4 ppm. The carbonyl stretching frequency in the IR spectrum is also atypically low for a ketone and the O–H stretch is weak. Both these features are consistent with the OH group forming an intramolecular hydrogen bond to the neighbouring C=O. These features are in fact very characteristic of *ortho* OH groups in aromatic systems, which should be noted for future reference. The hydrogen bonding also contributes to the quite large chemical shift of the C(O)CH_2 at 3.0 ppm. Without it, such a CH$_2$ would typically be found around 2.5 ppm.

The OH attached to the aromatic ring also accounts for the deshielded carbon resonance at 162.5 ppm—further evidence for the phenolic functional group.

8.12 **Example 12**

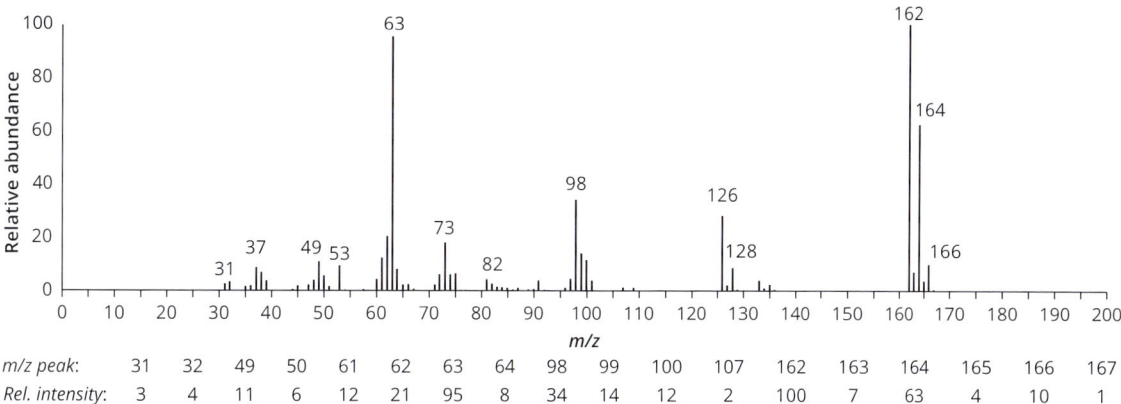

m/z peak:	31	32	49	50	61	62	63	64	98	99	100	107	162	163	164	165	166	167
Rel. intensity:	3	4	11	6	12	21	95	8	34	14	12	2	100	7	63	4	10	1

Mass spectrum

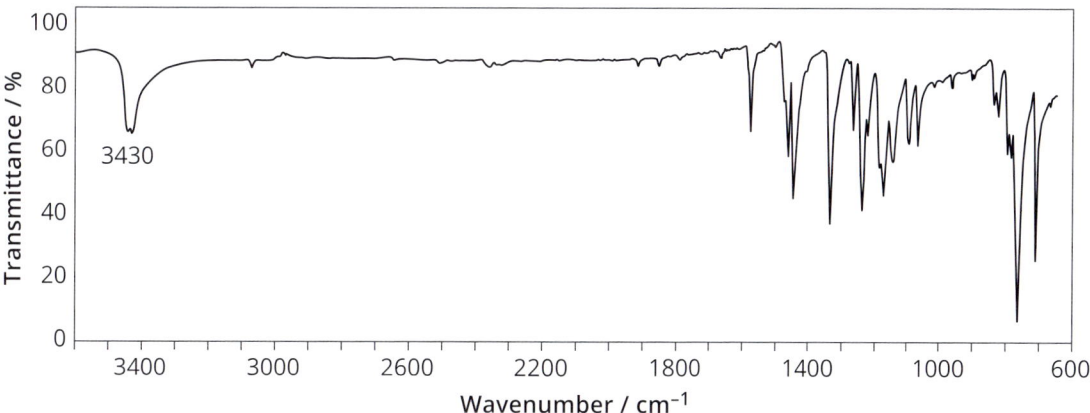

Infrared spectrum

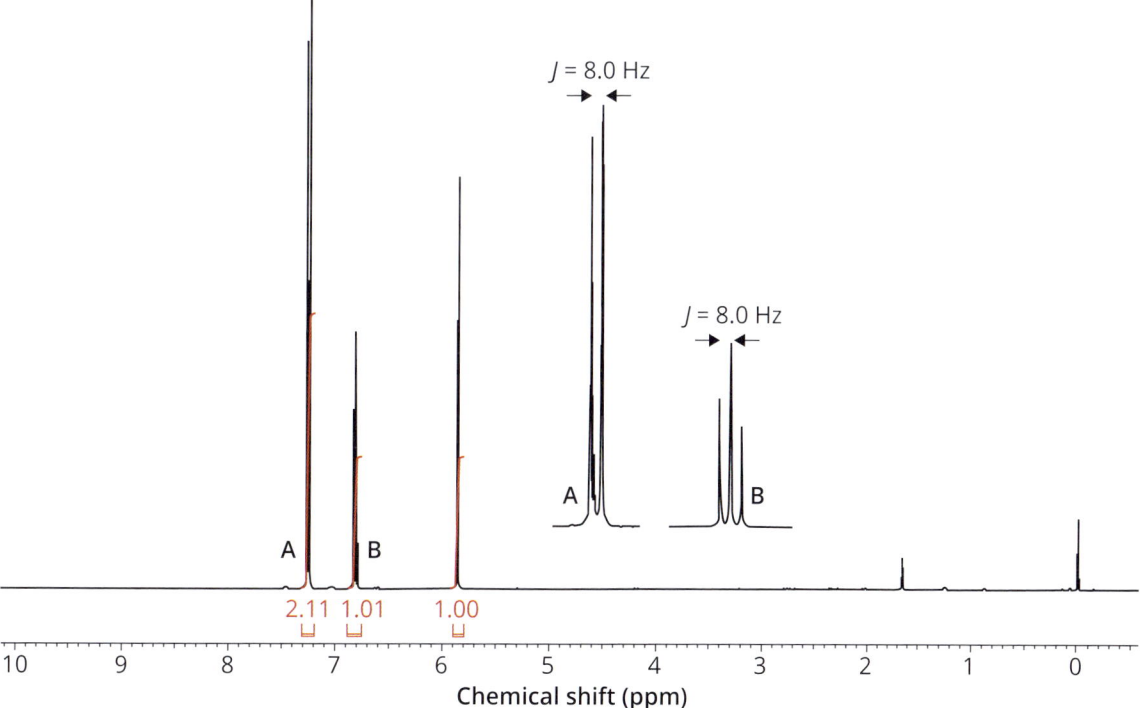

¹H NMR spectrum (400 MHz, CDCl₃)

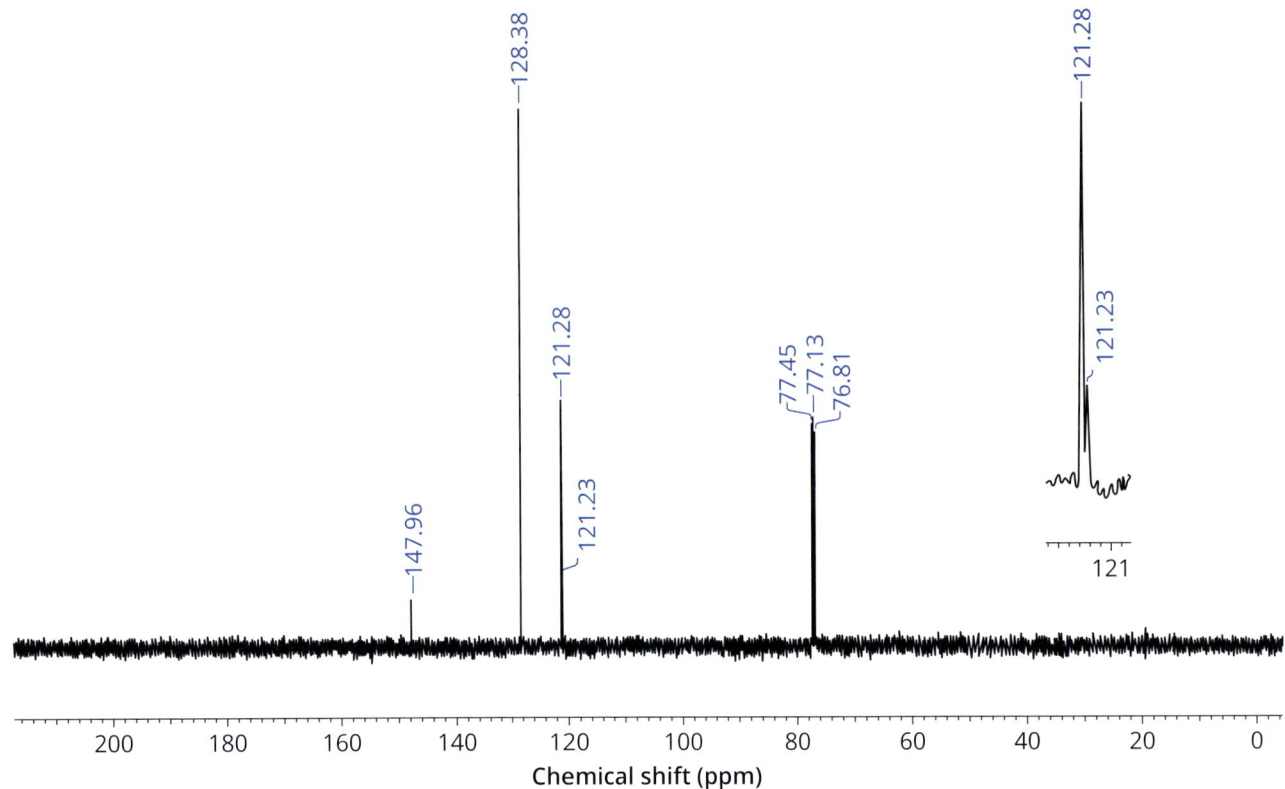

¹³C NMR spectrum (100 MHz, CDCl₃)

8.12.1 Initial analysis of Example 12

(a) The peaks at m/z 162, m/z 164, and m/z 166 in the mass spectrum are likely to represent isotopes of the molecular ion (M⁺•).

(b) The cluster of isotope peaks separated by two mass units each (m/z 162, m/z 164, and m/z 166) are in the ratio (approximately) 9:6:1, which is typical of the presence of *two* chlorine atoms.

(c) An even m/z value for the M⁺• peaks indicates that either no nitrogen atoms or an even number (e.g. 2, 4, etc) are present in the compound.

(d) The high abundance of the molecular ion suggests that the compound is likely to be aromatic.

(e) The number of carbon atoms in the molecular ion is predicted to be ~6 based on the relative abundance of its M and M+1 peaks in the mass spectrum.

(f) The IR shows a smooth, broad absorption at 3450 cm⁻¹, which is typical of an alcohol group (OH stretch).

(g) The ¹H NMR shows three signals. We should ignore the trace CHCl₃ at 7.26 ppm, water at 1.65 ppm, and TMS at 0.00 ppm.

(h) The ratio of signal integrals is 2:1:1, which indicates the presence of four hydrogen atoms (or a multiple).

(i) ¹³C NMR shows four signals (remember not to count the CDCl₃ 1:1:1 triplet at 77 ppm), and the chemical shift ranges indicate unsaturated carbon environments for all of them. We need to consider the possibility of symmetry within an aromatic ring here, which would reduce the number of signals observed.

(j) The composition accounted for so far is $C_4H_4Cl_2O$, which adds up to a mass of 138 (for ^{35}Cl).

(k) From conclusions (a) and (j), we can deduce that the difference in mass is 24 (162–138), and we can predict that two extra carbon atoms account for this based on (e). A tentative molecular formula is therefore $C_6H_4Cl_2O$.

(l) For $C_6H_4Cl_2O$, we can calculate that the DU = 4. Together with signals around 7–8 ppm in the ^1H NMR, this value indicates the likely presence of an aromatic ring.

8.12.2 Detailed analysis of Example 12

(m) ^1H NMR: This shows a 2H doublet at 7.3 ppm and a 1H triplet at 6.8 ppm sharing a 8.0 Hz coupling. The signals are in the aromatic region, and the 1H triplet signal is consistent with a symmetrical =CH–C*H*=CH– system. A C*H*=CH$_2$ system would lead to a more complex splitting pattern due to the *cis* and *trans* relationships of the coupling hydrogen atoms, and the *J* values would be much larger for alkenes, especially *trans* couplings that would be in excess of 10 Hz. Furthermore, the DU indicates that an aromatic ring is present. Thus, the inner CH gives rise to the 1H triplet, and, if symmetrical, the adjacent two equivalent CHs would give rise to the 2H doublet.

(n) ^1H NMR: The aromatic region of the spectrum is consistent with a 1,2,3-trisubstituted aromatic ring. Due to the splitting patterns discussed above we can also conclude that the ring must be symmetrical.

(o) ^1H NMR: A 1H singlet is seen at 5.8 ppm. This signal could correspond to an unsaturated CH environment, as it would be rather high to be a saturated CH environment. However, the IR has already indicated the presence of an alcohol, and the chemical shift is consistent with an OH assignment (noting that OH shifts can be highly variable).

(p) ^{13}C NMR: The carbon NMR spectrum shows four aromatic signals, two of which are of weaker intensity, one of medium intensity, and one of greater intensity. The evidence from the tentative molecular formula indicated elements of symmetry such that there are fewer signals than carbon atoms. Thus, the large signal at 128.4 ppm could point to the presence of more than one carbon atom in the same environment.

The analysis of the data indicates that the following structural elements are present, and these account for all of the observed proton and carbon signals in the NMR spectra.

8.12.3 Conclusion and final structure of Example 12

The structural elements identified match the predicted molecular formula and weight of $C_6H_4Cl_2O$ and 162/164/166 g/mol respectively. Adding these structural elements together seems straightforward, considering that there are three groups and a trisubstituted ring. However, do not overlook making sure that the ring has the requisite symmetry to account for the splitting pattern in the ^1H NMR spectrum and the carbon environments in the ^{13}C NMR spectrum. The molecular structure can be pieced together to give 2,6-dichlorophenol (**EG12**).

EG12

8.12.4 Further rationalization of Example 4

This fragmentation spectrum represents multiple fragmentation mechanisms. The peak at m/z 126 in the mass spectrum corresponds to a loss of HCl with the isotope abundance pattern at m/z 126/128 indicative of a radical cation containing the remaining single chlorine atom. There is evidence of the loss of the second chlorine atom via the isotope abundance pattern at m/z 91, although this is a minor peak. The base peak represents a fragment at m/z 63 and a second prominent fragment peak is found at m/z 98, which represents a loss of HCl followed by CO [M–HCl–CO], leaving a radical cation fragment. This complex mechanism is represented in the diagram following this paragraph. The loss of both chlorine atoms to form the base peak at m/z 63 is supported by the fact that the isotope pattern at m/z 63 represents no M+2 peak indicative of chlorine. Note that only neutral molecules have been lost in both cases, as indicated by the even m/z value of these fragments (radical loss leads to odd cation fragments when the molecular ion is even, and vice versa). Loss of CO is common for aromatic phenols, which leads to ring opening. In addition, when more than one functionality exists on the aromatic ring, an 'ortho effect' often occurs, which promotes a rearrangement to expel a neutral molecule from the radical cation (as observed here). Furthermore, the ortho effect is pronounced when the second substituent of a phenol is adjacent, but not an alkyl group, as is the case here. This would suggest that at least one chlorine and a hydroxyl group are adjacent substituents on the aromatic ring. Given the mass constraints (the presence of an aromatic ring with a hydroxyl and chlorine substituent), the second chlorine is likely to be a third substituent on the ring, but it is not clear where this is located from the mass spectra alone. The base peak at m/z 63 contains no chlorine and is a cation with a mass difference from 98 of 35. It therefore represents the loss of a chlorine radical.

| m/z 161.96 | m/z 126 | m/z 126 | m/z 126 | m/z 98 |

From the NMR, the two carbon CH groups *ortho* to the Cl give rise to the intense signal at 128.4 ppm, and the signal at 121.3 ppm is the single CH *para* to the OH. The carbon bearing the OH can be identified as having the largest shift at 148.0 ppm due to the strong oxygen electron-withdrawing effect. It is worth noting that the C–Cl and C–OH signals at 121.2 and 148.0 ppm respectively are not too different in intensity, despite there being a 2:1 ratio in the carbon atom count. It is therefore timely to advise caution with this type of analysis when making early deductions about the structure without having corroborating evidence. Finally, note that the two protons *ortho* to the chlorine atoms do not show $^4J_{HH}$ coupling to each other because they have identical environments, owing to symmetry.

8.13 **Example 13**

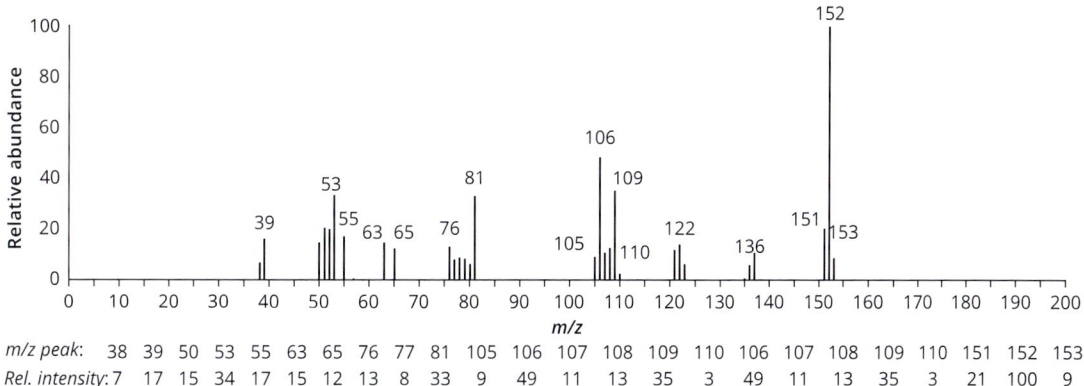

m/z peak: 38 39 50 53 55 63 65 76 77 81 105 106 107 108 109 110 106 107 108 109 110 151 152 153

Rel. intensity: 7 17 15 34 17 15 12 13 8 33 9 49 11 13 35 3 49 11 13 35 3 21 100 9

Mass spectrum

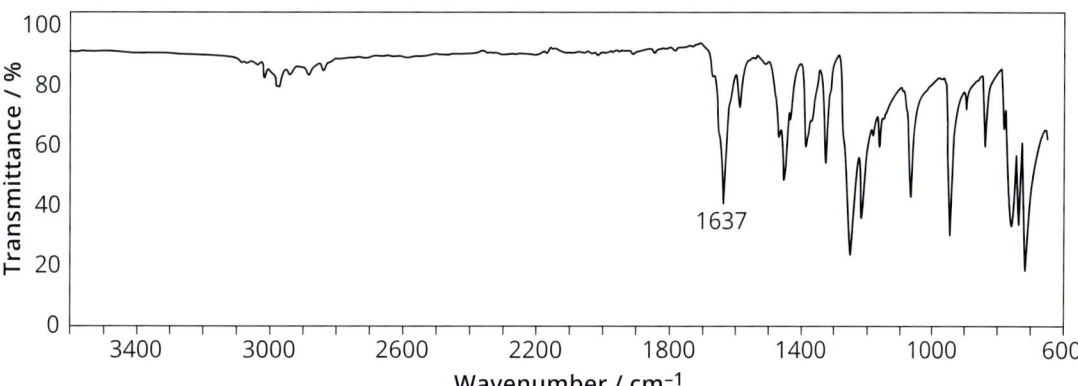

Infrared spectrum

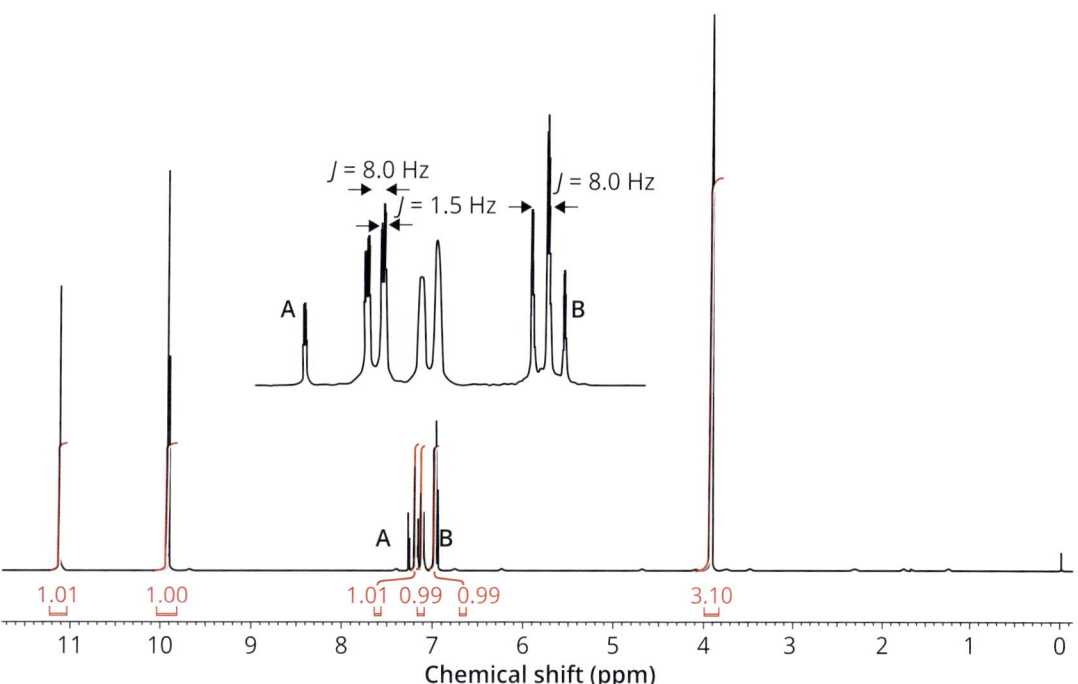

¹H NMR spectrum (400 MHz, CDCl₃)

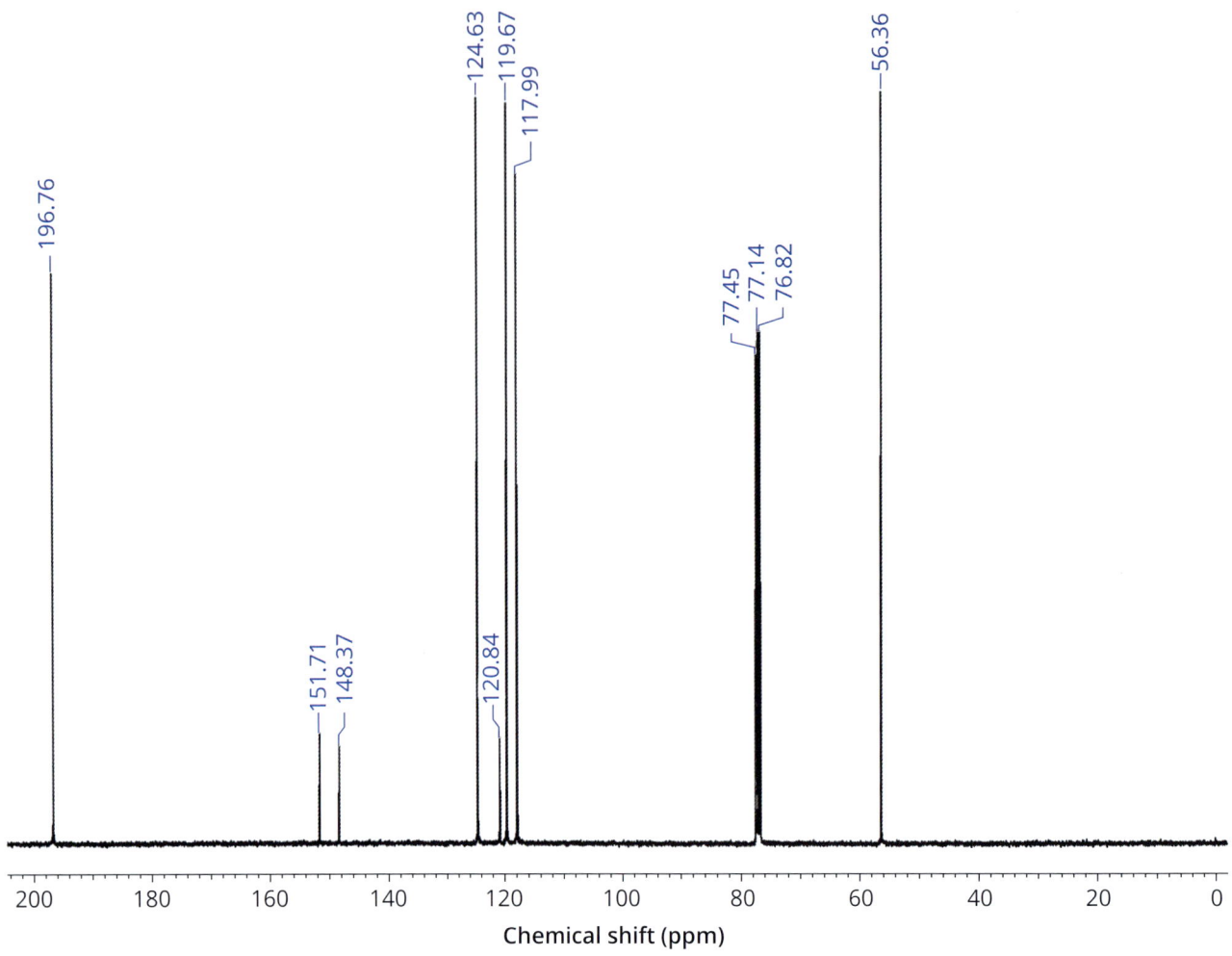

13C NMR spectrum (100 MHz, CDCl₃)

^{13}C NMR spectrum (100 MHz, CDCl$_3$)

8.13.1 Initial analysis of Example 13

(a) The peak at m/z 152 in the mass spectrum is likely to represent the molecular ion (M$^{+\bullet}$).

(b) The isotopes of the molecular ion peak do not indicate the presence of either chlorine or bromine.

(c) An even M$^{+\bullet}$ indicates that either no nitrogen atoms or an even number (e.g. 2, 4, etc) are present in the compound.

(d) The base peak is the molecular ion, which suggests that the compound may be aromatic.

(e) The number of carbon atoms in the molecular ion is predicted to be ~8 based on the relative abundance of its M and M+1 peaks in the mass spectrum.

(f) The infrared spectrum indicates the presence of a carbonyl group due to the strong absorption at ~1640 cm^{-1}. However, as this is quite a low value for a carbonyl group, we cannot confirm with certainty that this is indeed a carbonyl group by relying on the IR alone. A cross-check with the ^{13}C NMR reveals that this spectrum shows a signal at 196.8 ppm, which is typical of a carbonyl group, and therefore confirms the general assignment.

(g) The region around 3000 cm^{-1} in the IR spectrum shows quite a broad absorption, but it is neither very intense nor over a very great range.

(h) The ^{1}H NMR shows six signals. We should ignore the trace CHCl$_3$ at 7.26 ppm and TMS at 0.00 ppm.

(i) The ratio of signal integrals is 1:1:1:1:1:3, which indicates the presence of eight hydrogen atoms (or a multiple).

(j) ^{13}C NMR shows eight signals, and the chemical shift ranges indicate one saturated and seven unsaturated carbon environments. Remember that we should not count the CDCl$_3$ 1:1:1 triplet at 77 ppm.

(k) Of the seven unsaturated carbon signals, six are in the aromatic region and one is in the carbonyl region.

(l) The estimated molecular composition so far is C$_8$H$_8$O, which adds up to a mass of 120.

(m) From conclusion (a) and (l), we arrive at a difference in mass of 32 (152–120). This difference is most likely due to two extra oxygen atoms, although it could also be due to a sulfur atom.

(n) For C$_8$H$_8$O$_3$ (or the alternative C$_8$H$_8$OS), we can calculate that the DU = 5. Our initial analysis of the NMR and MS spectra suggested the presence of an aromatic ring and carbonyl group, and therefore all the elements of unsaturation have been identified.

8.13.2 Detailed analysis of Example 13

(o) ^{1}H NMR: The signal at 11.1 ppm is a singlet 1H. As the signal is above 10 ppm, it arises from a deshielded and uncoupled nucleus—e.g. an –OH group. A carboxylic acid immediately springs to mind; however, this signal is sharp whereas acids typically give a very broadened signal. A carboxylic acid would also give a jagged IR absorption over a wide range (2500–3200 cm^{-1}, O–H stretch). This is not clearly evident in the IR spectrum, where the absorption is not that intense.

(p) ^{1}H NMR: The signal at 9.9 ppm is a singlet 1H. The chemical shift of this uncoupled signal is typical of an aldehyde. This observation can be supported further by the presence of the ^{13}C signal at 196.8 pm and the C=O stretch at 1640 cm^{-1}.

(q) A cluster of three peaks (1:1:1) in the ^{1}H aromatic region is seen at 7.2, 7.1, and 6.9 ppm. The expansion primarily reveals two double doublets (each with an 8.0 Hz coupling plus a smaller, long-range 1.5 Hz coupling between the protons at 7.2 and 7.1 ppm, although this small coupling is less well resolved for the 7.1 ppm signal) and one triplet with only an 8.0 Hz coupling. This pattern is typical of a 1,2,3-trisubstituted aromatic ring.

(r) The ^{13}C NMR spectrum supports the idea of the trisubstituted aromatic ring in that there are three more intense signals (the protonated carbons) and three weaker ones (the substituted carbons)

(s) ^{1}H NMR: The signal at 3.9 ppm is a 3H singlet, which is indicative of an isolated methyl group, with the deshielding typical of a neighbouring oxygen atom CH$_3$–O. This assignment is supported by the presence of an aliphatic resonance at 56.4 ppm in the ^{13}C spectrum.

The analysis of the data indicates that the following structural elements are present, and these account for all of the observed proton and carbon signals in the NMR spectra.

1   2   3   4

8.13.3 Conclusion and final structure of Example 13

The structural elements identified match the predicted molecular formula and weight of $C_8H_8O_3$ and 152 g/mol respectively. Our analysis has revealed a trisubstituted ring and three substituents. This gives three possible options (positional isomers or regio-isomers), 5–7, for the unknown **EG13**.

To distinguish between these three, we must consider some of the finer details of the spectra. Firstly, the OH group gave a sharp signal in the ^1H NMR spectrum yet is also very deshielded. Secondly, the carbonyl stretching frequency in the IR spectrum is atypically low for an aldehyde and the OH stretch is weak. Both of these features are consistent with the OH group forming an intramolecular hydrogen bond to the neighbouring C=O. This would rule out isomer **7**, but distinguishing between isomers **5** and **6** is harder and would need further spectroscopic investigation. At this stage, all we can say is that **EG13** is either compound **5** or **6**.

8.13.4 Further rationalization of Example 13

Here, as in other examples, it is worth keeping in mind that radical cation formation, at different functional group sites within a molecule, is possible. This can lead to multiple different fragmentation mechanisms represented by the same mass spectrum. This is likely the case here where a number of different isobaric radical cation charge sites are likely due to the presence of aldehyde, phenol, and ether groups all containing oxygen atoms that can form the radical cation. The radical cation has an even integer value at m/z 152. We can see that a number of fragment peaks also have even integer values, but fragments are usually odd when the radical cation is even and vice versa. There are three possibilities for even fragments from an even radical cation: i) loss of a neutral species, ii) certain types of rearrangement (e.g. McLafferty), or iii) a change in the number of nitrogen atoms between radical cation and its fragments. The latter is less likely here as the nitrogen rule suggests that no nitrogen or an even number of nitrogen atoms (e.g. 2 or 4, which seems unlikely) are present. The first thing to note when looking at the fragment peaks is that the peak at m/z 151 ($[M–H]^+$) corresponds to loss of an H radical. This is common in aliphatic and aromatic aldehydes, and these also often lose a neutral CO molecule corresponding to the fragment peak at m/z 123 (M–H–CO), as shown in the following diagram.

The more abundant fragment at *m/z* 121 is 31 mass units less than the radical cation, which is characteristic of aromatic ethers cleaving at the bond α to the ring to lose CH_3O, as shown below.

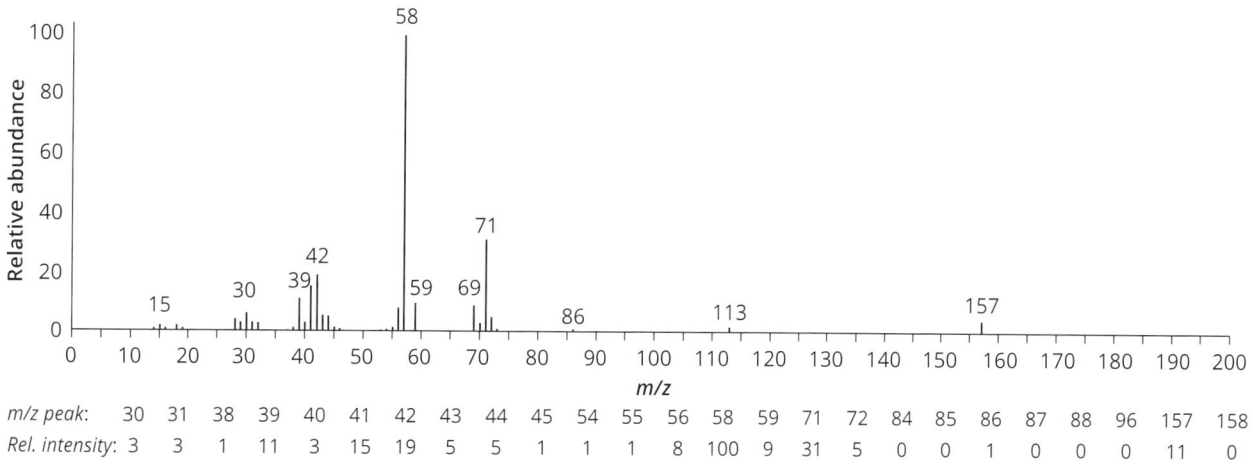

m/z 152   *m/z* 121

Aromatic ethers can also lose a methyl radical, followed by loss of CO. This is a total loss of 43 mass units from the radical cation which fits the fragment peak at *m/z* 109 (note this is an odd integer value because the methyl radical is lost).

m/z 152   *m/z* 109

The even fragment at *m/z* 106 is indicative of loss of one or more neutral species rather than a radical and corresponds with loss of H_2O followed by CO. This is particularly indicative of aromatic phenols with adjacent non-alkyl substituents (see also Example 12) due to the *ortho* effect and can indicate adjacent hydroxyl and aldehyde constituents on the aromatic ring which is in fact the case here.

Whilst differentiation of structures **5** and **6** cannot be reliably performed with the data presented thus far, this can be achieved through the use of the two-dimensional (2D) NMR techniques introduced in Chapter 10 (section 10.2.4). We can recognize that the large carbon shifts at 147 and 151 ppm arise from the deshielding caused by the directly attached oxygen atoms. Assigning which of these carbon centres carries the OH and which the OMe is not possible due to the small shift difference, but here again we could employ 2D correlation experiments to define these assignments reliably, as is also illustrated in section 10.2.4.

8.14 Example 14

m/z peak:	30	31	38	39	40	41	42	43	44	45	54	55	56	58	59	71	72	84	85	86	87	88	96	157	158
Rel. intensity:	3	3	1	11	3	15	19	5	5	1	1	1	8	100	9	31	5	0	0	1	0	0	0	11	0

Mass spectrum

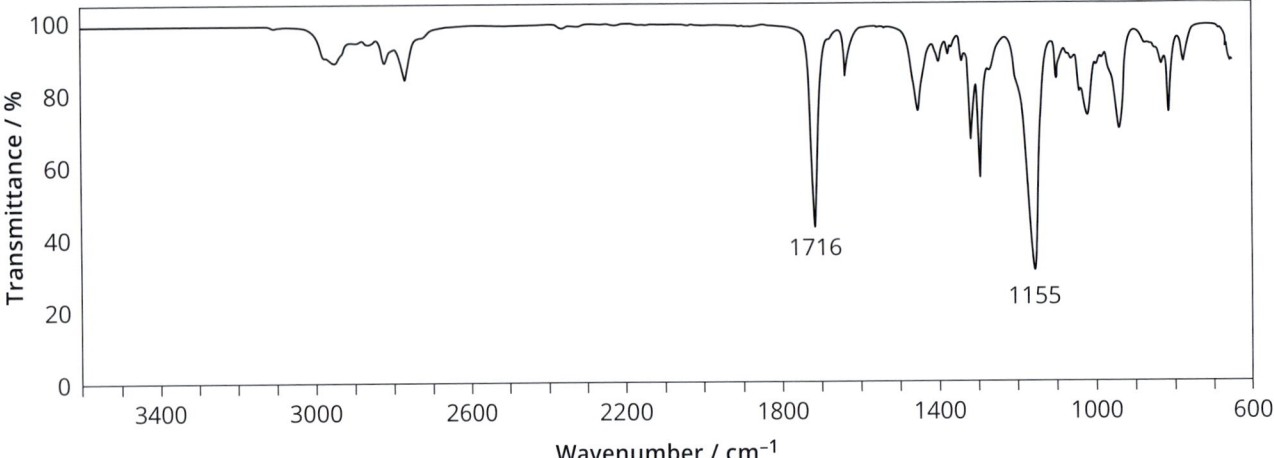

Infrared spectrum

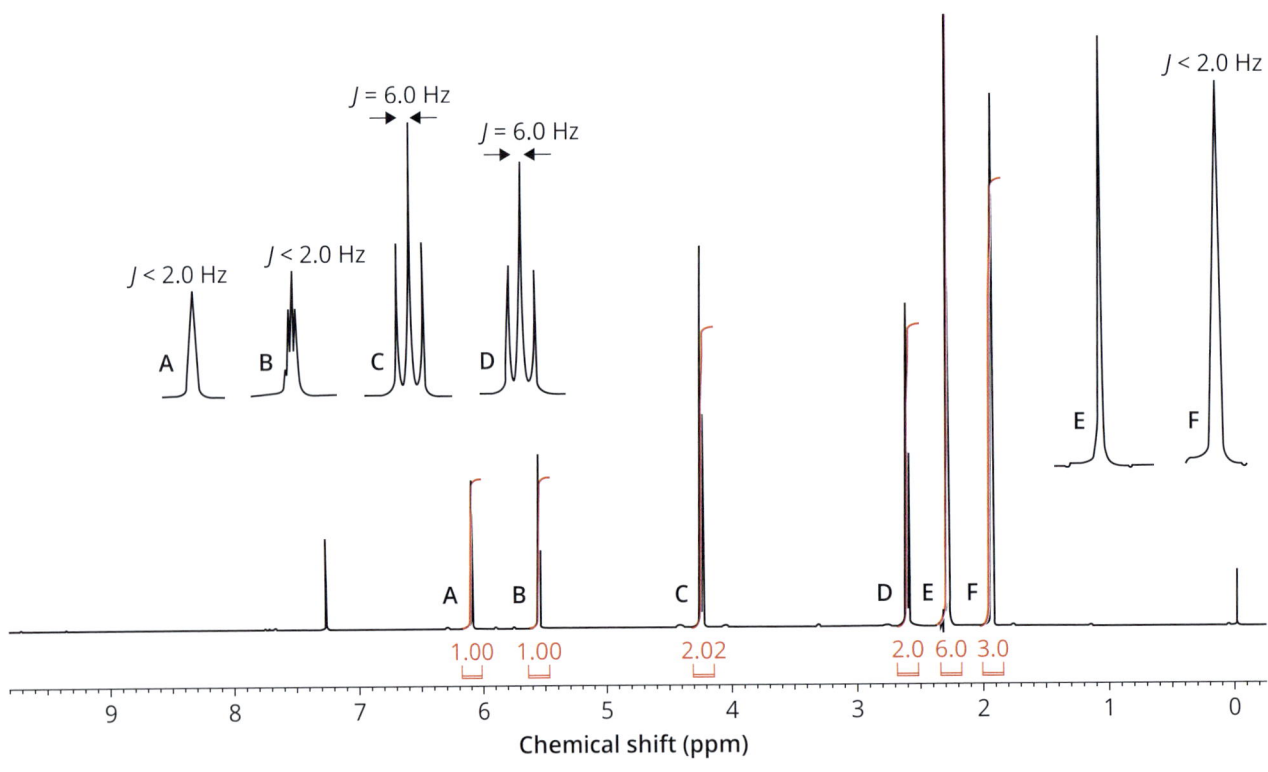

¹H NMR spectrum (400 MHz, CDCl₃)

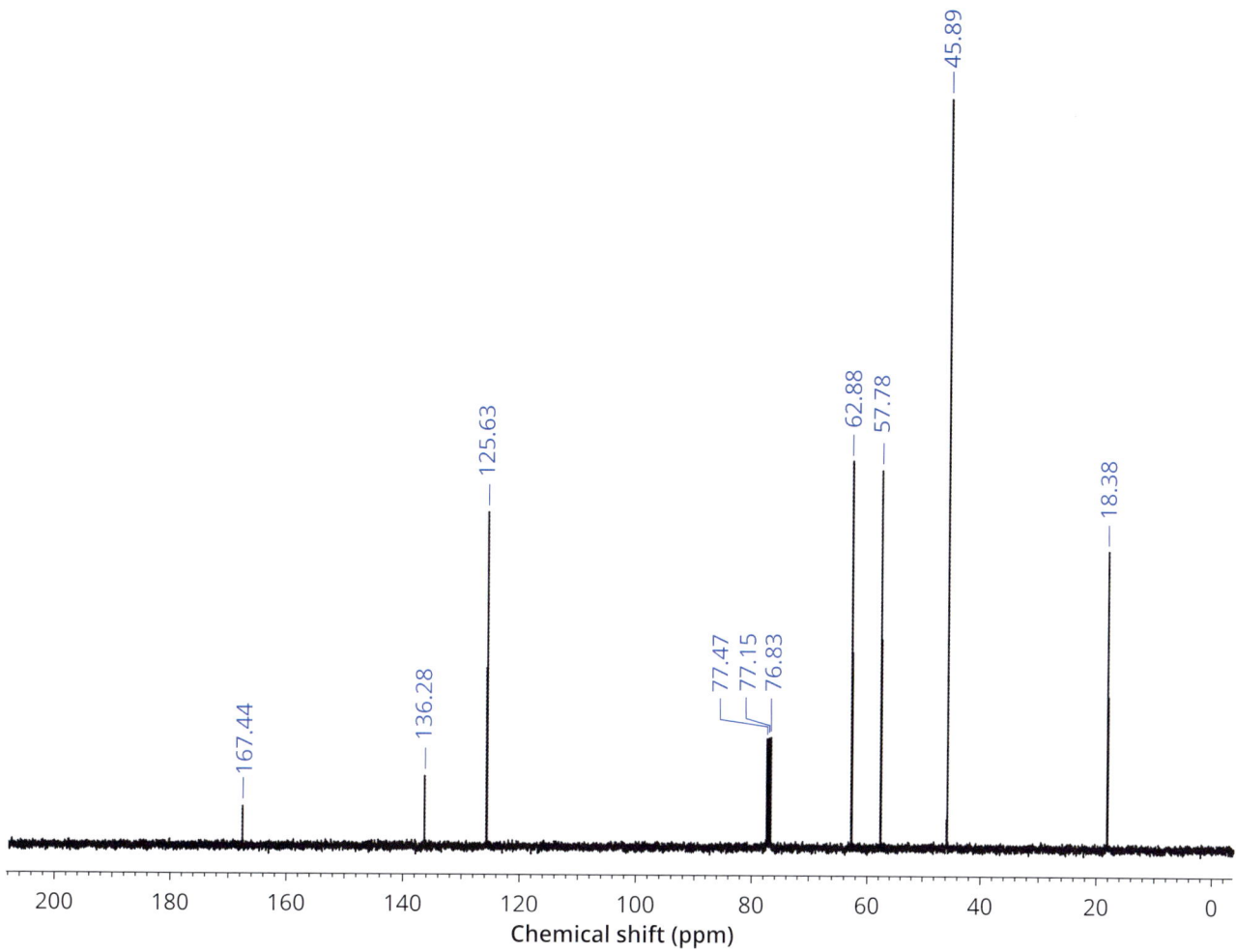

¹³C NMR spectrum (100 MHz, CDCl₃)

8.14.1 Initial analysis of Example 14

(a) The peak at *m/z* 157 in the mass spectrum is likely to represent the molecular ion (M⁺˙).

(b) The isotopic abundances of the molecular ion suggest that neither chlorine nor bromine a present.

(c) An odd M⁺˙ indicates that at least one nitrogen atom is present in the structure, or any odd multiple of them (e.g. 3,5 etc).

(d) The low abundance of the molecular ion suggests that the compound may not be aromatic.

(e) The abundance of the molecular ion and its M+1 isotope peak are too low to be used to predict the number of carbon atoms.

(f) The infrared spectrum indicates the presence of a carbonyl group, but the type of carbonyl group cannot be determined for certain at this stage. However, a cross-check with the ¹³C NMR for a signal between 165 and 220 ppm, which is typical of a carbonyl group, is quickly done and confirms the general assignment.

(g) The ¹H NMR shows six signals. We should ignore the trace CHCl₃ at 7.26 ppm and TMS at 0.00 ppm.

(h) The ratio of signal integrals is 1:1:2:2:6:3, which indicates the presence of fifteen hydrogen atoms (or a multiple).

(i) ^{13}C NMR shows seven signals, and the chemical shift ranges indicate four saturated and three unsaturated carbon environments (but note that there are not necessarily *only* seven carbon atoms present). Remember to ignore the CDCl$_3$ 1:1:1 triplet at 77 ppm.

(j) The estimated molecular composition of the compound is C$_7$H$_{15}$NO, which adds up to a mass of 129.

(k) From conclusions (a) and (j), we can deduce that the difference in mass is 28 (157–129). This is likely due to C+O or 2×N. If a C+O is present, then the lack of an eighth signal in the ^{13}C means that two atoms are equivalent. If a 2×N is present, we would still have an odd number of nitrogen atoms in total and therefore an odd molecular ion.

(l) Whether for C$_8$H$_{15}$NO$_2$ or for C$_7$H$_{15}$N$_3$O, the DU = 2 in each case. In (f), we deduced that a carbonyl group is present. Thus, the unknown must also contain another double bond or a saturated ring (an unsaturated ring would add two more DBEs).

8.14.2 Detailed analysis of Example 14

(m) ^{1}H NMR: The signal at 6.12 ppm is a broad unresolved multiplet 1H. The signal is broad, but not 'smooth', which indicates small couplings to other H atoms (that are not fully resolved).

(n) ^{1}H NMR: The signal at 5.57 ppm is a multiplet 1H. The signal is not a true quintet as the ratio of the lines is not the expected 1:4:6:4:1. Hence the assignment of a multiplet is safer, but the small splittings ($J < 2$ Hz) indicate that there will be coupling for this hydrogen atom, likely across at least 4 bonds.

(o) Taken together, the chemical shifts in (m) and (n) are indicative of the presence of two unsaturated =CH signals. There is no evidence of the J couplings indicative of E or Z displacement of the CH groups (i.e., $J > 8$ Hz). A terminal =CH$_2$ is suggested (with a geminal $^2J_{HH} \sim 1.5$ Hz), which also fits with the presence of a double bond indicated by the DU calculation.

(p) ^{1}H NMR: The signal at 4.25 ppm is a 2H triplet with $J = 6.0$ Hz. It shows deshielding typical of a neighbouring oxygen atom and splitting that indicates a neighbouring CH$_2$ group.

(q) ^{1}H NMR: The signal at 2.60 ppm is a 2H triplet with $J = 6.0$ Hz. In this case, the deshielding is typical of a mild neighbouring electron-withdrawing group, but not oxygen. The splitting indicates a neighbouring CH$_2$ group.

(r) Taken together (p) and (q) are indicative of the presence of a –CH$_2$–CH$_2$–O structural element.

(s) ^{1}H NMR: The signal at 2.30 ppm is a 6H singlet. The chemical shift indicates that the protons are attached to a moderately electron-withdrawing group.

(t) The 6H integral would indicate two identical methyl groups with no couplings. The presence of N is indicated by MS, so an NMe$_2$ group is suggested by the evidence.

(u) ^{1}H NMR: The signal at 1.95 ppm is a 3H multiplet. Whist this signal is triplet-like, it does not have the 1:2:1 ratio of a 'true' triplet. The size of the coupling(s) ($J \sim 1.5$ Hz) is unlikely to be a vicinal (3-bond) coupling; it is probably longer range. The data suggests a methyl group (3H) attached to a slightly electron-withdrawing group, with the possibility of longer-range couplings.

(v) ^{13}C NMR shows seven signals. Three of these are unsaturated and the other four saturated carbons.

(w) ^{13}C NMR: The signal at 167.4 ppm indicates a carbonyl group of some sort—this interpretation is supported by IR.

(x) ^{13}C NMR: There are only two signals with chemical shifts around 100–140 ppm indicating an alkene, which is consistent with deduction in (o).

(y) ^{13}C NMR: The three signals with chemical shifts of 46, 58, and 63 ppm are consistent with the CH_3–N and –CH_2–CH_2–O structural elements identified in conclusions (t) and (r).

The evidence indicates the presence of the following structural elements:

| 1 | 2 | 3 | 4 | 5 |

A quick check sum shows that the structural elements add to the formula $C_8H_{15}NO_2$, which matches the predicted molecular ion peak in the mass spectrum (*m/z* of 157). All atoms have therefore been found and accounted for and DU=2 satisfied.

8.14.3 Conclusion and final structure of Example 14

Now that the pieces have been identified, we just need to arrange the molecular jigsaw pieces and complete the picture, leaving no valencies unfilled. Structural elements **4** and **5**, each with one unfilled valency, can be seen as edge pieces, and so the core of the molecule must therefore be made from structural elements **1–3** in some order. At this point, considering secondary evidence will help us start the process of connecting the pieces.

In this example, two pieces of evidence are most useful to get started. Firstly, the chemical shift of the second CH_2 in fragment **3** is 2.60 ppm. This is very close to the shift of the NMe_2 protons at 2.30 ppm, which suggests that **4** is connected to **3**. Secondly, the small couplings < 2 Hz, which were not easily distinguished, are found in both structural elements **2** and **5**, suggesting that these two are connected.

This leaves two larger structural elements, which can be connected by fragment **1** to give the final structure **EG14**.

At this stage it is prudent to double check that all structural elements have been used, that the functional groups look sensible, and that all assignments are consistent with the solution structure.

EG14 is an important methacrylate ester which, when polymerized, forms a water-soluble polymer due to the basic amine group.

8.14.4 Further rationalization of Example 14

The mass spectrum in this example is challenging to interpret without additional spectroscopic information. The reason for this is that multiple functional groups are present (amine and ester), alongside a single nitrogen, which means that there is potential for rearrangements, single-bond fragmentation, and fragments with and without a nitrogen atom. However, esters and amines typically undergo α-cleavage, and esters also often undergo McLafferty rearrangements. The most prominent peak in the mass spectrum is the base peak at m/z 58, which represents the product of α-cleavage and retention of the charge on the nitrogen-containing tertiary amine fragment. Often when even fragments at m/z 58 and m/z 86 are present with an odd m/z molecular ion, this indicates a McLafferty rearrangement of an aliphatic ketone or the α-cleavage of an amine (secondary or tertiary ones are most prominent). In this spectrum the appearance of m/z 30 and m/z 44 support the presence of a tertiary amine. These are rationalized via primary and secondary fragmentation reactions shown in the following diagrams.

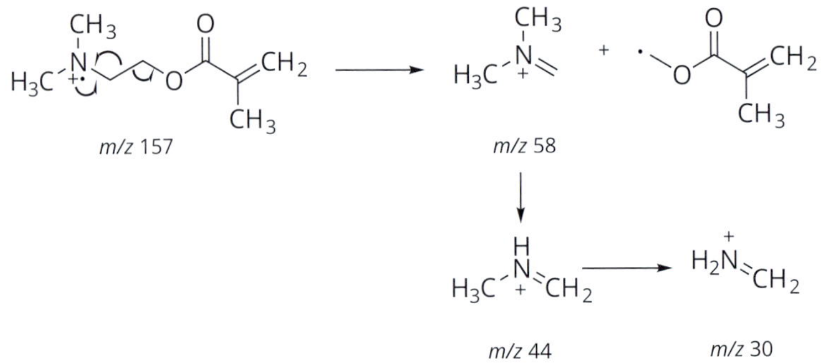

Note that we cannot use the nitrogen rule to differentiate the carbonyl or amine fragments as the amine fragment is an even-electron fragment containing a single nitrogen and the McLafferty rearrangement for the carbonyl is an odd-electron fragment in which the nitrogen is lost; in both cases the m/z value will be even given that they are derived from an odd molecular ion.

Other peaks are likely the result of fragmentation initiated by the carbonyl group. For example, α-cleavage around the carbonyl results in the peak at m/z 69. Note that inductive cleavage of the ether radical cation also produces a peak at m/z 69.

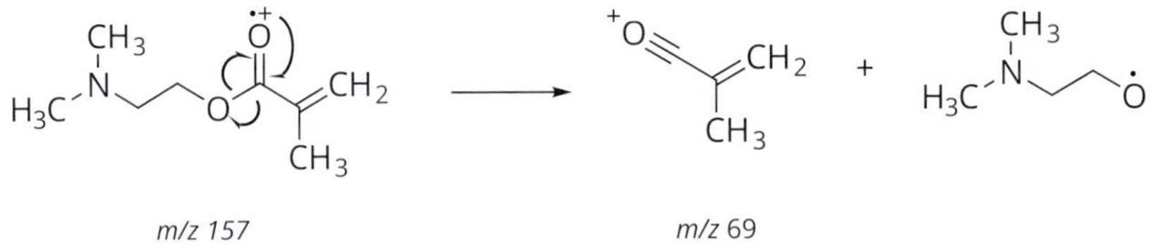

A peak at m/z 41 is likely the result of inductive cleavage of the carbonyl. There is no nitrogen in this cation, so its odd m/z is commensurate with this mechanism. The peak at m/z 71 corresponds to the cation formed by a McLafferty rearrangement.

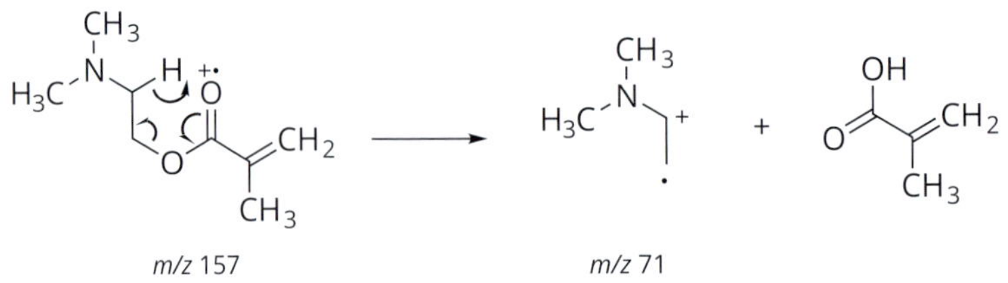

Considering the NMR spectra, the two carbon resonances of the alkene group at 136.3 and 125.6 ppm show quite different intensities. Recall that it is quite common to observe non-protonated carbon atoms displaying lower intensity signals, allowing the more intense 125.6 ppm resonance to be assigned to the terminal alkene CH_2 and the weaker 136.3 ppm peak to that without an attached proton. The very intense carbon peak at 45.9 ppm can be attributed to the two equivalent methyl groups of the $N-(CH_3)_2$ structural element, the shift also being consistent with this. The 62.9 and 57.8 ppm peaks are quite similar in shift, so one needs to be cautious in assigning these as the OCH_2 or NCH_2 carbons. Whilst it is reasonable to expect the $O-CH_2$ to be of larger shift, the use of 2D correlation techniques would distinguish these by using the already assigned proton chemical shifts of these groups (see Chapter 10).

Finally, we note the very intense and obvious absorption in the fingerprint region of the infrared spectrum at 1155 cm^{-1}. This arises from the stretch of the C–O single bond of the ester functionality and provides an example of where a specific absorption in the fingerprint region can be assigned to a specific structural feature.

8.15 Example 15

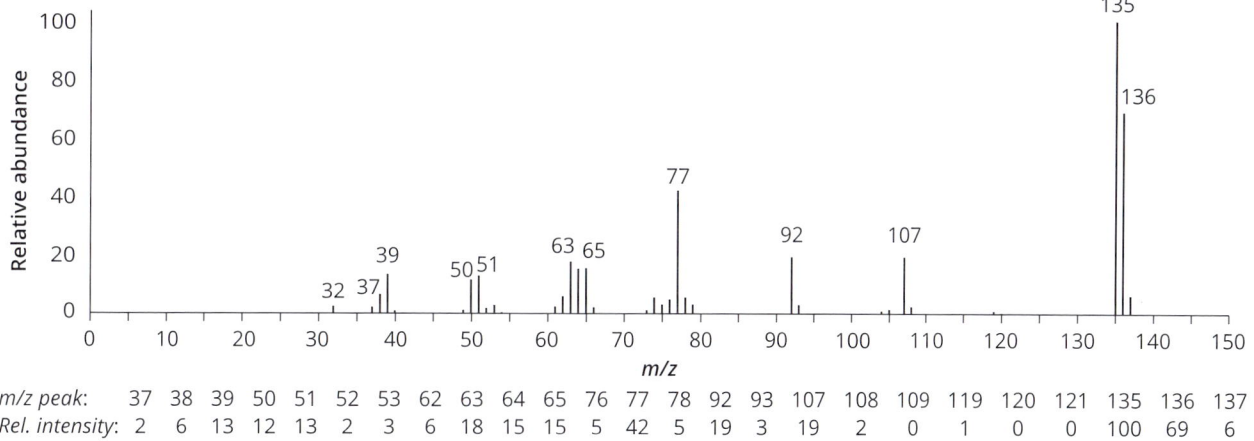

m/z peak:	37	38	39	50	51	52	53	62	63	64	65	76	77	78	92	93	107	108	109	119	120	121	135	136	137
Rel. intensity:	2	6	13	12	13	2	3	6	18	15	15	5	42	5	19	3	19	2	0	1	0	0	100	69	6

Mass spectrum

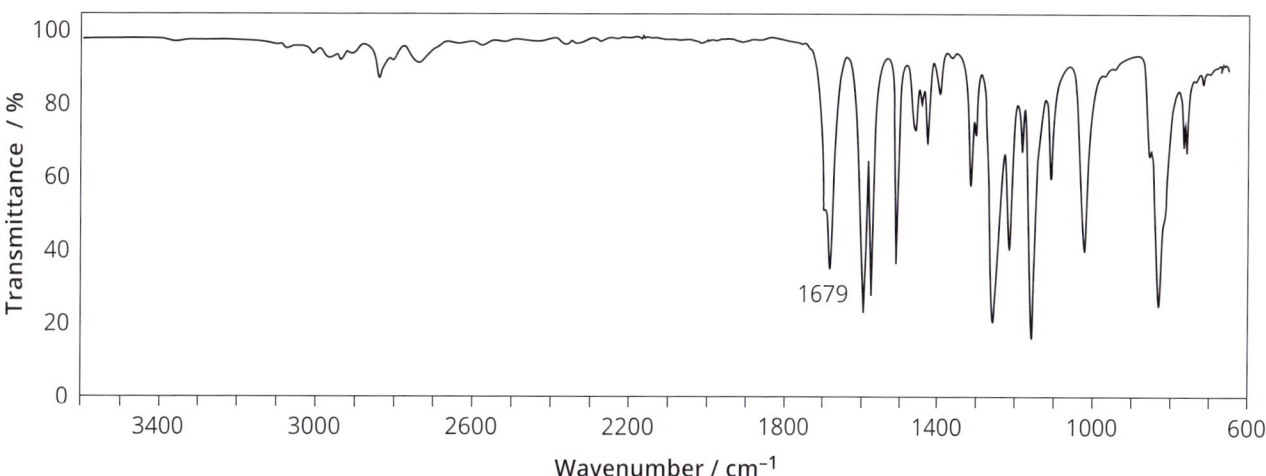

Infrared spectrum

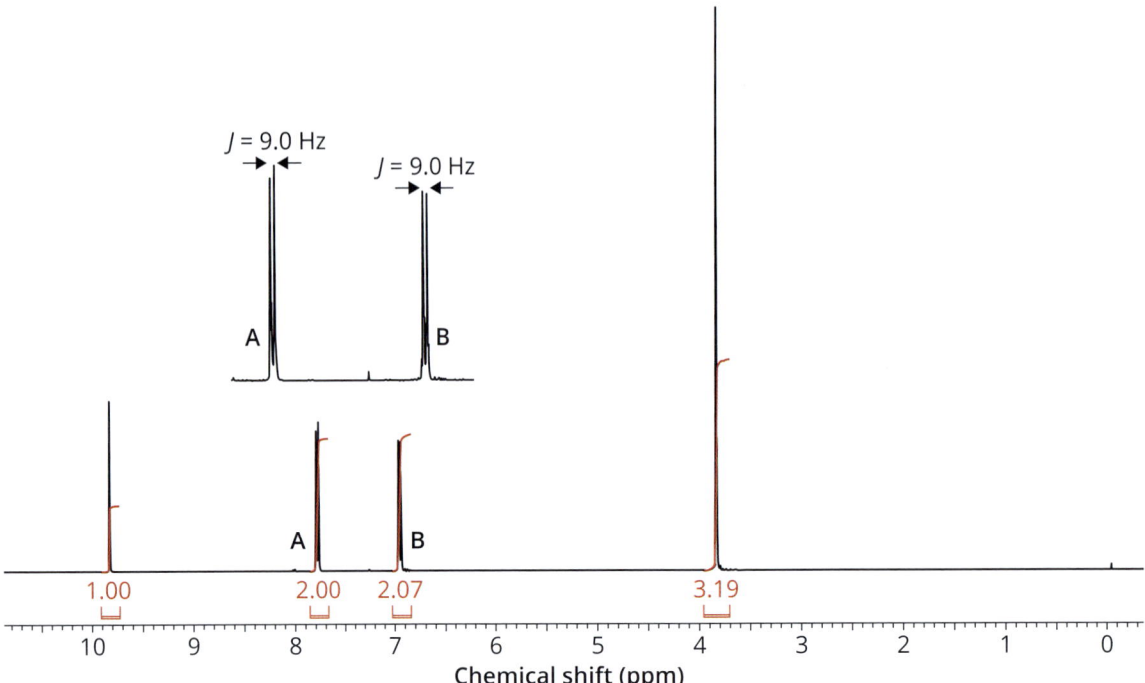

¹H NMR spectrum (400 MHz, CDCl₃)

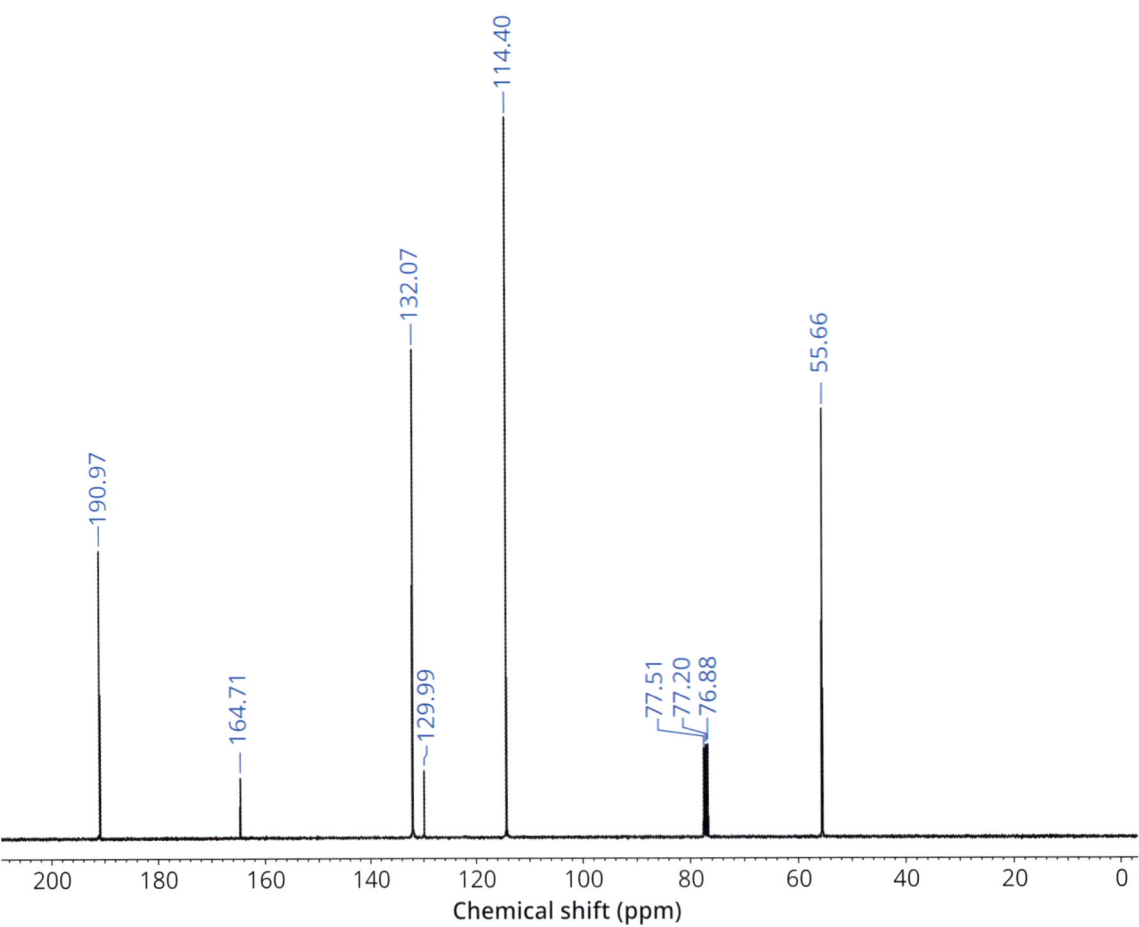

¹³C NMR spectrum (100 MHz, CDCl₃)

8.15.1 Initial analysis of Example 15

(a) The peak at *m/z* 136 in the mass spectrum likely represents the molecular ion ($M^{+\bullet}$).

(b) An even $M^{+\bullet}$ indicates that either no nitrogen atoms or an even number of them (e.g. 2, 4, etc) are present in the compound.

(c) The isotope pattern for the proposed $M^{+\bullet}$ indicates that neither chlorine nor bromine is present.

(d) The high intensity of the molecular ion suggests that the compound is likely to be aromatic.

(e) The base peak at *m/z* 135 indicates that at least one functional group that readily loses hydrogen is present, for example an aldehyde.

(f) The number of carbon atoms in the molecular ion is predicted to be ~8 based on the relative abundance of its M and M+1 peaks in the mass spectrum.

(g) The infrared spectrum shows an absorption in the carbonyl region at ~1700 cm^{-1}.

(h) The IR shows no clear-cut absorptions in the range 2500–3600 cm^{-1} other than the ubiquitous CH stretch at 3000 cm^{-1}.

(i) If we ignore the trace $CHCl_3$ at 7.26 ppm and TMS at 0.00 ppm, the ^1H NMR shows four signals, two of which are singlets and the other two doublets.

(j) The ratio of signal integrals is 1:2:2:3, which indicates the presence of eight hydrogen atoms (or a multiple).

(k) The ^{13}C NMR shows six signals—remember not to include the 1:1:1 triplet at 77 ppm due to the solvent $CDCl_3$. The chemical shift ranges indicate five unsaturated and one saturated carbon environments. The infrared already indicated in (g) that one of the signals will be a carbonyl group; the small peak at 191.0 ppm is the obvious candidate for this.

(l) The composition accounted for at this stage is C_6H_8O based on the NMR data, which adds up to a mass of 96 g/mol. The difference between this and the observed mass of the molecular ion is 40 mass units. Two additional carbon atoms are predicted from the isotope patterns in the mass spectrum data (f), which leaves 16 mass units that are likely to be an oxygen. The estimated molecular formula is $C_8H_8O_2$.

8.15.2 Detailed analysis of Example 15

(m) ^1H NMR: The singlet at 9.8 ppm has a large chemical shift and so must be in proximity to an electron-withdrawing group. Being on the left-hand side of the spectrum, it would have to be attached to an unsaturated carbon (and most likely an aldehyde), but it could also be from an O–H deshielded by intramolecular H-bonding. What is sure is that it is uncoupled and therefore must be isolated from neighbouring hydrogen atoms. It would be safest to describe the fragment as just '–H' at this stage.

(n) The base peak at *m/z* 135 in the mass spectrum is one mass unit less than the molecular ion, suggesting a hydride loss which is characteristic of aldehydes.

(o) ^1H NMR: There are two signals at 7.8 and 6.9 ppm, each of which integrates to 2H. These signals are in the aromatic region of the spectrum, and the centrosymmetric pattern of the signals is indicative of the classic roofing pattern of a 1,4-disubstituted aromatic ring.

(p) ^1H NMR: There is a 3H singlet at 3.9 ppm. The 3H singlet is classically identified with an isolated methyl group. The chemical shift is indicative of a CH_3 group in an electron-withdrawing environment, and δ values around 4 ppm point towards it being an OCH_3.

(q) ^{13}C NMR: The peak at 55.7 ppm is in the aliphatic region of the spectrum, and the chemical shift indicates a carbon attached to an electron-withdrawing atom. A cross-check with the conclusion from the ^{1}H NMR spectrum suggests that this signal is most likely associated with the methoxy group inferred in (p) above.

(r) ^{13}C NMR: Our conclusion that a 1,4-disubstituted aromatic ring is present [see (o)] means that we can re-evaluate the ^{13}C spectrum and also check the estimated molecular formula. A 1,4-disubstituted system would be expected to provide four carbon environments (164.7, 132.1, 130.0, 114.4 ppm) for a six-carbon aryl ring. An additional oxygen atom is indicated by the methoxy group, and thus our molecular formula can now be confirmed as $C_8H_8O_2$, consistent with the mass spectrum.

(s) For $C_8H_8O_2$, we can calculate the DU = 5, which matches with the structural elements suggested thus far: one aromatic ring and one carbonyl group.

The analysis of the data indicates that the following structural elements are present, and these account for all of the observed proton and carbon signals in the NMR spectra.

| 1 | 2 | 3 | 4 |

8.15.3 Conclusion and final structure of Example 15

The structural elements identified match the predicted molecular formula and predicted mass of 136 Da. There are different ways in which you can go about connecting these structural elements, remembering that there must be no unfilled valencies at the end. In this example the characteristic singlet at 9.8 ppm is strongly indicative of an aldehyde CH and so this would be a logical place to start. But if you could not remember this chemical shift, you could still add these pieces together to make a sensible structure. The simple illustration below shows the importance of *checking that your final structure is still consistent with the spectra*. The uppermost structure is not a 1,4-disubstituted aromatic system whereas the molecular structure of **EG15** (4-methoxybenzaldehyde) is consistent with this key aspect and all the other data.

EG15

8.15.4 Further rationalization of Example 15

The fragment peak at *m/z* 135 represents loss of a hydrogen radical (M–H) and the peak at *m/z* 107 represents further loss of CO, which is characteristic of aromatic aldehydes (this behaviour was also seen in Example 13). The aromatic nature of the compound is indicated by the prominent molecular ion peak and the presence of the fragment at *m/z* 77, indicative of a phenyl cation (with its characteristic ring opening decomposition to give *m/z* 51).

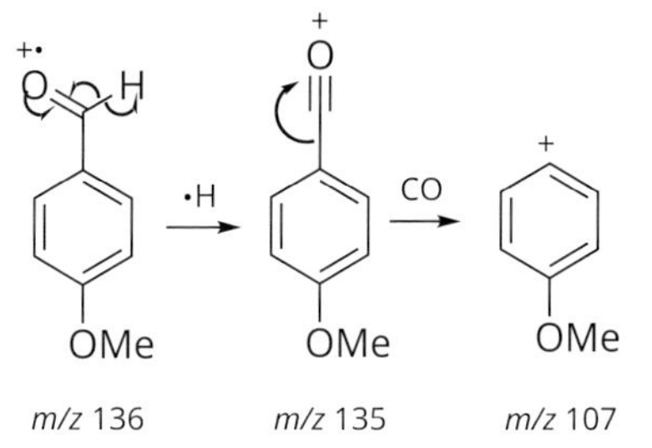

m/z 136   *m/z* 135   *m/z* 107

The mass difference from the molecular ion to benzaldehyde (the simplest aromatic aldehyde) is 31, which typically represents OCH_3 or CH_2OH. As this is most likely an alkyl substituent on the ring, it corresponds with a substituted methoxy group, for example OCH_3 (rather than the alternative CH_2OH). The methoxy substituent can cleave at the bond between the ring and the oxygen atom to form a benzaldehyde cation although this is clearly less favoured as there is only a very minor peak at *m/z* 105 in the mass spectrum (unlabelled). Further loss of CHO would lead to the formation of a phenyl cation.

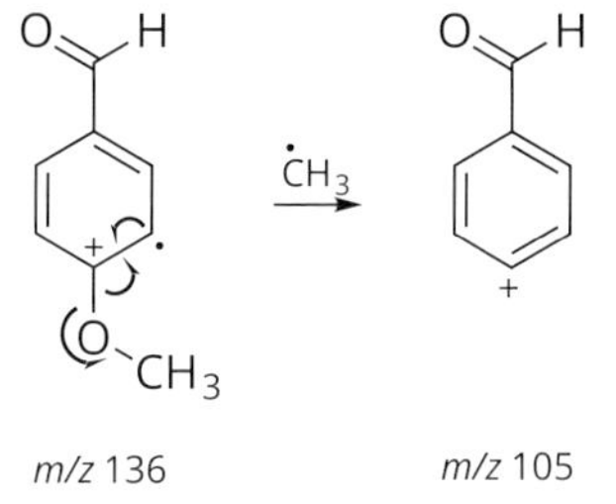

m/z 136   *m/z* 105

A favoured mechanism for methoxy group-driven fragmentation is when the oxygen atom is the site of the radical and charge. This leads to β-bond fragmentation with loss of a methyl radical. This is followed by loss of CO leading to a characteristic five-membered ring structure. There is a small peak at *m/z* 93 corresponding to this structure. Further loss of the aldehyde CHO group leads to *m/z* 65, also present in the mass spectrum.

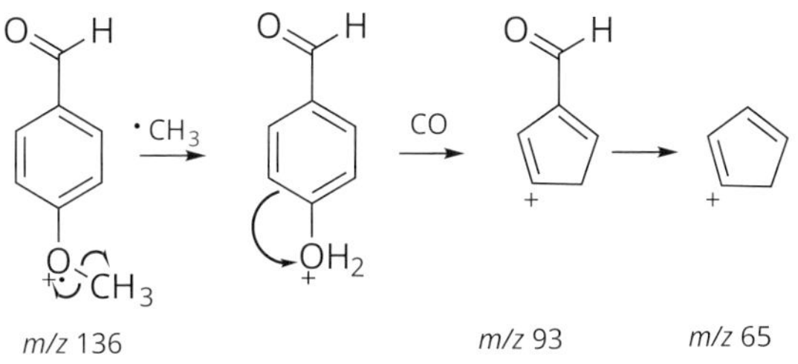

m/z 136   *m/z* 93   *m/z* 65

In the NMR spectra, we can understand the differences in the carbon chemical shifts for the aromatic ring by considering the influence of the attached groups. The largest shift at 164.7 ppm is caused by the strong electron-withdrawing (deshielding) effect of the directly attached oxygen, causing this carbon to stand away from the others. Conversely, the adjacent carbon is significantly shielded and so appears at only 114.4 ppm due to the resonance delocalization of the oxygen lone pair into the aromatic ring, as shown in the following diagram. The same resonance shielding also explains the smaller shift of the corresponding proton resonance at 6.9 ppm.

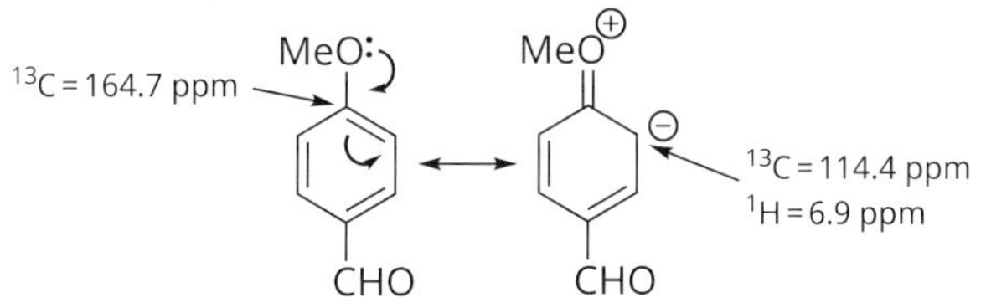

$^{13}C = 164.7$ ppm

$^{13}C = 114.4$ ppm
$^{1}H = 6.9$ ppm

CHO   CHO

The aldehyde, on the other hand, allows electron density to move out of the aromatic ring and so leads to relative deshielding of the adjacent centres, causing the carbons and protons to resonate at 132.1 ppm and 7.8 ppm respectively.

Summary of worked example solutions

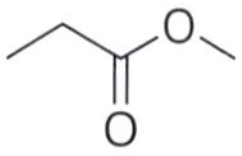

EG1

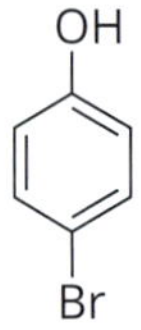

EG2

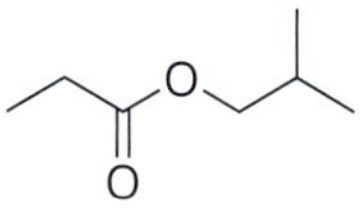

EG3

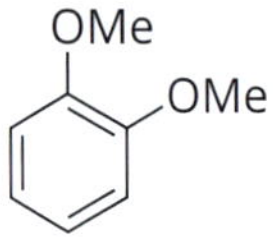

EG4

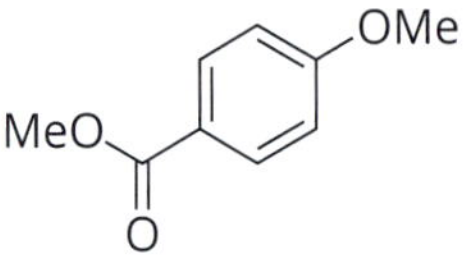

EG5

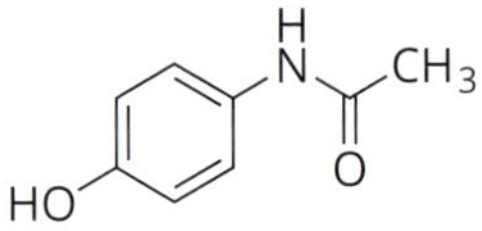

EG6

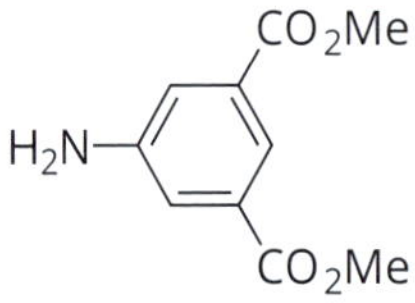

EG7

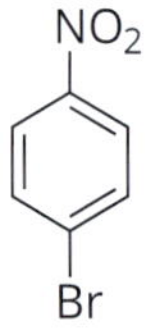

EG8

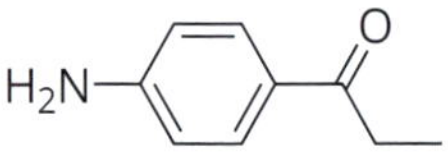

EG9

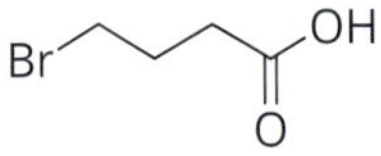

EG10

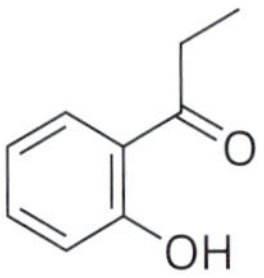

EG11

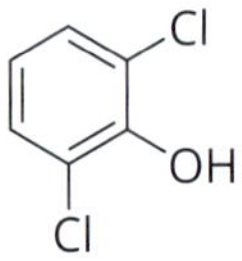

EG12

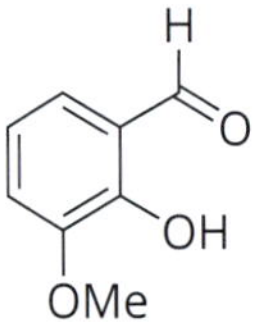

EG13

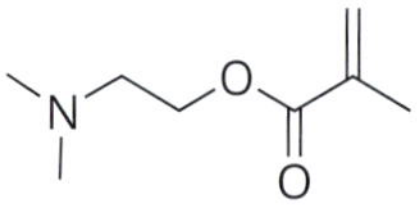

EG14

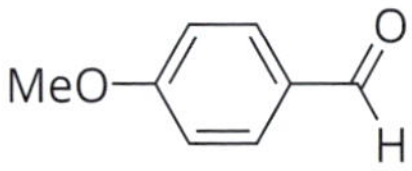

EG15

CHAPTER 9

Problems

In this chapter fifteen sets of spectra are provided for you to work through independently. These are presented approximately in increasing order of difficulty. Worked answers are not provided in this case, as we hope that you will by now have started to develop your own systematic approach to solving problems of this kind. Regardless of the exact way in which you tackle these problems, we highly recommend that you approach the task in a logical order, recording clearly the conclusions you can draw at each stage, whether they be firm deductions or just possibilities at that point—just as we modelled in Chapter 8. The identities of the compounds are provided in Appendix 2 so that you can check your answers.

Note that in some cases multiple structures, specifically positional isomers (regioisomers), can be considered consistent with the available spectroscopic data. These are also listed in Appendix 2 and may be considered acceptable answers to the problems. Such ambiguity is not uncommon in real-world situations and highlights how important it is to remain open to alternative structures when problem-solving. Regioisomers can often be differentiated through more advanced techniques, such as those presented in subsequent chapters.

9.1 Problem 1

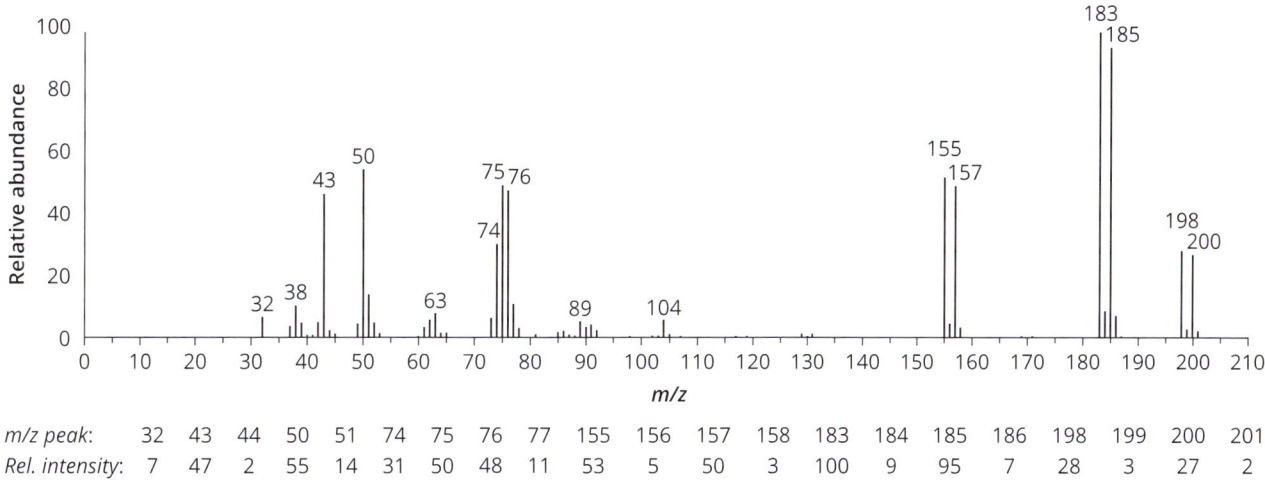

m/z peak:	32	43	44	50	51	74	75	76	77	155	156	157	158	183	184	185	186	198	199	200	201
Rel. intensity:	7	47	2	55	14	31	50	48	11	53	5	50	3	100	9	95	7	28	3	27	2

Mass spectrum

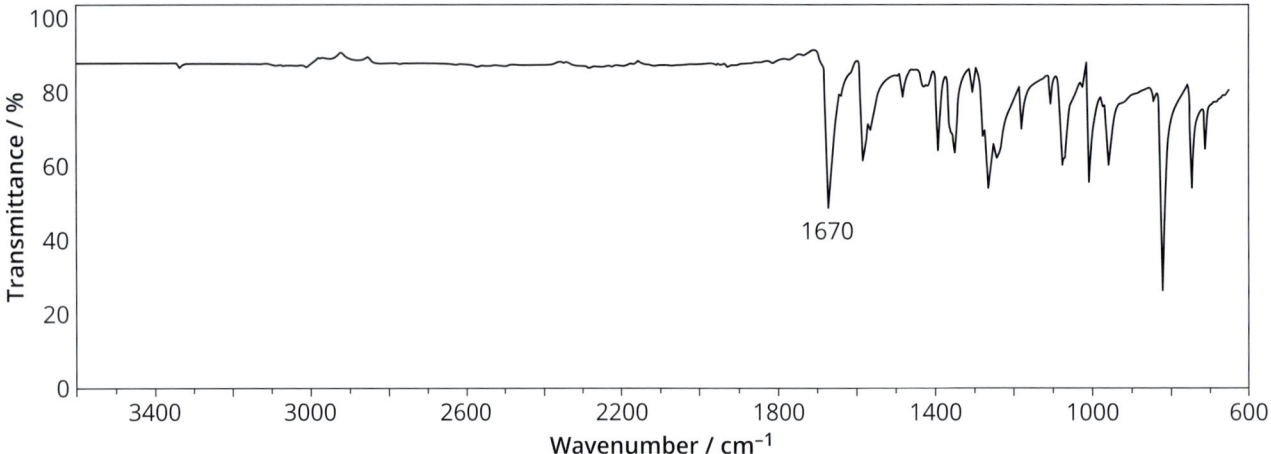

Infrared spectrum

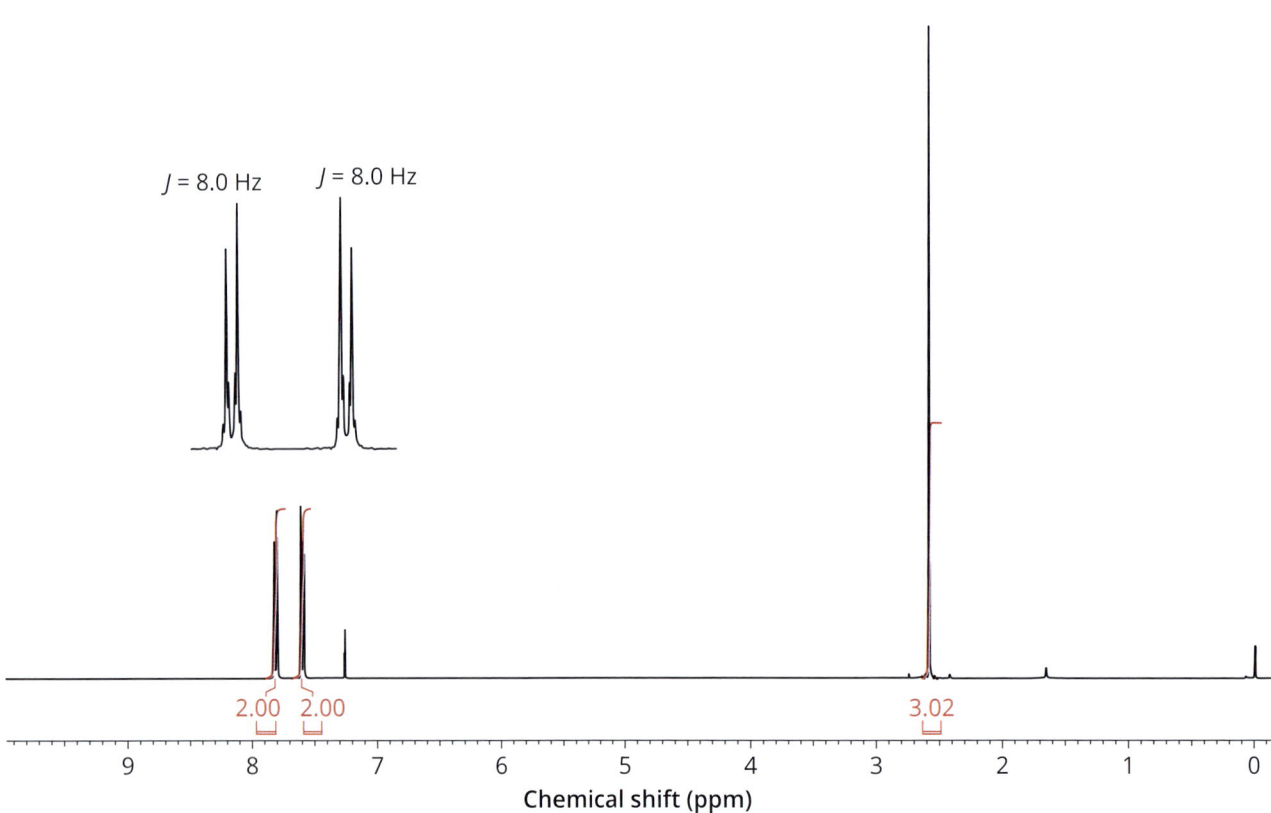

¹H NMR spectrum (400 MHz, CDCl₃)

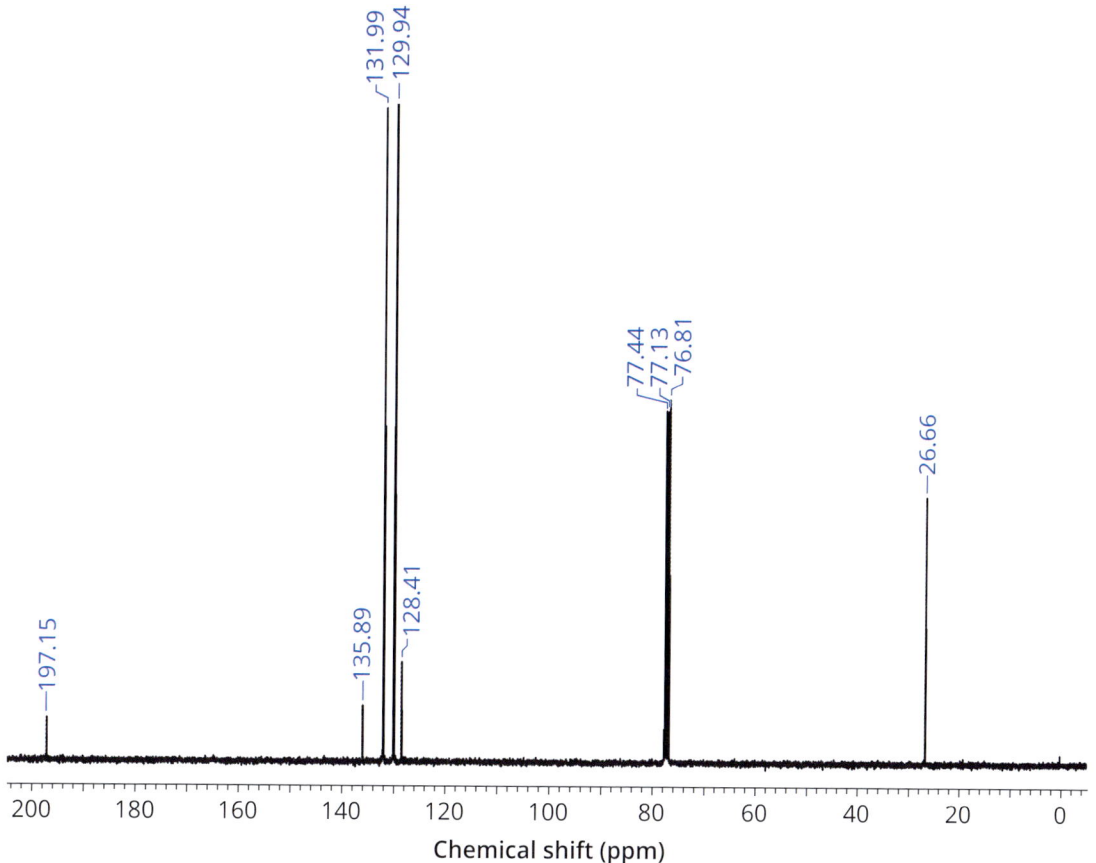

^{13}C NMR spectrum (100 MHz, CDCl$_3$)

9.2 **Problem 2**

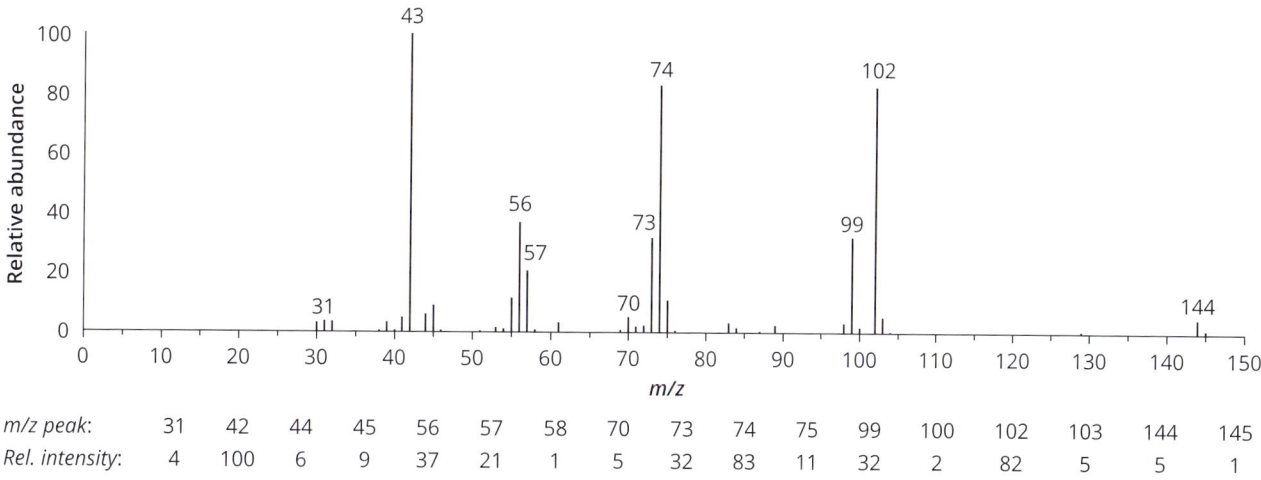

m/z peak:	31	42	44	45	56	57	58	70	73	74	75	99	100	102	103	144	145
Rel. intensity:	4	100	6	9	37	21	1	5	32	83	11	32	2	82	5	5	1

Mass spectrum

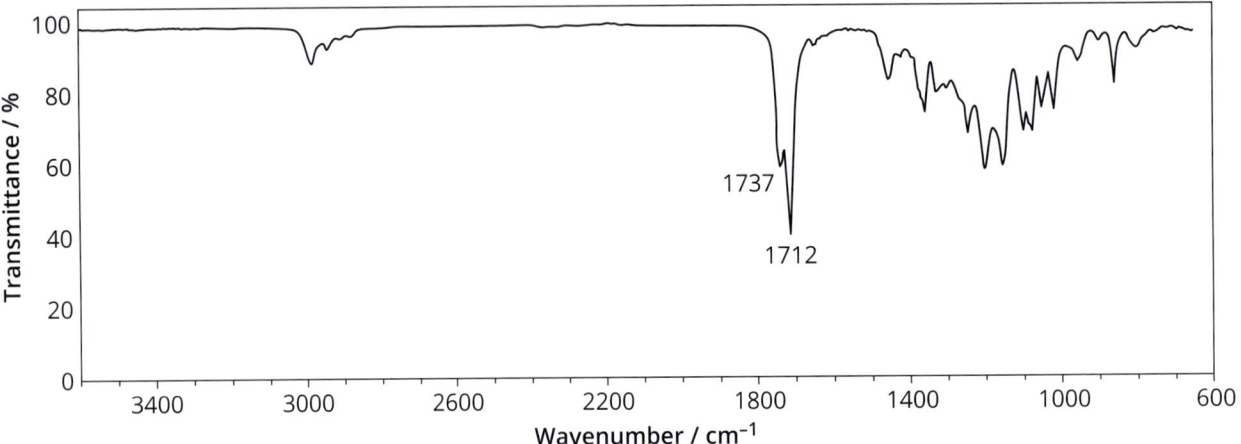

Infrared spectrum

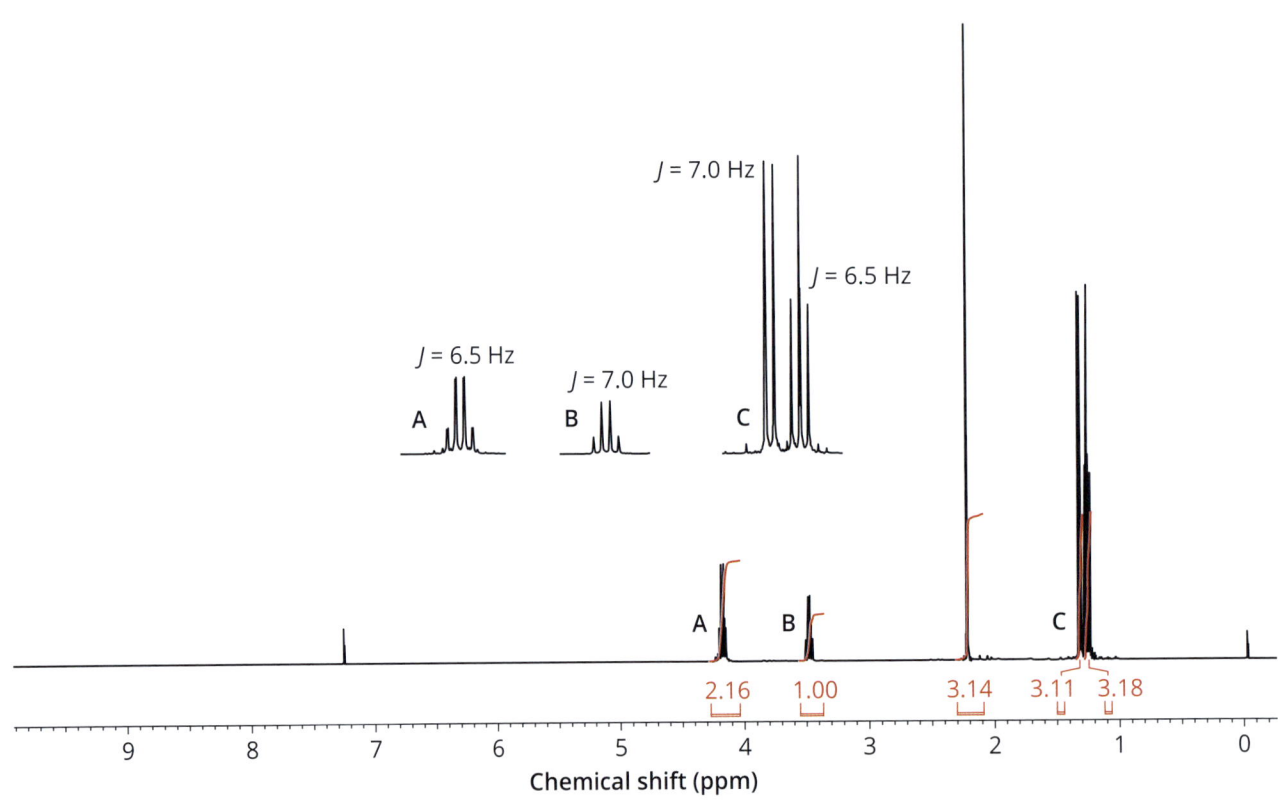

¹H NMR spectrum (400 MHz, CDCl₃)

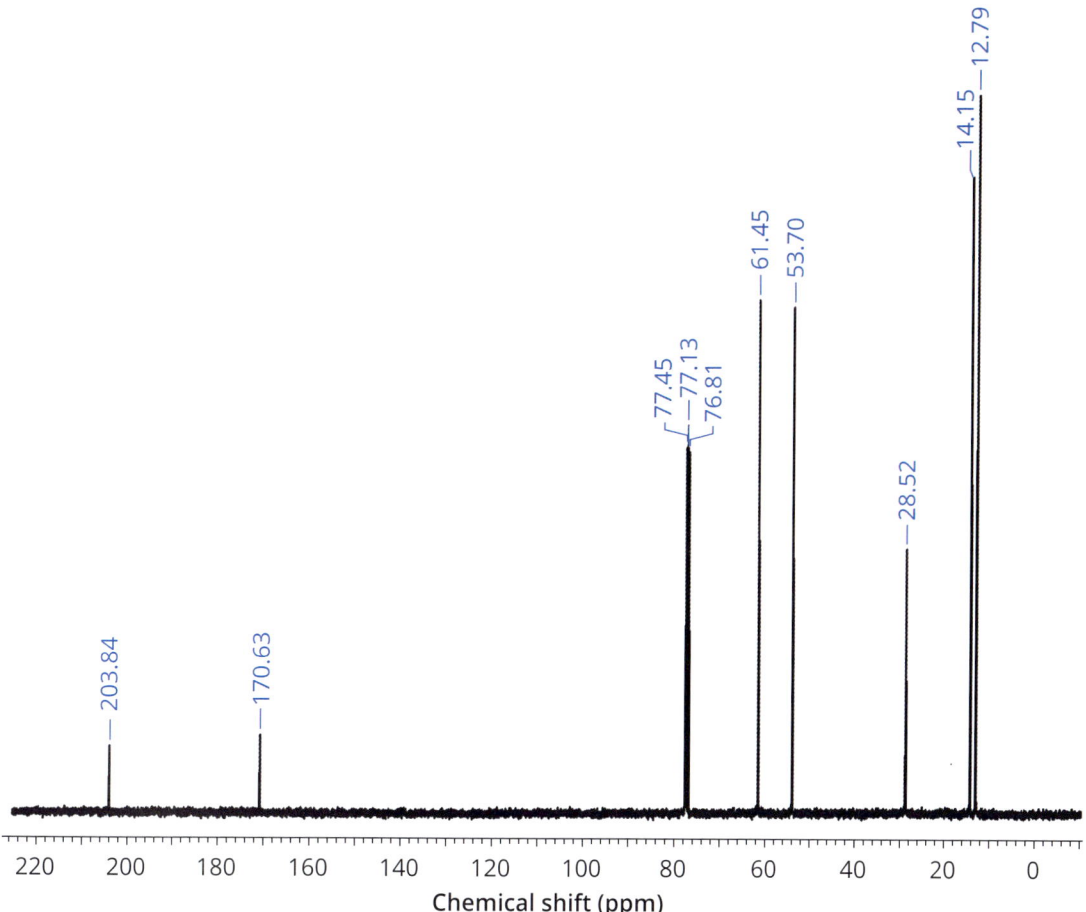

^{13}C NMR spectrum (100 MHz, CDCl$_3$)

9.3 Problem 3

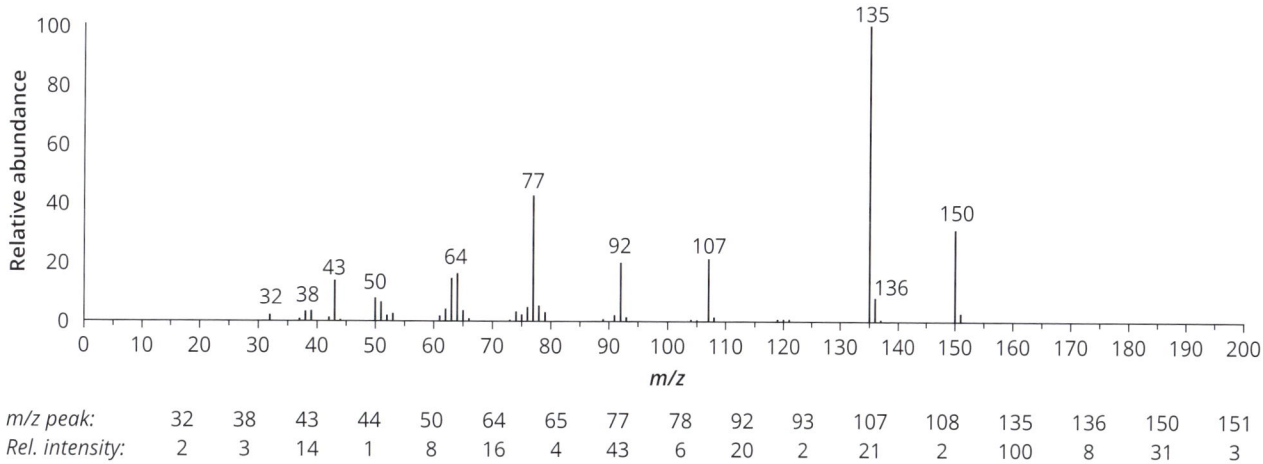

m/z peak:	32	38	43	44	50	64	65	77	78	92	93	107	108	135	136	150	151
Rel. intensity:	2	3	14	1	8	16	4	43	6	20	2	21	2	100	8	31	3

Mass spectrum

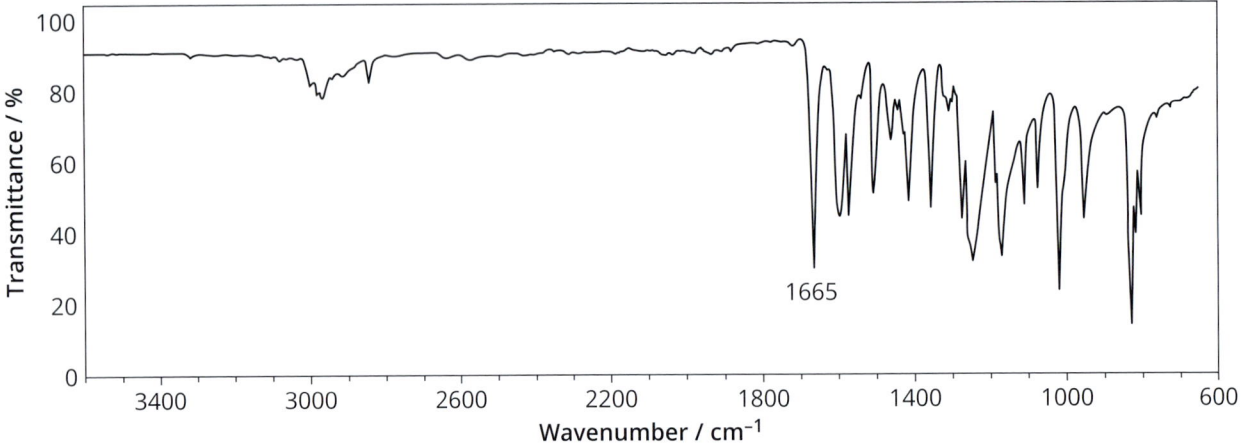

Infrared spectrum

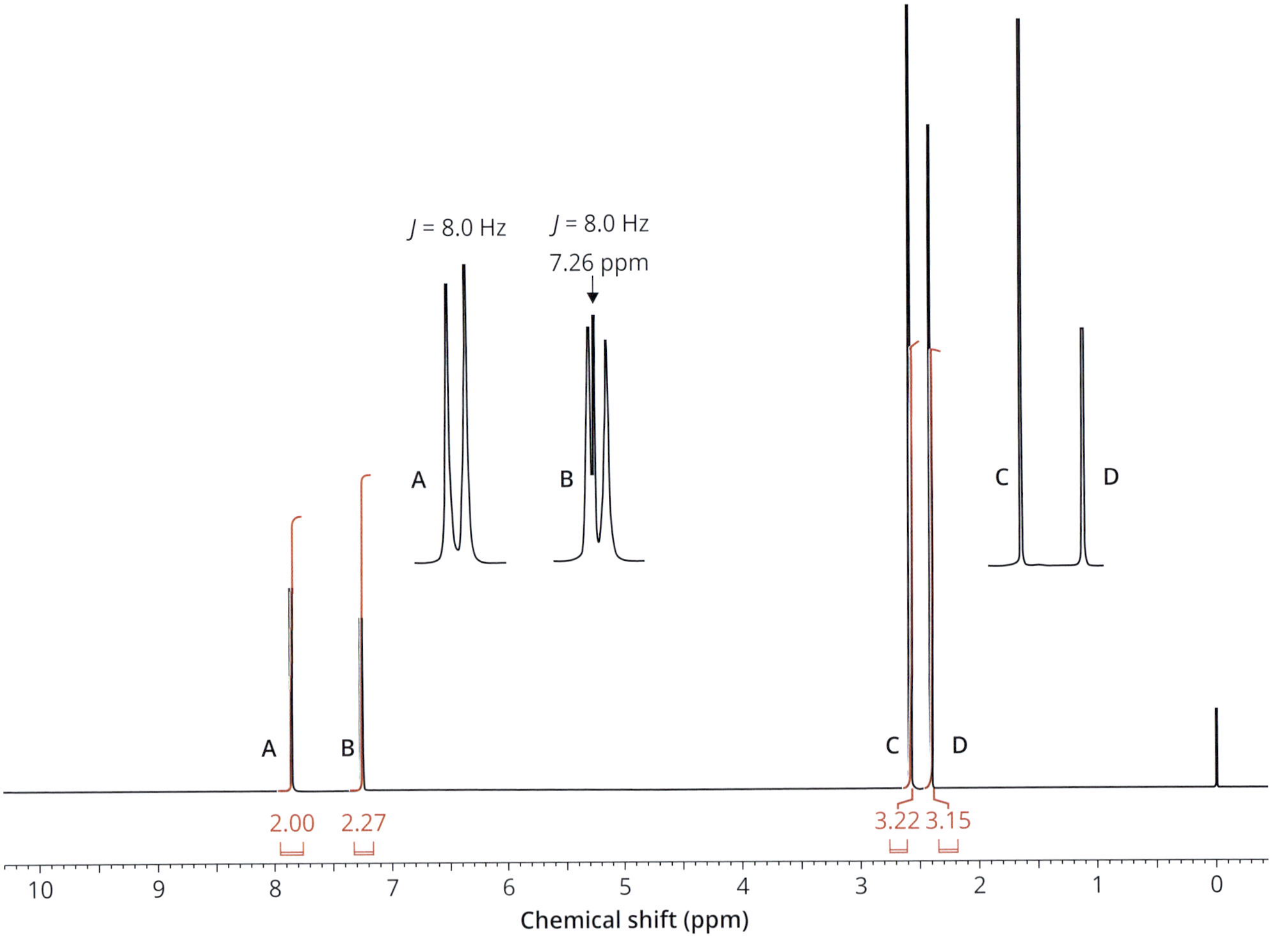

1H NMR spectrum (400 MHz, CDCl₃)

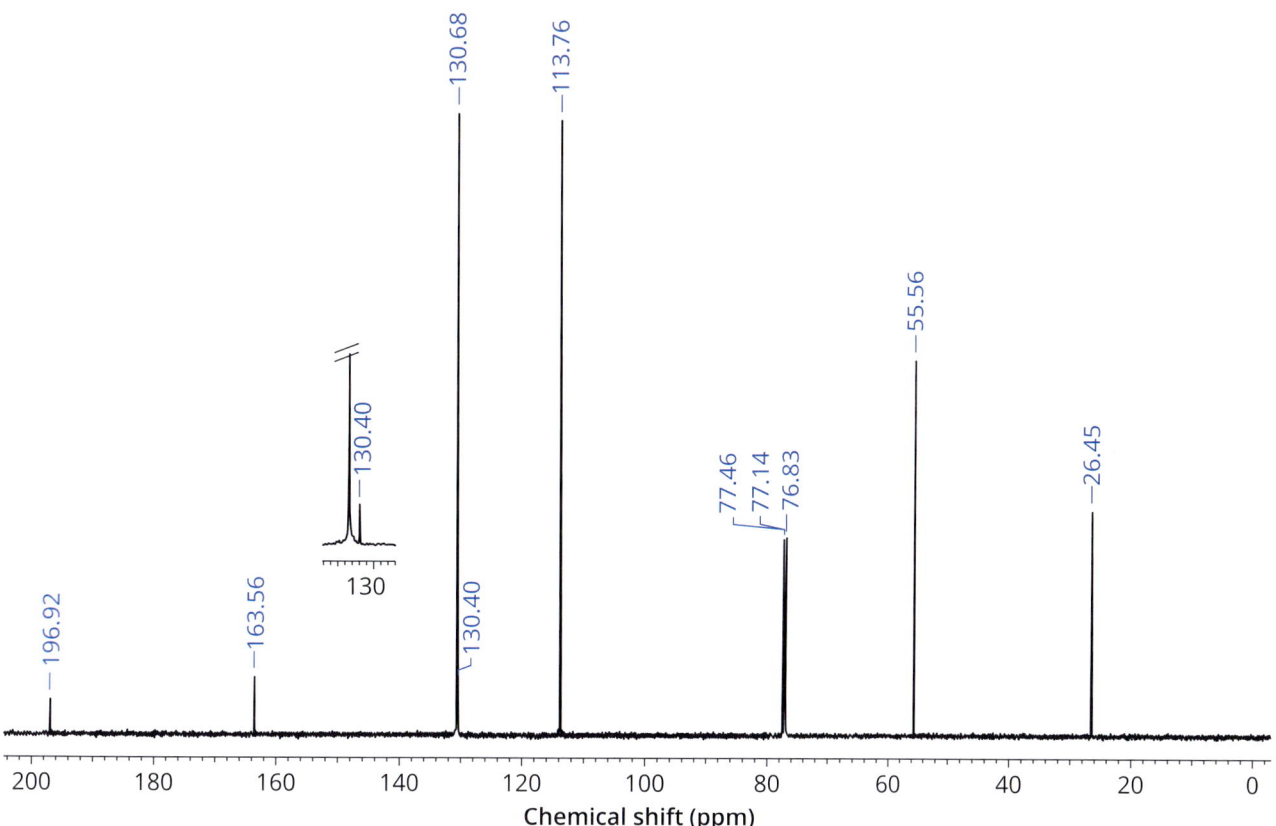

^{13}C NMR spectrum (100 MHz, CDCl$_3$)

9.4 Problem 4

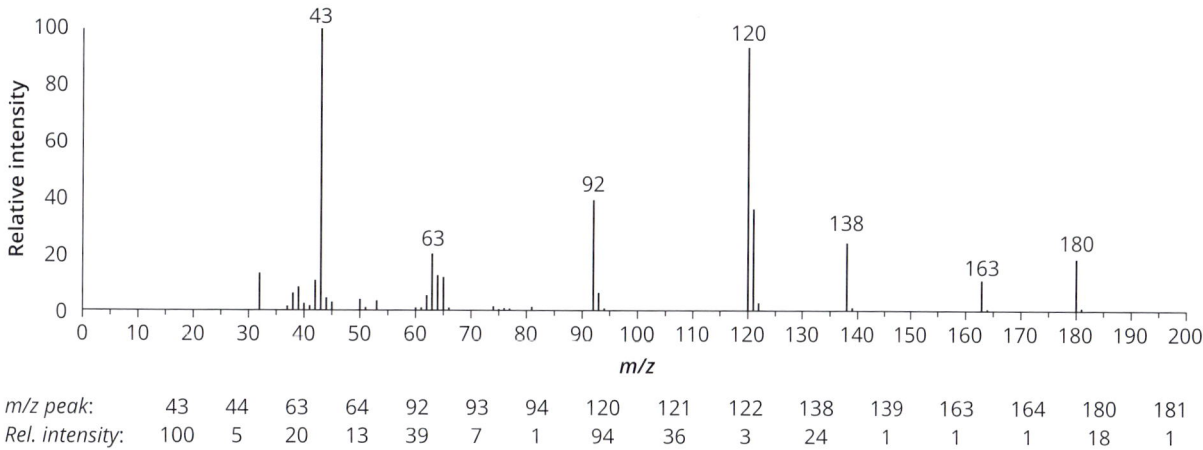

m/z peak:	43	44	63	64	92	93	94	120	121	122	138	139	163	164	180	181
Rel. intensity:	100	5	20	13	39	7	1	94	36	3	24	1	1	1	18	1

Mass spectrum

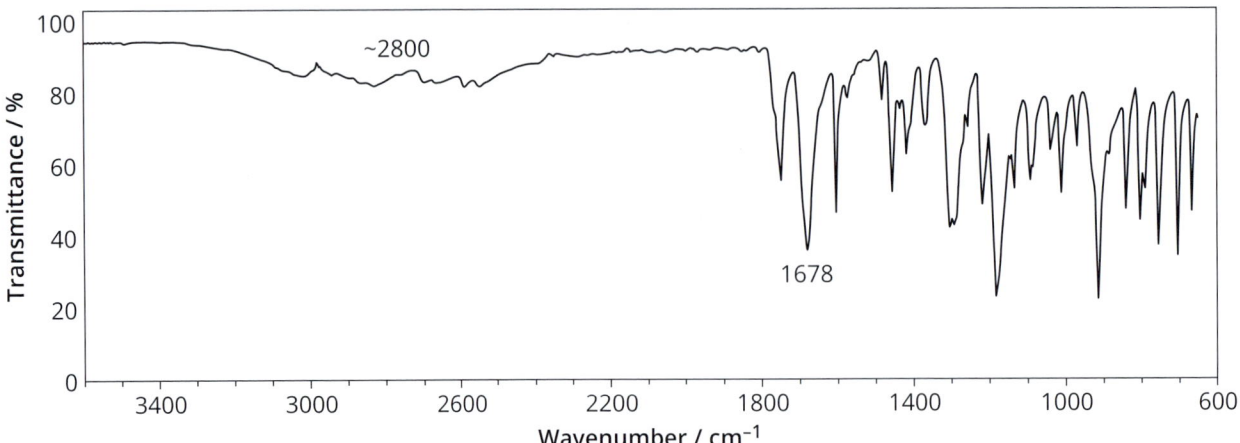

Infrared spectrum

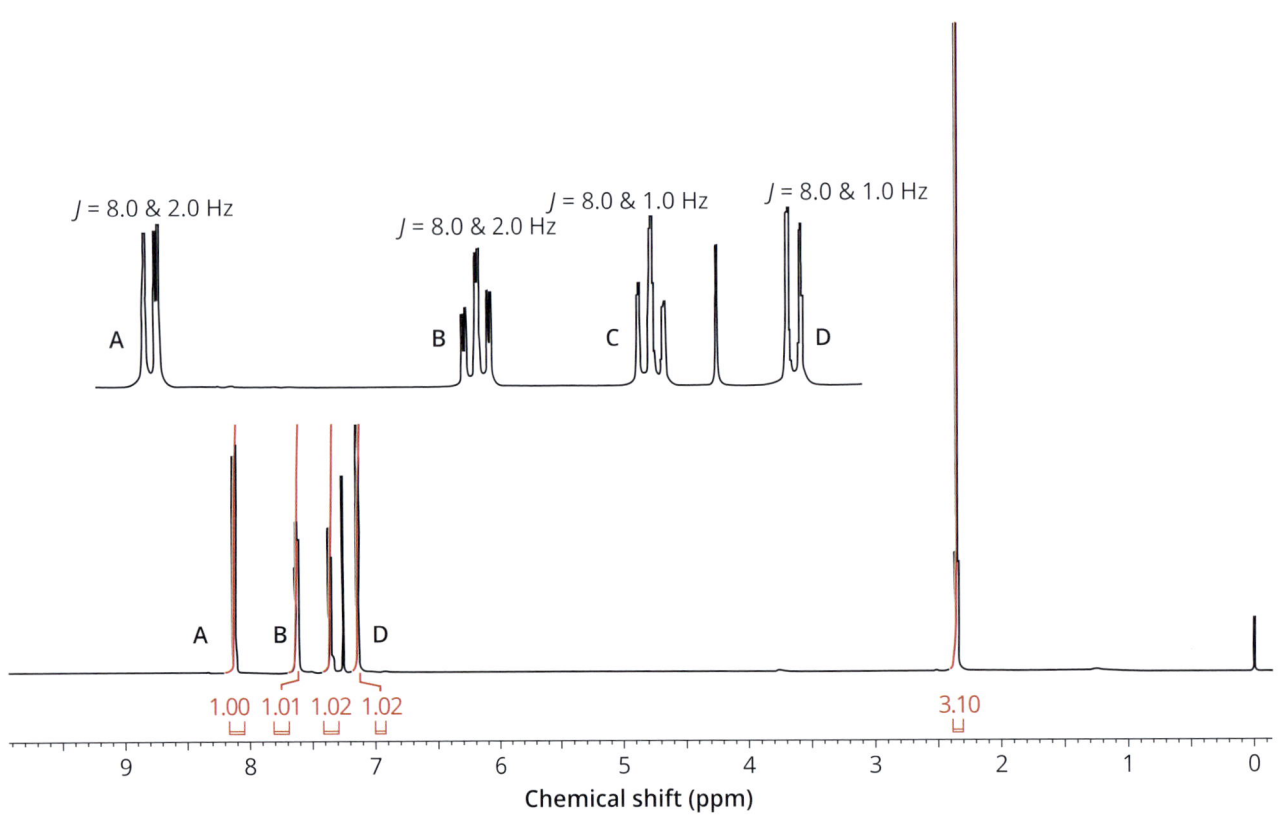

¹H NMR spectrum (400 MHz, CDCl₃)

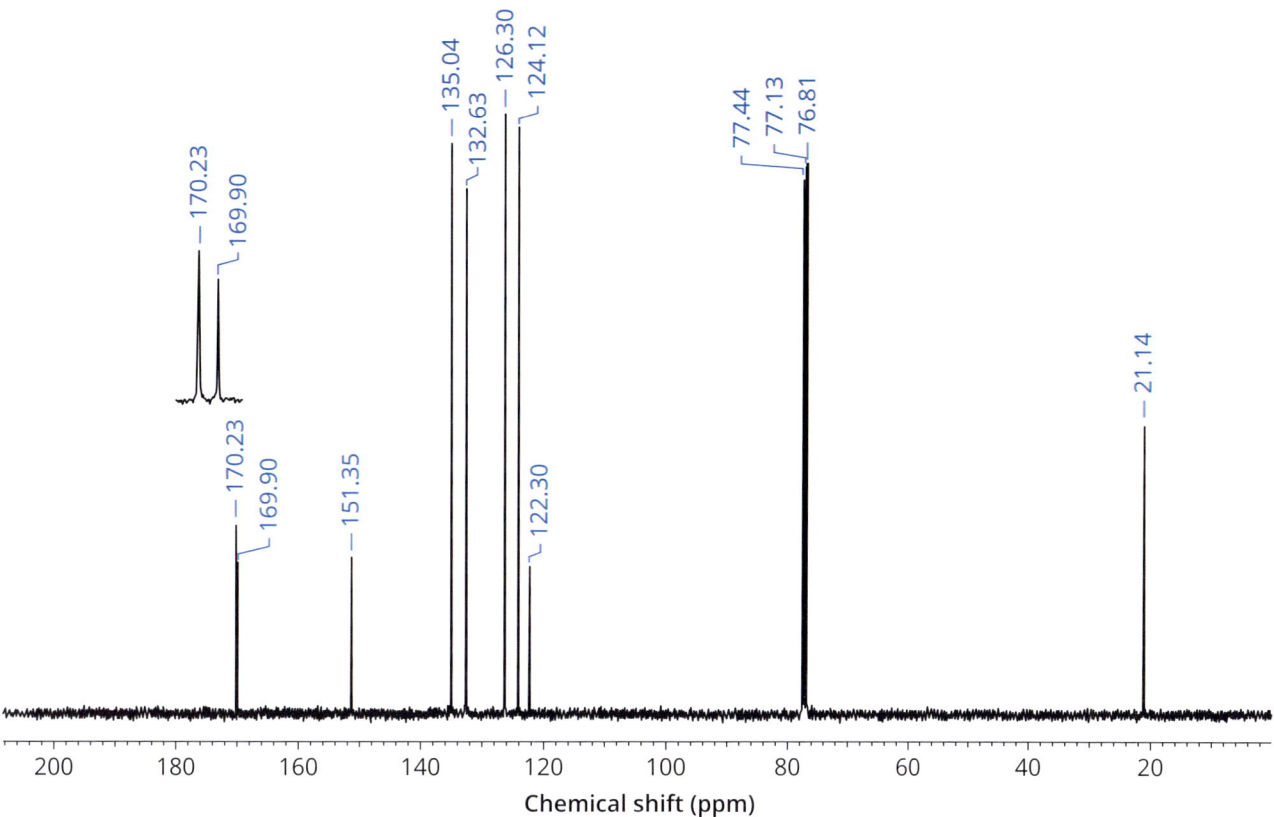

^{13}C NMR spectrum (100 MHz, CDCl$_3$)

9.5 **Problem 5**

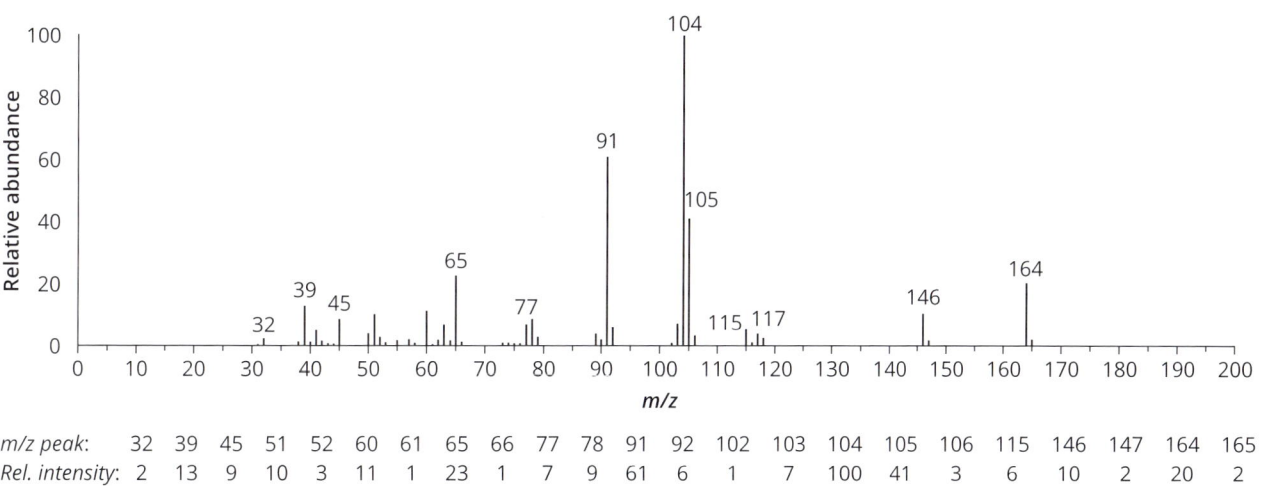

m/z peak:	32	39	45	51	52	60	61	65	66	77	78	91	92	102	103	104	105	106	115	146	147	164	165
Rel. intensity:	2	13	9	10	3	11	1	23	1	7	9	61	6	1	7	100	41	3	6	10	2	20	2

Mass spectrum

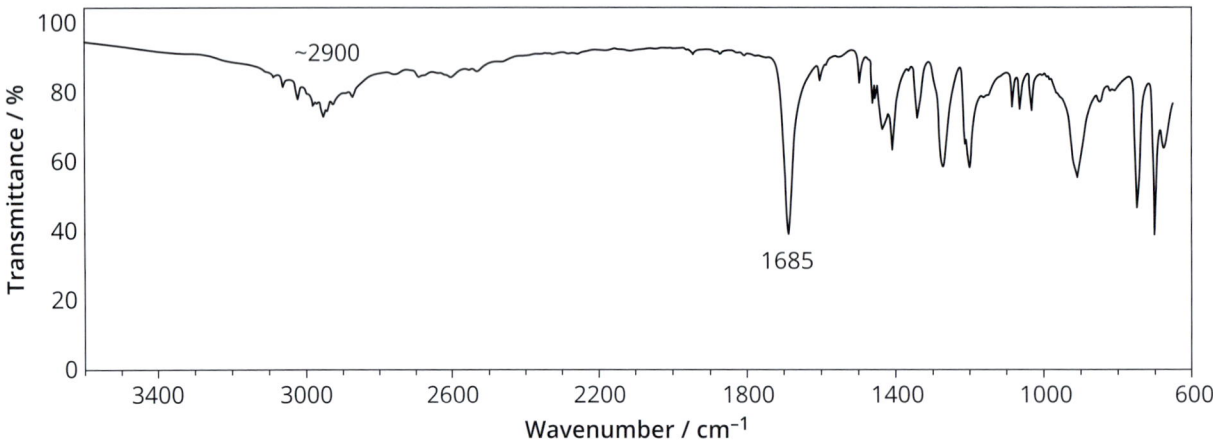

Infrared spectrum

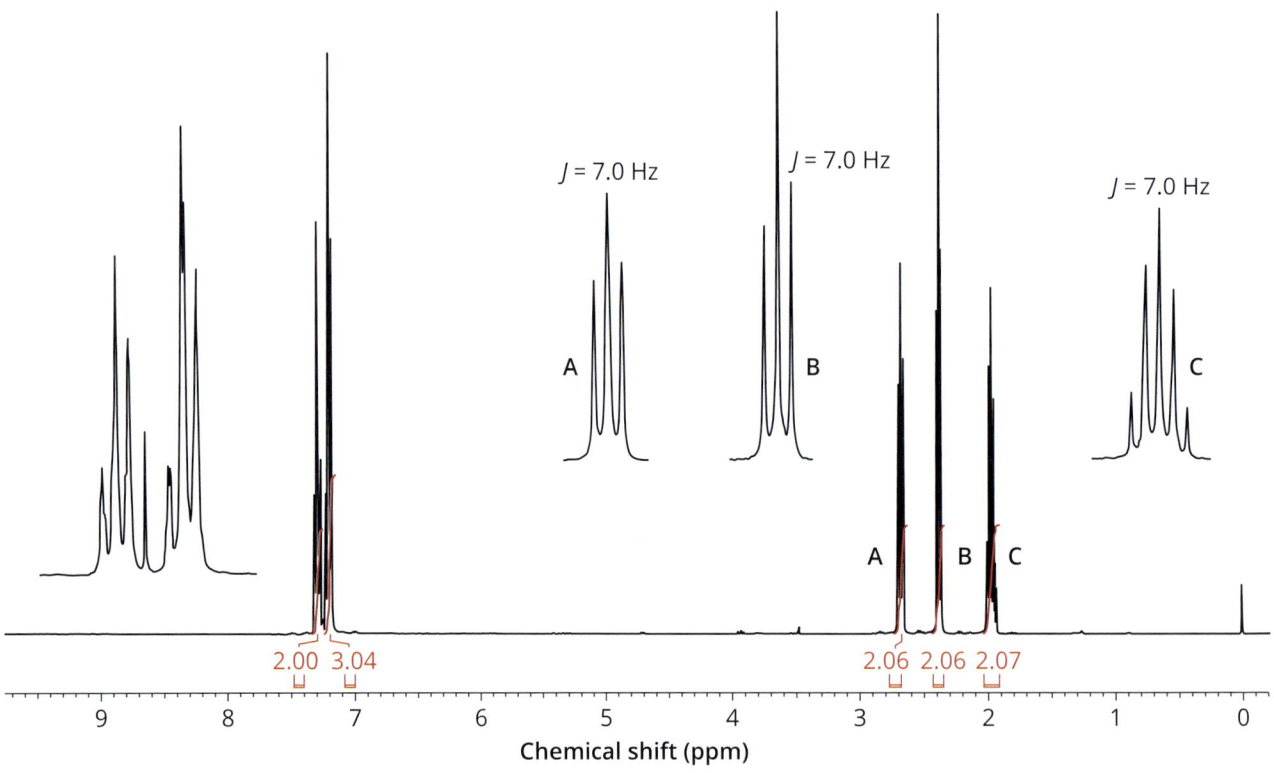

¹H NMR spectrum (400 MHz, CDCl₃)

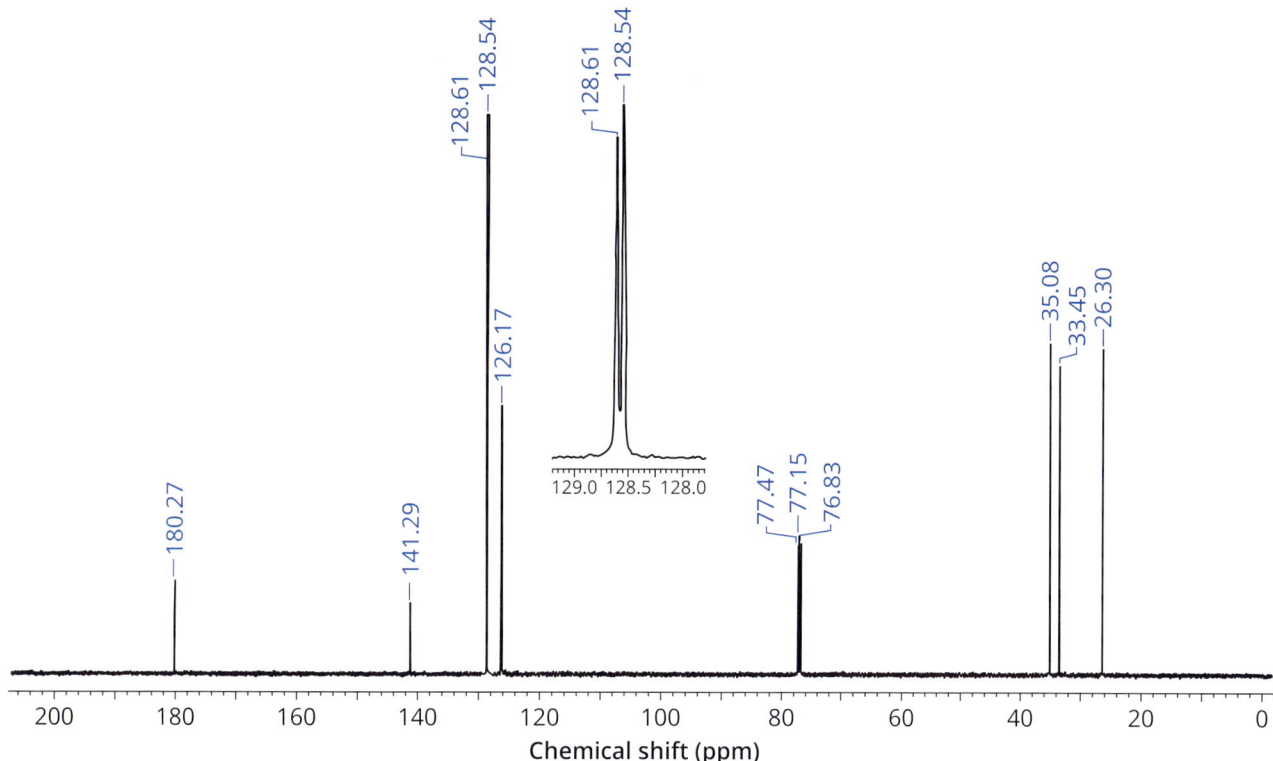

¹³C NMR spectrum (100 MHz, CDCl₃)

9.6 **Problem 6**

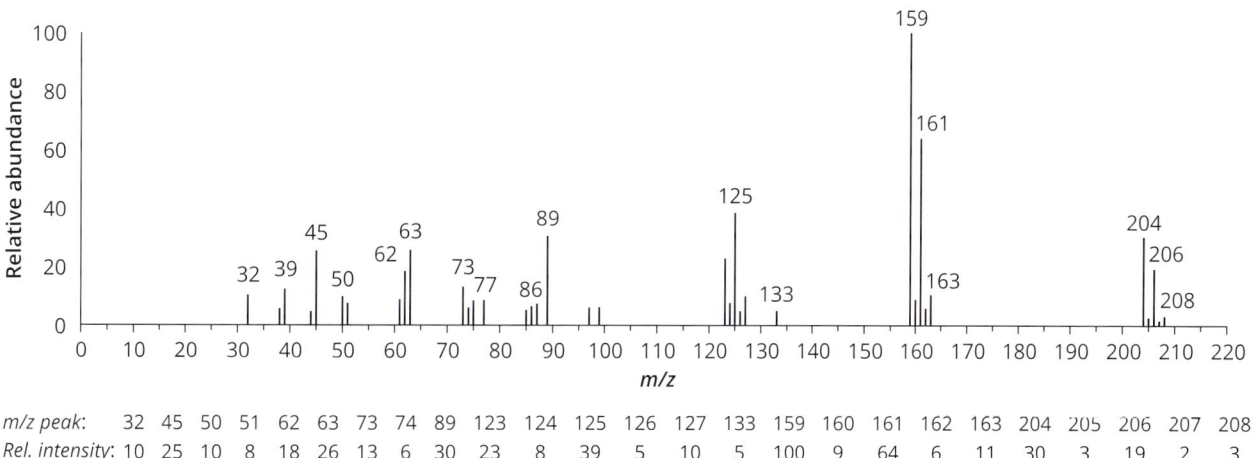

m/z peak:	32	45	50	51	62	63	73	74	89	123	124	125	126	127	133	159	160	161	162	163	204	205	206	207	208
Rel. intensity:	10	25	10	8	18	26	13	6	30	23	8	39	5	10	5	100	9	64	6	11	30	3	19	2	3

Mass spectrum

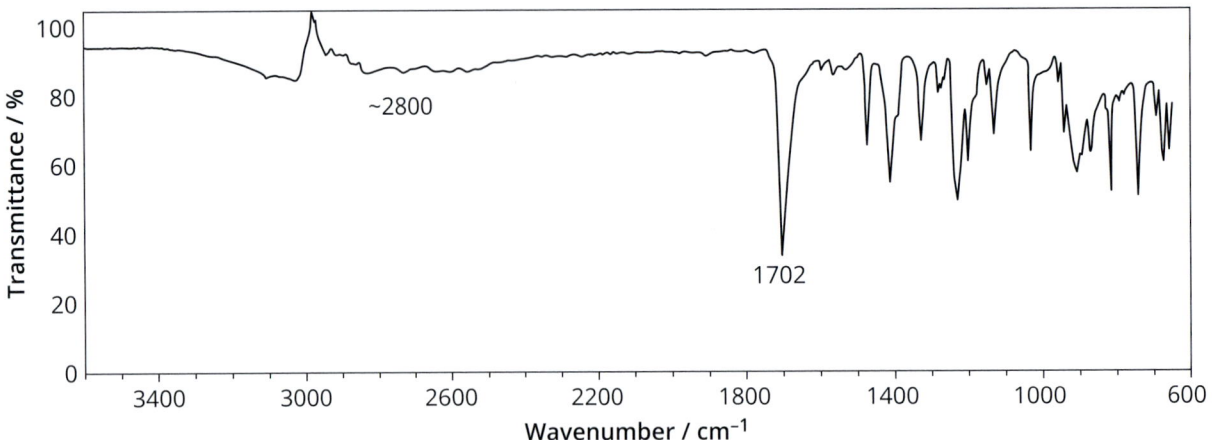

Infrared spectrum

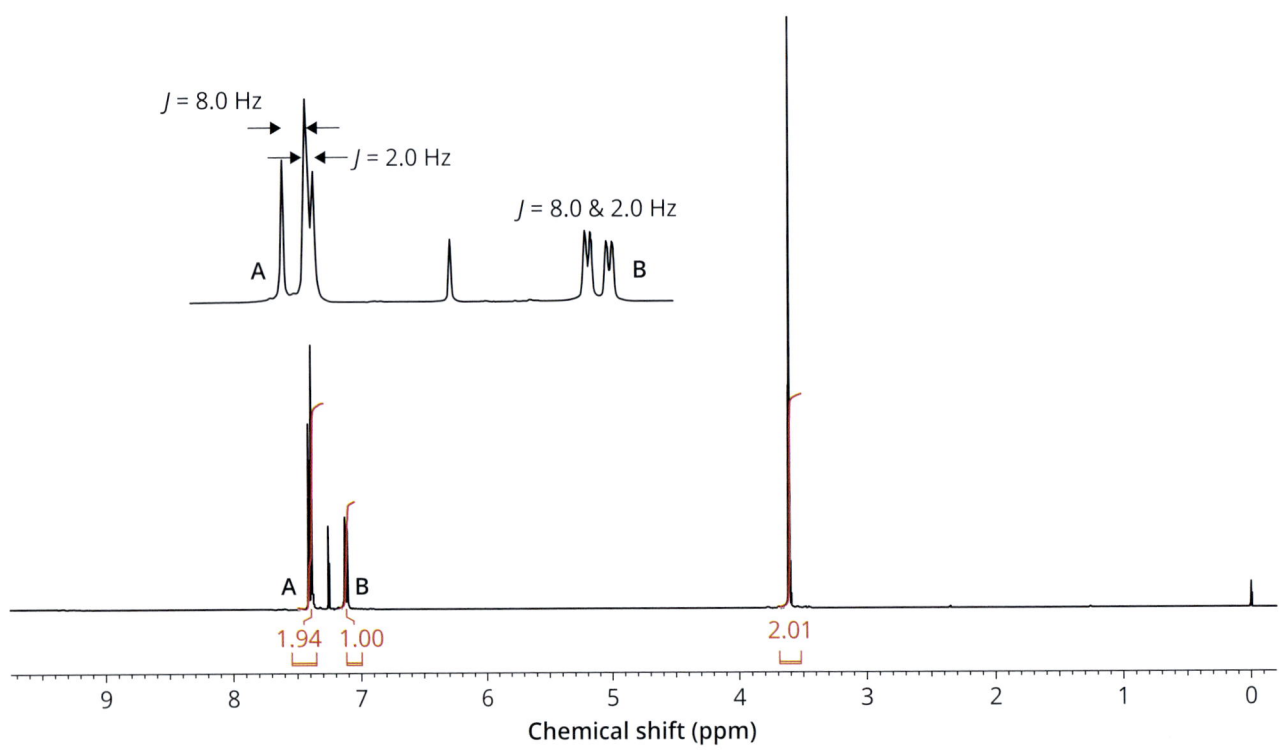

¹H NMR spectrum (400 MHz, CDCl₃)

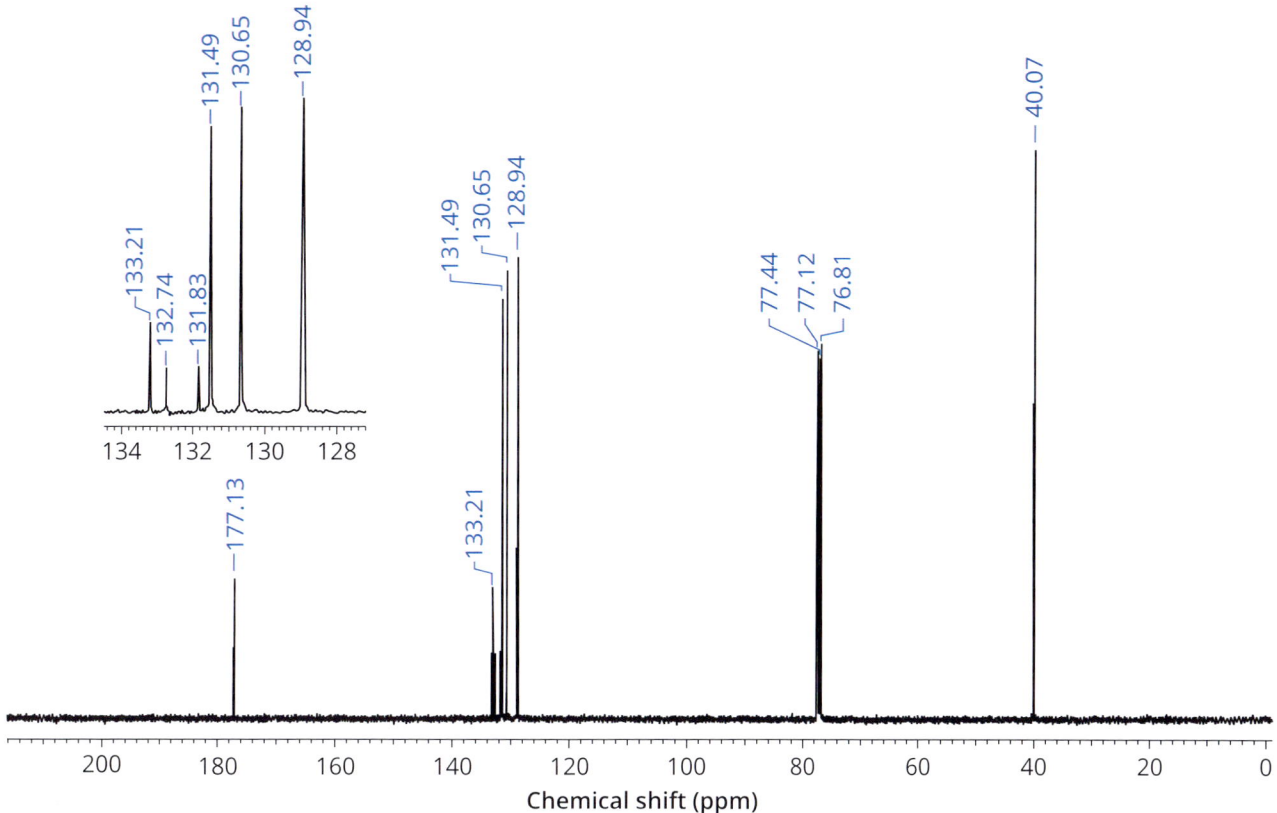

^{13}C NMR spectrum (100 MHz, CDCl$_3$)

9.7 **Problem 7**

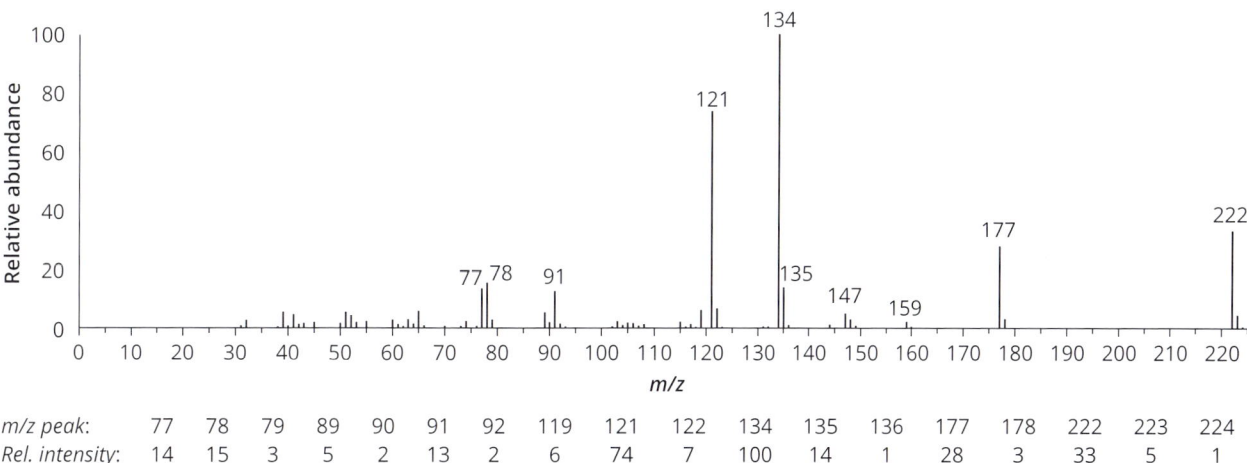

m/z peak:	77	78	79	89	90	91	92	119	121	122	134	135	136	177	178	222	223	224
Rel. intensity:	14	15	3	5	2	13	2	6	74	7	100	14	1	28	3	33	5	1

Mass spectrum

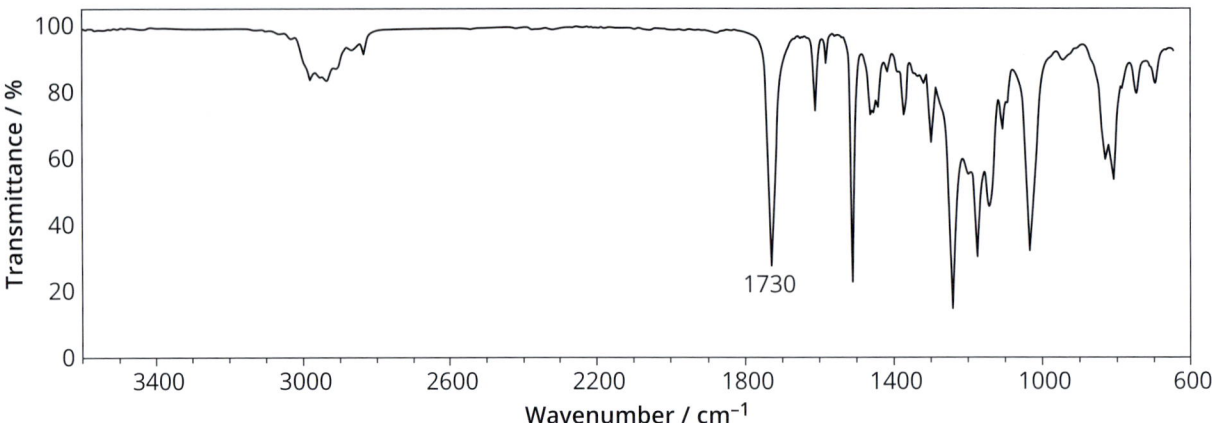

Infrared spectrum

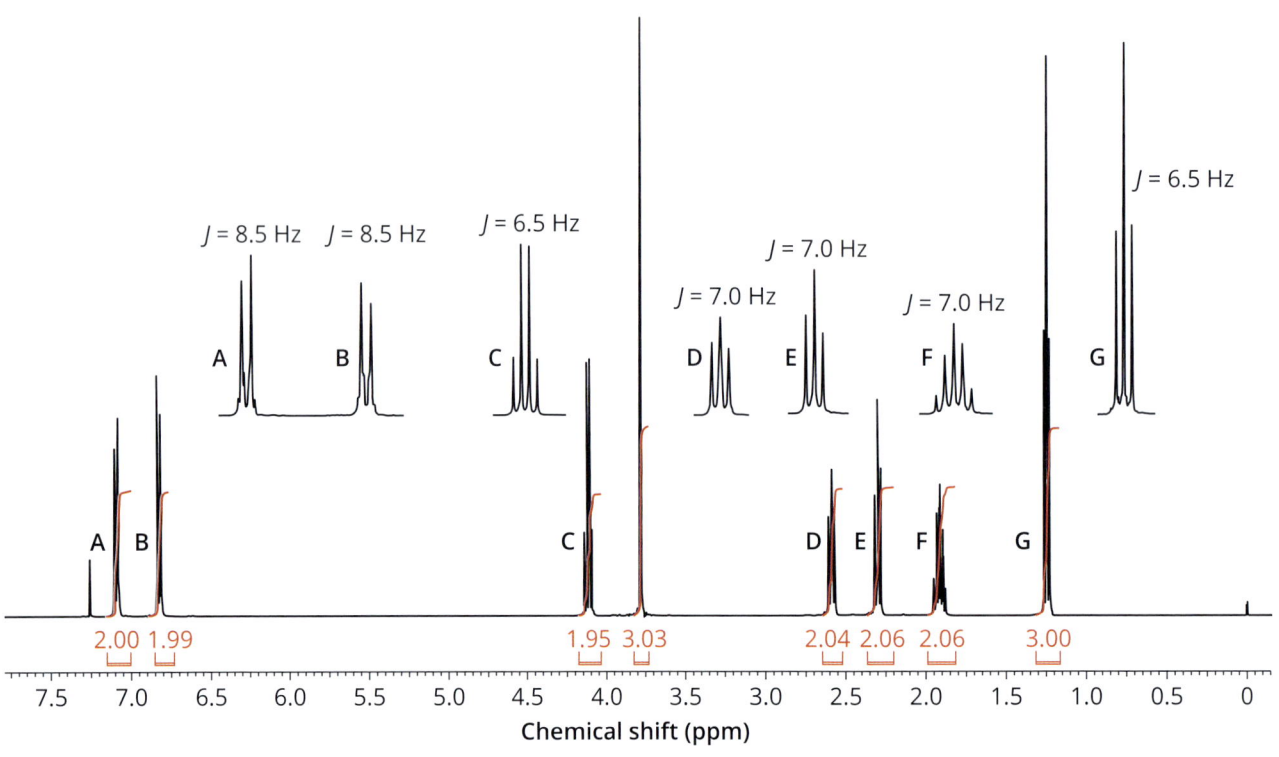

¹H NMR spectrum (400 MHz, CDCl₃)

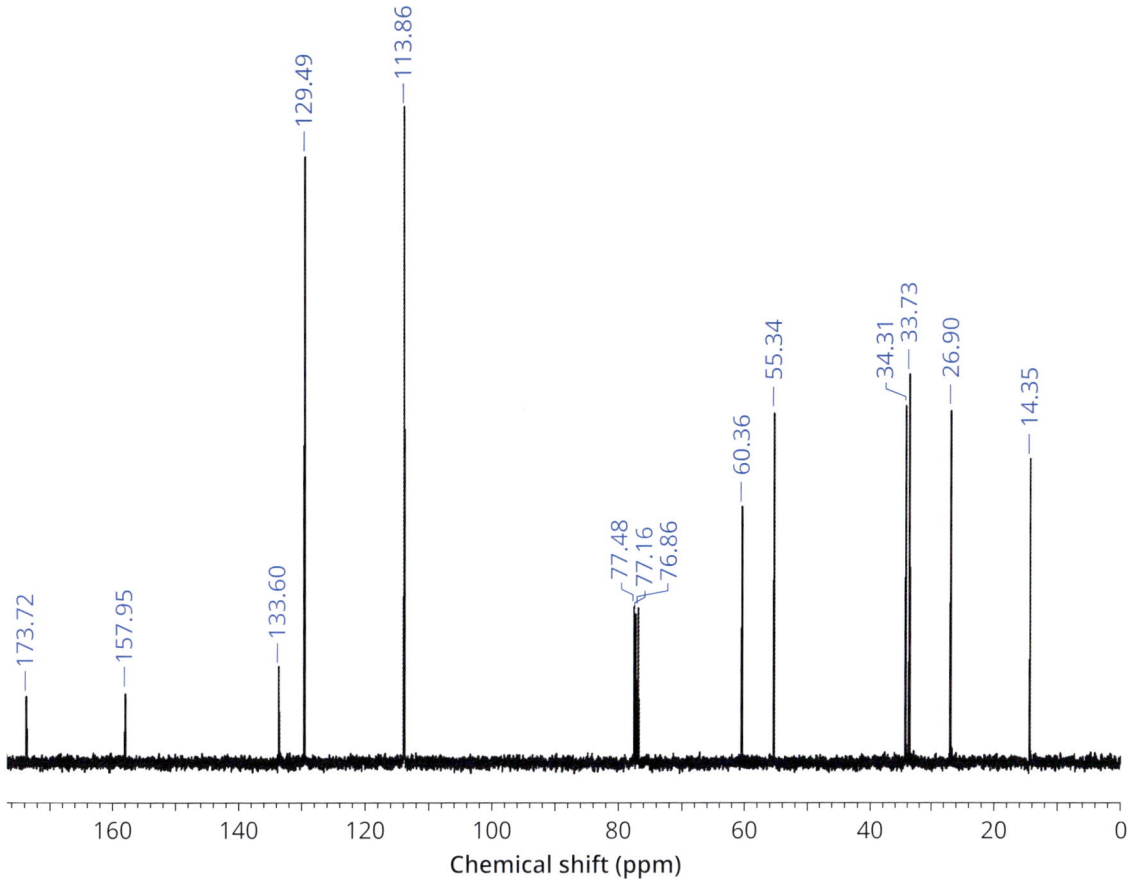

13C NMR spectrum (100 MHz, CDCl₃)

9.8 **Problem 8**

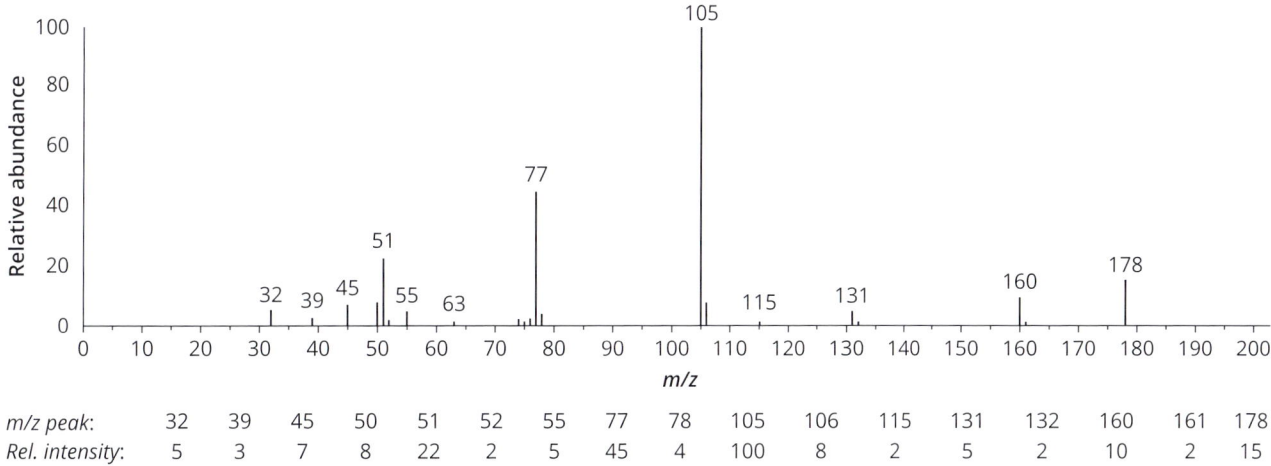

m/z peak:	32	39	45	50	51	52	55	77	78	105	106	115	131	132	160	161	178
Rel. intensity:	5	3	7	8	22	2	5	45	4	100	8	2	5	2	10	2	15

Mass spectrum

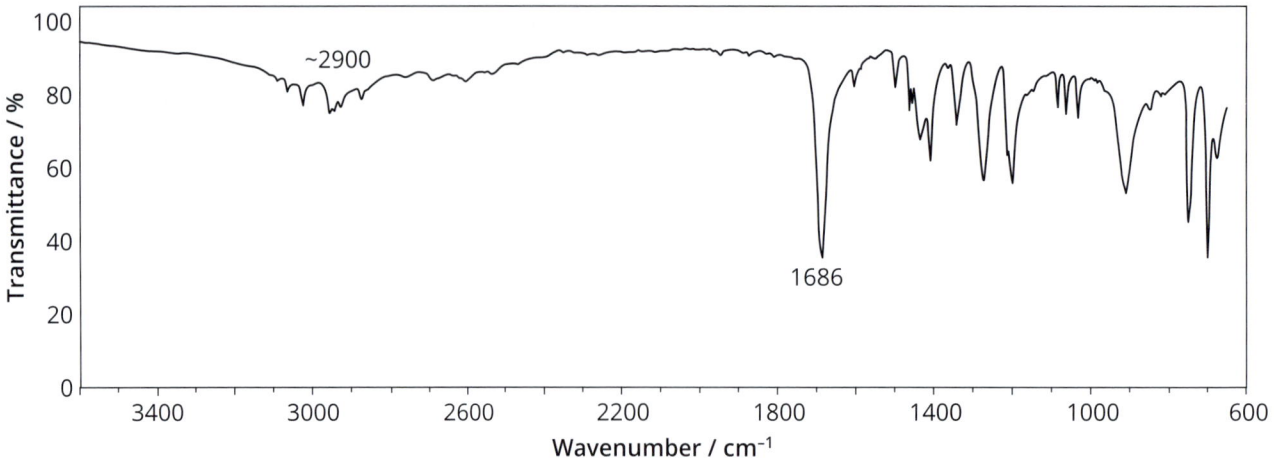

Infrared spectrum

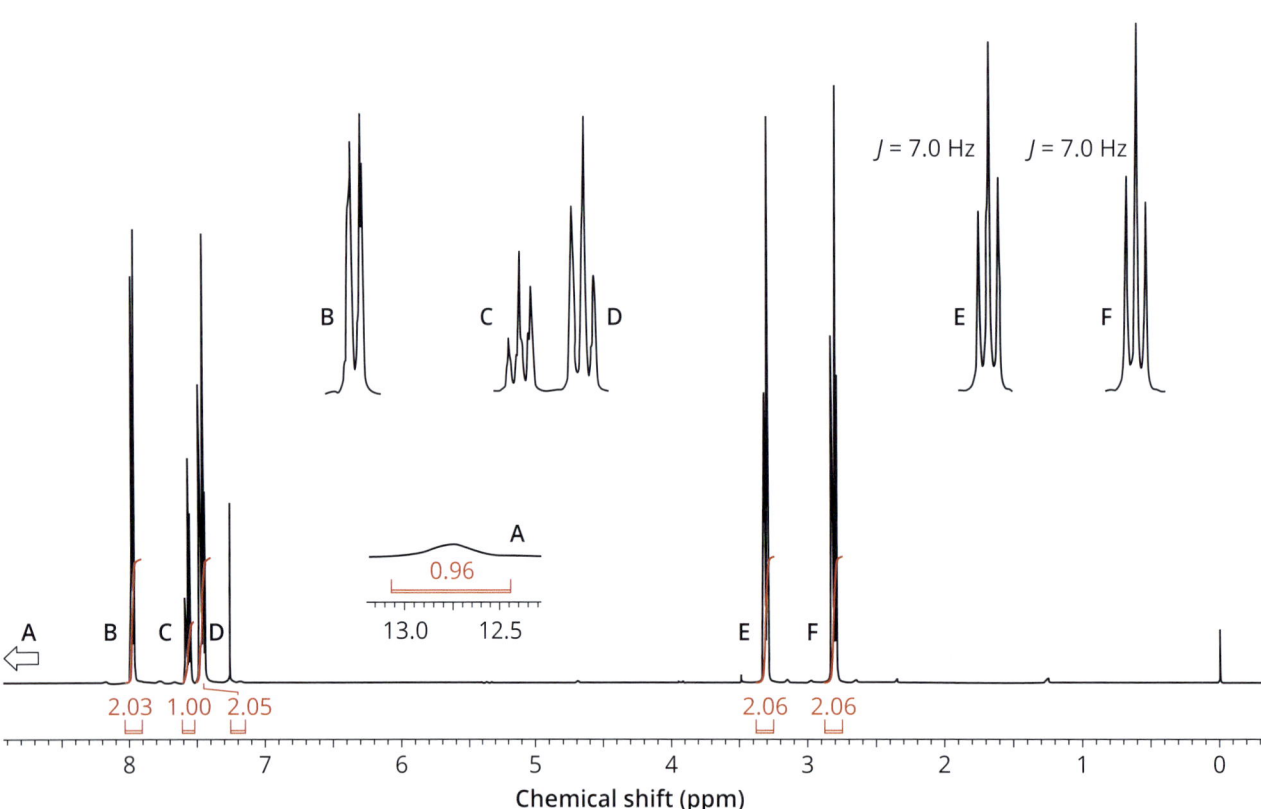

¹H NMR spectrum (400 MHz, CDCl₃)

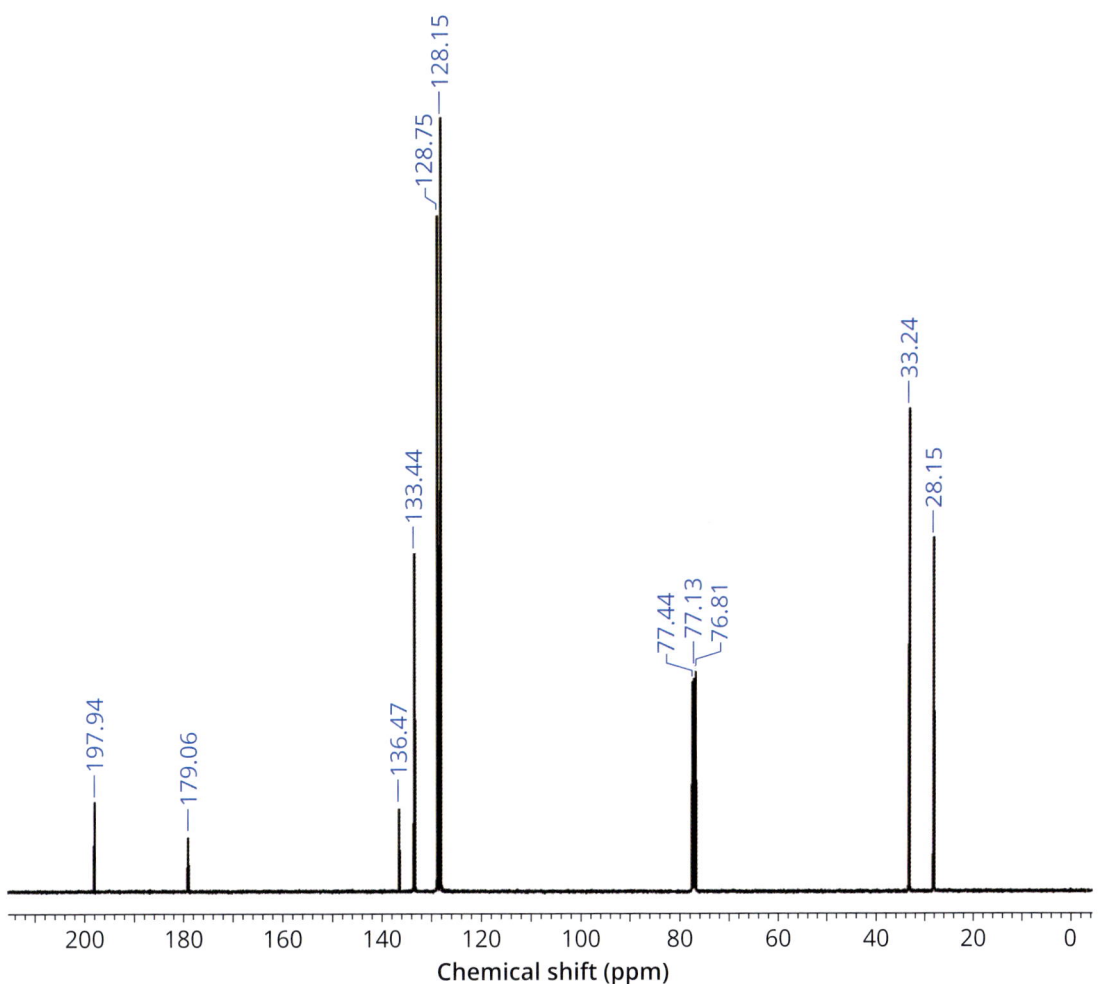

^{13}C NMR spectrum (100 MHz, CDCl$_3$)

9.9 Problem 9

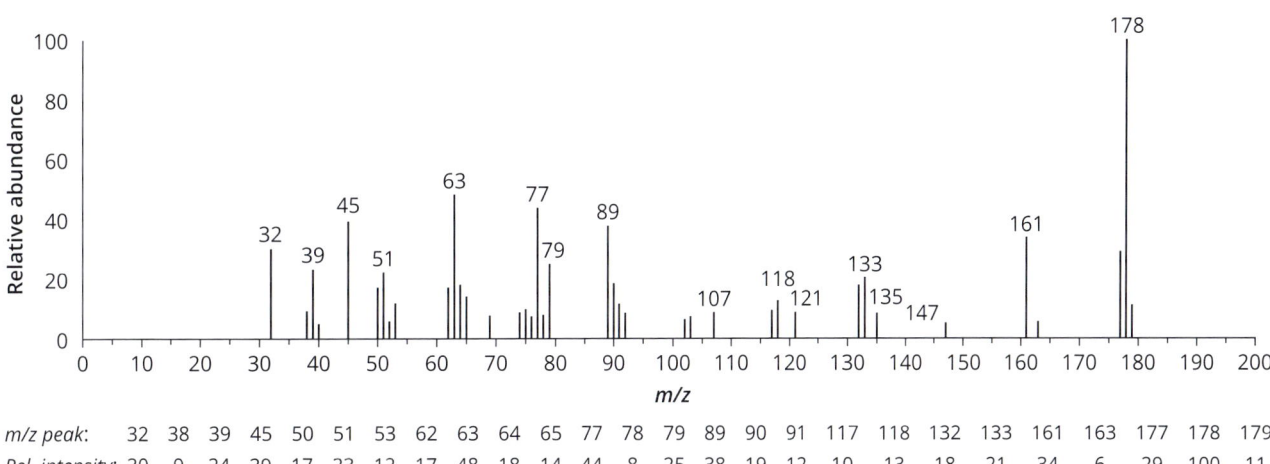

m/z peak:	32	38	39	45	50	51	53	62	63	64	65	77	78	79	89	90	91	117	118	132	133	161	163	177	178	179
Rel. intensity:	30	9	24	39	17	23	12	17	48	18	14	44	8	25	38	19	12	10	13	18	21	34	6	29	100	11

Mass spectrum

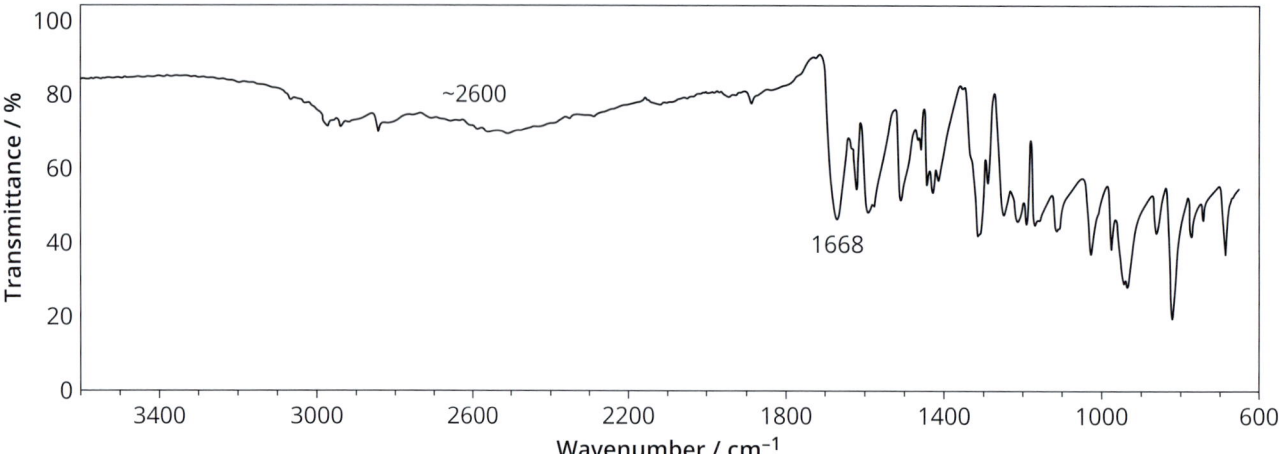

Infrared spectrum

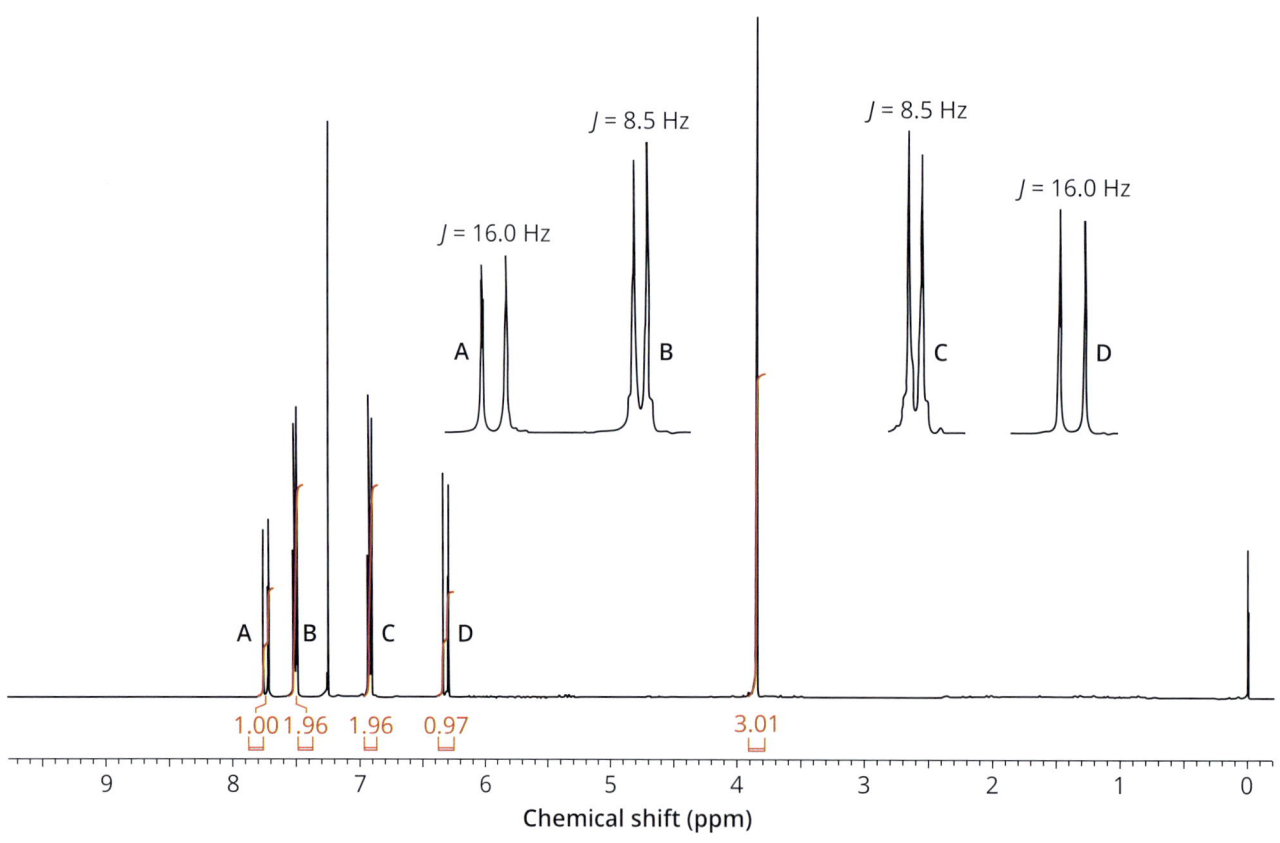

¹H NMR spectrum (400 MHz, CDCl₃)

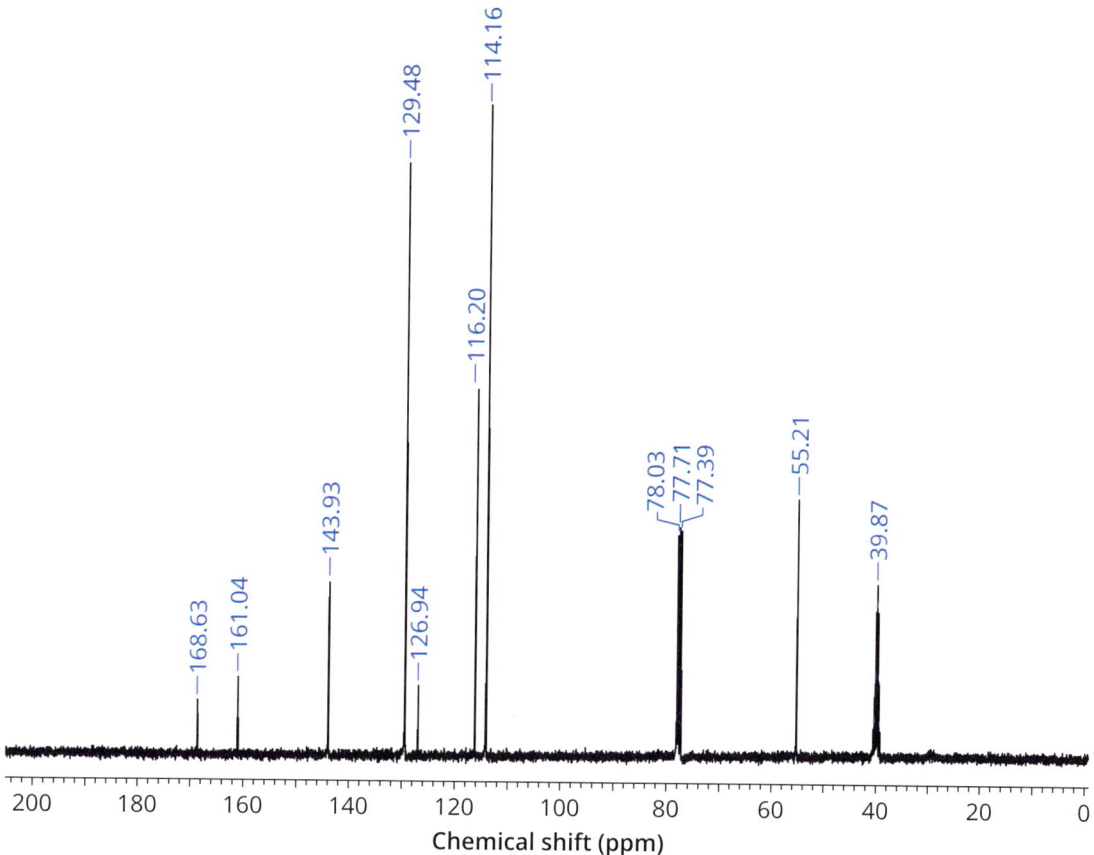

^{13}C NMR spectrum (100 MHz, CDCl$_3$ + d_6-DMSO)

> **HINT**
>
> Note the use of a mixed deuterated solvent system here (to aid solubility).

9.10 Problem 10

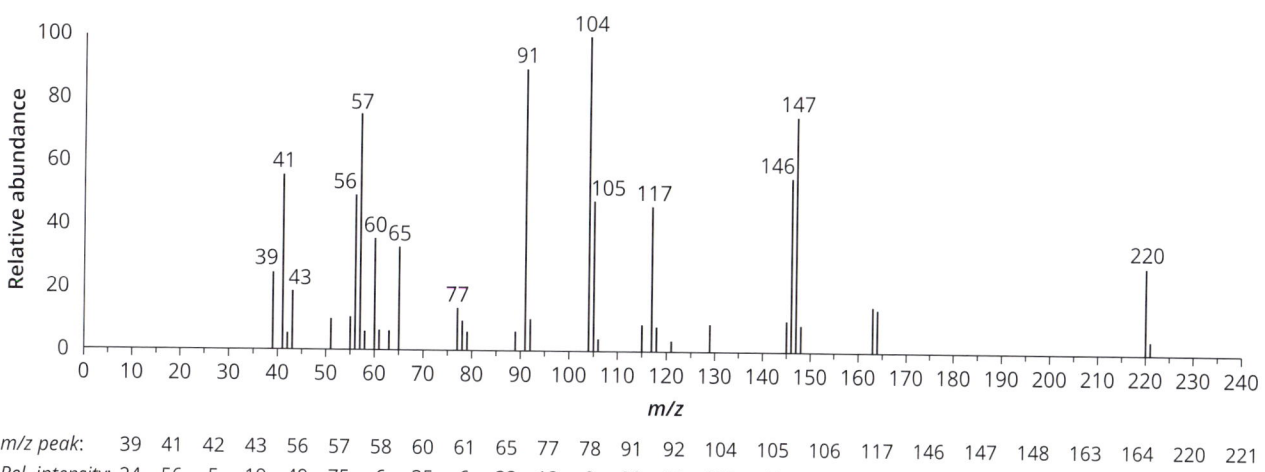

m/z peak:	39	41	42	43	56	57	58	60	61	65	77	78	91	92	104	105	106	117	146	147	148	163	164	220	221
Rel. intensity:	24	56	5	19	49	75	6	35	6	33	13	9	90	10	100	48	4	46	55	75	8	14	14	28	4

Mass spectrum

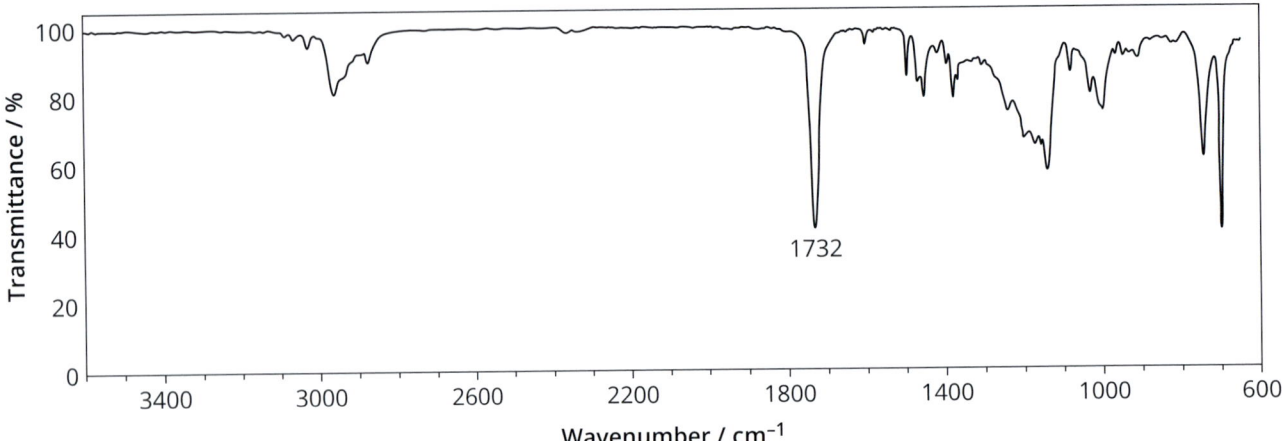

Infrared spectrum

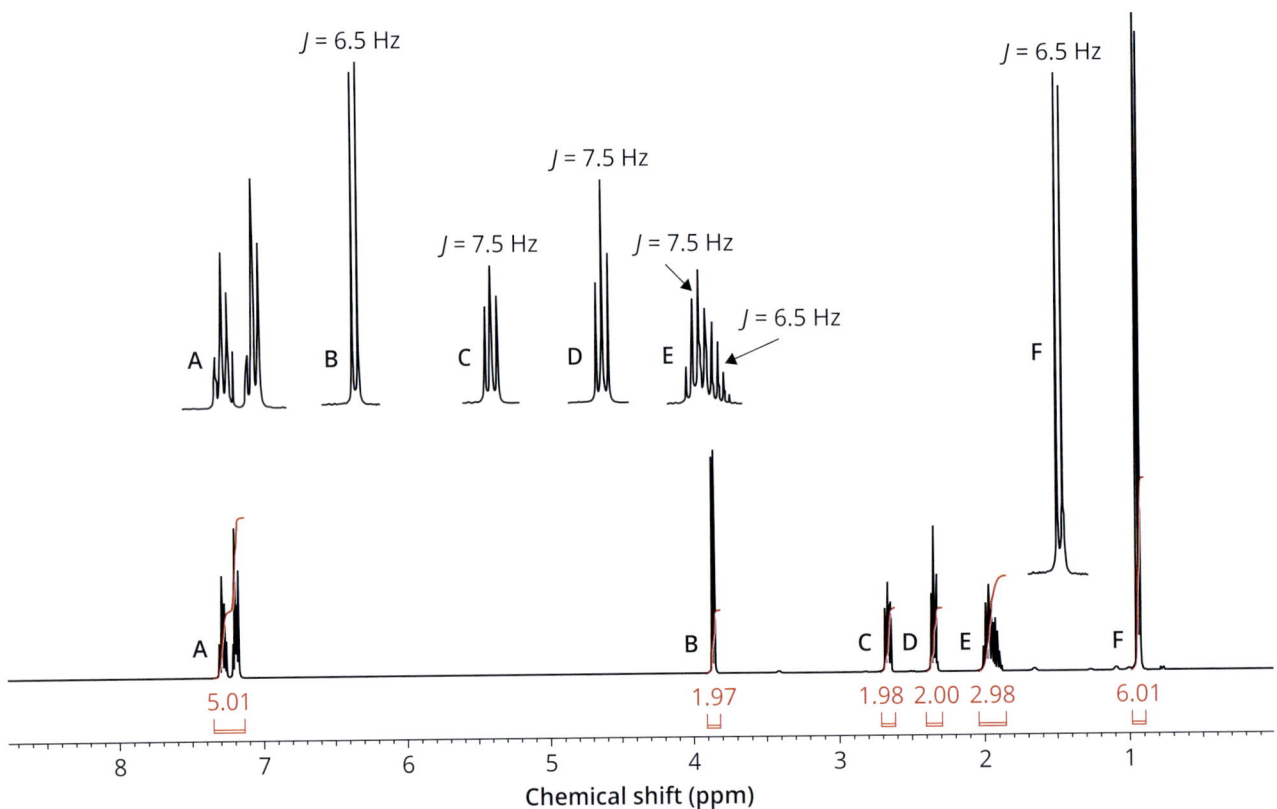

¹H NMR spectrum (400 MHz, CDCl₃)

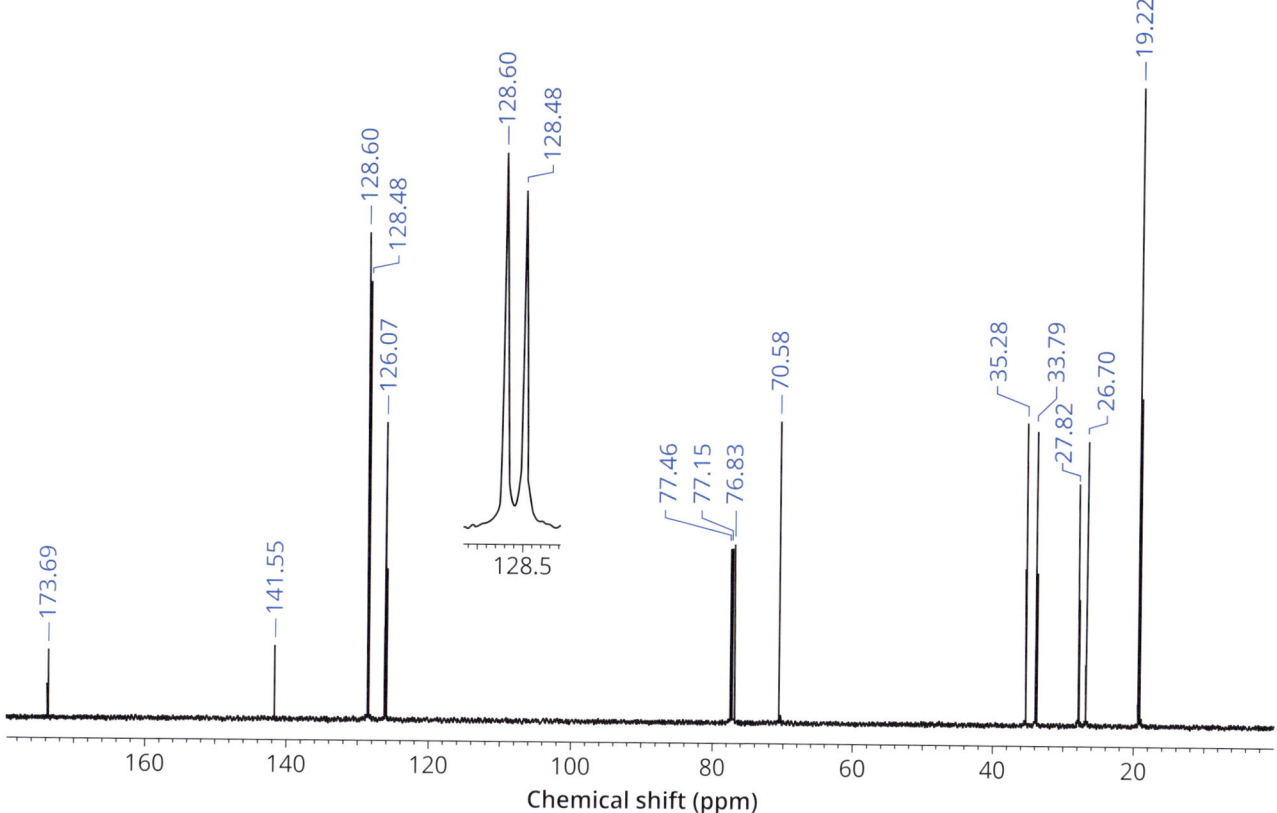

¹³C NMR spectrum (100 MHz, CDCl₃)

9.11 **Problem 11**

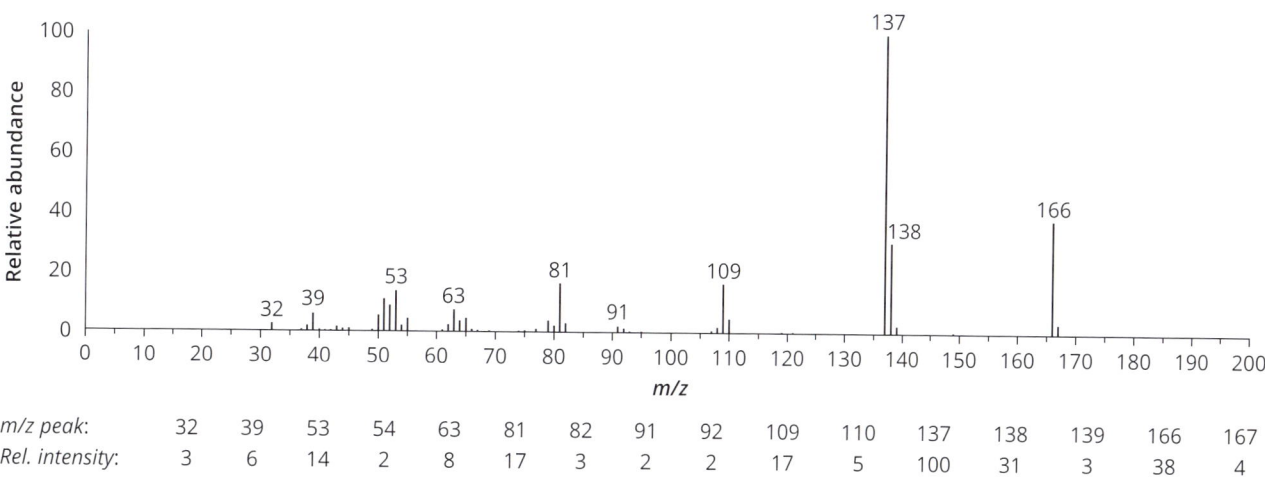

m/z peak:	32	39	53	54	63	81	82	91	92	109	110	137	138	139	166	167
Rel. intensity:	3	6	14	2	8	17	3	2	2	17	5	100	31	3	38	4

Mass spectrum

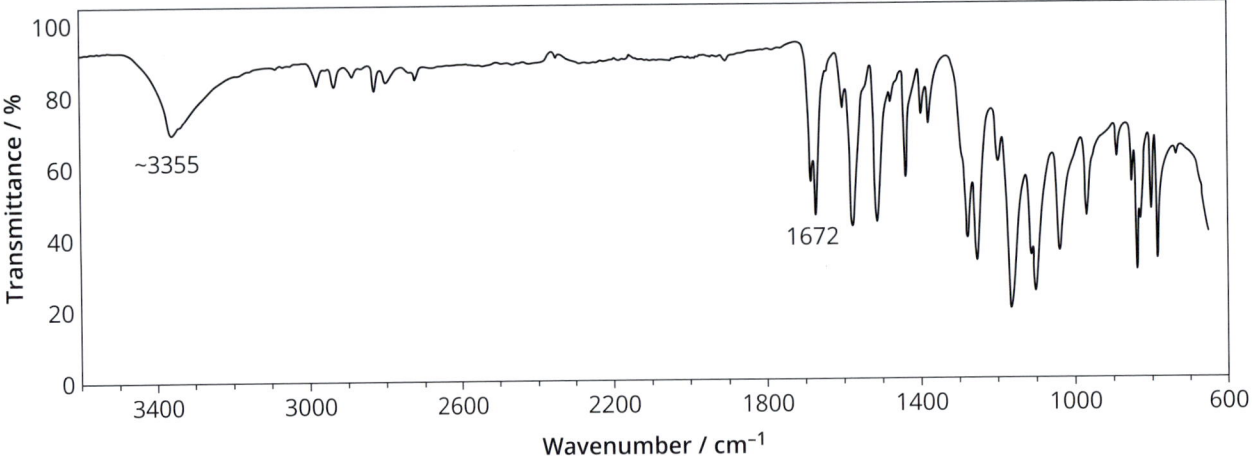

Infrared spectrum

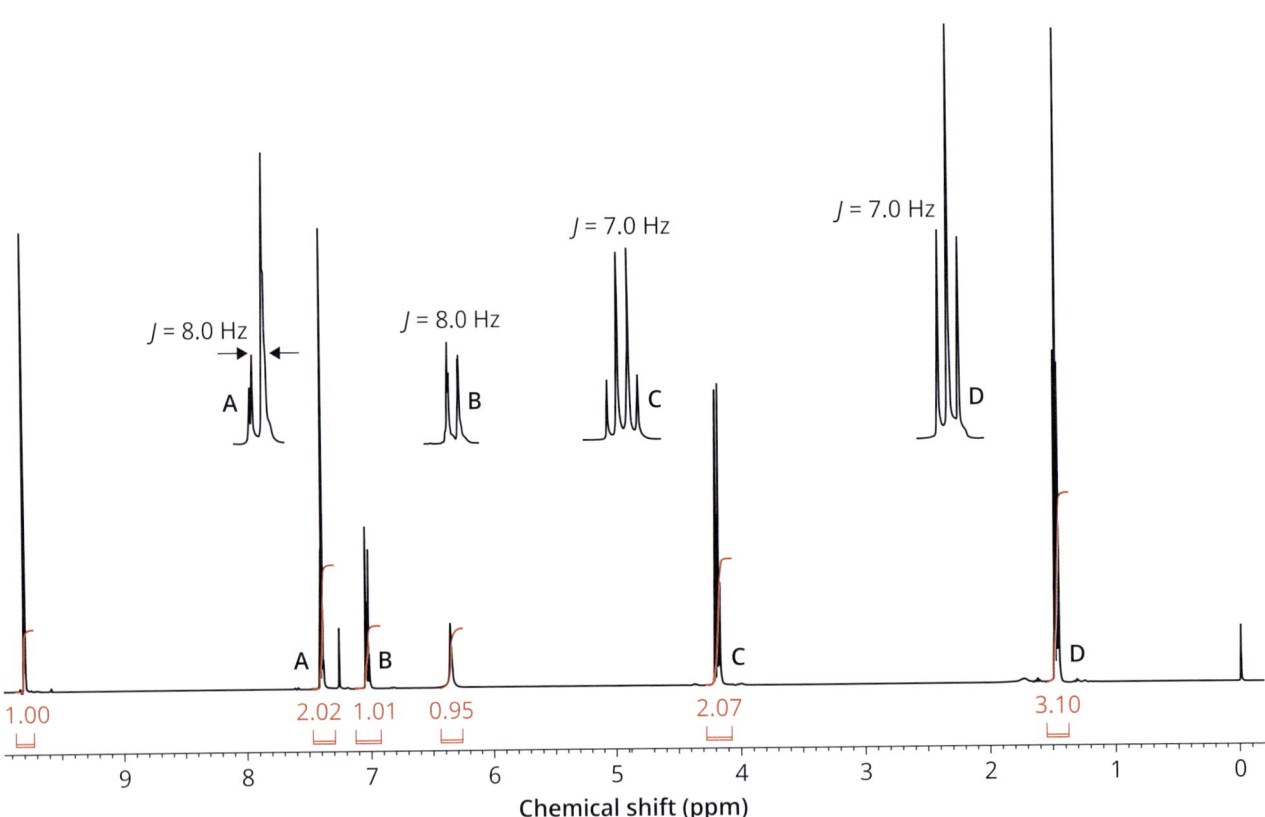

¹H NMR spectrum (400 MHz, CDCl₃)

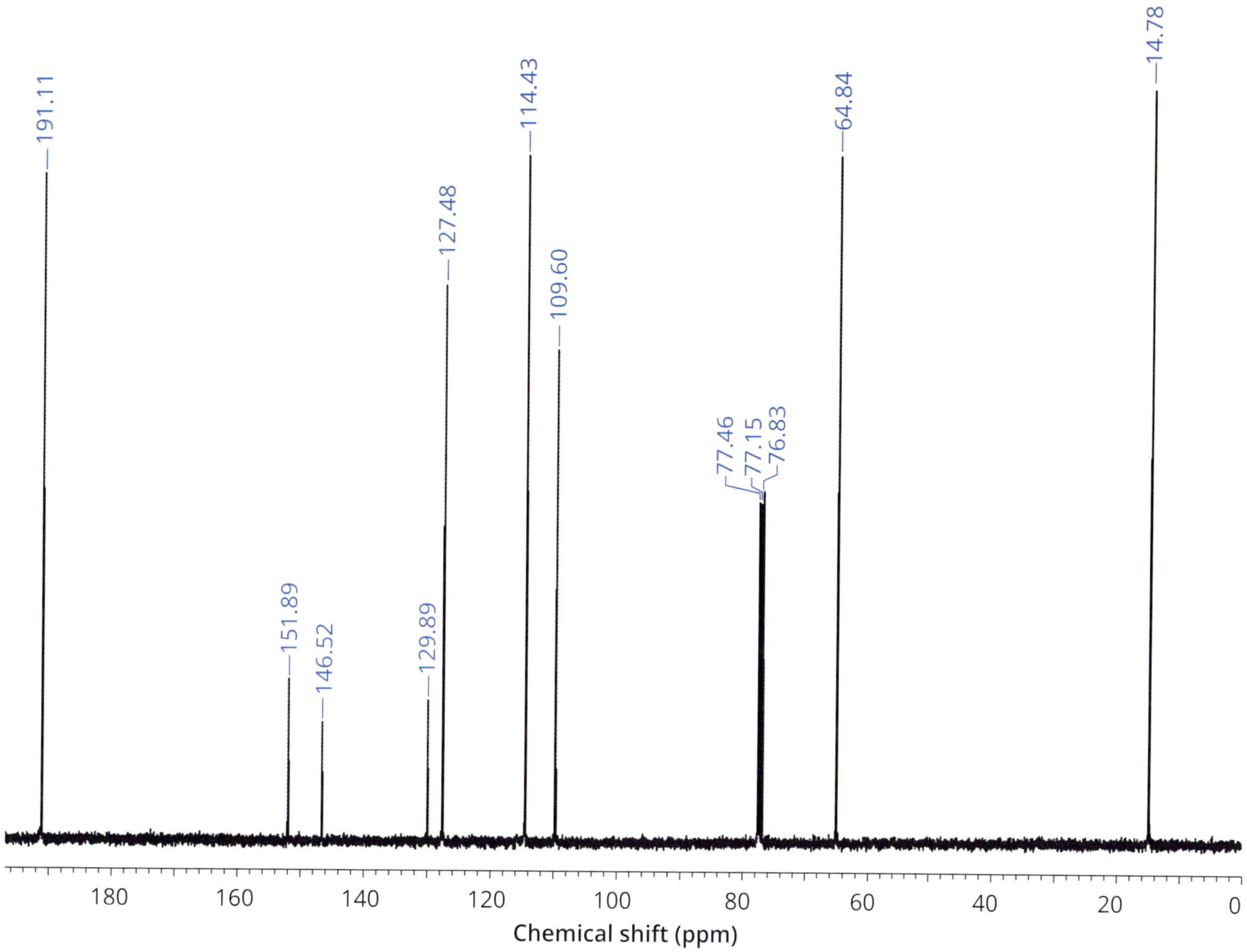

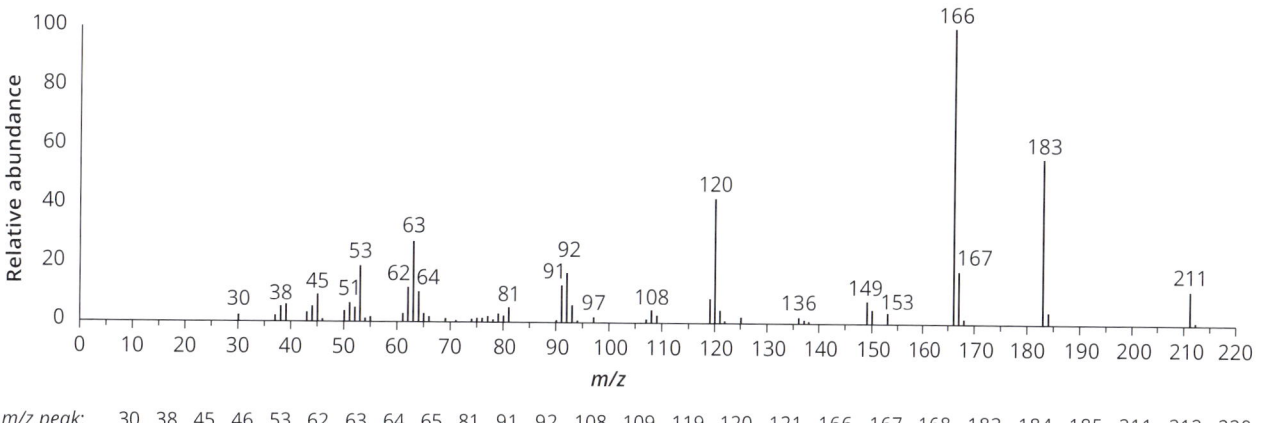

191.11
151.89
146.52
129.89
127.48
114.43
109.60
77.46
77.15
76.83
64.84
14.78
180
160
140
120
100
80
60
40
20
0
Chemical shift (ppm)

^{13}C NMR spectrum (100 MHz, $CDCl_3$)

9.12 Problem 12

Relative abundance
100
80
60
40
20
0
166
183
120
63
53
92
91
62
64
167
211
45
51
81
30
38
97
108
136
149
153
0 10 20 30 40 50 60 70 80 90 100 110 120 130 140 150 160 170 180 190 200 210 220
m/z

m/z peak:	30	38	45	46	53	62	63	64	65	81	91	92	108	109	119	120	121	166	167	168	183	184	185	211	212	220
Rel. intensity:	2	6	10	1	19	12	27	11	3	5	13	17	5	3	9	42	5	100	18	2	56	4	1	12	1	0

Mass spectrum

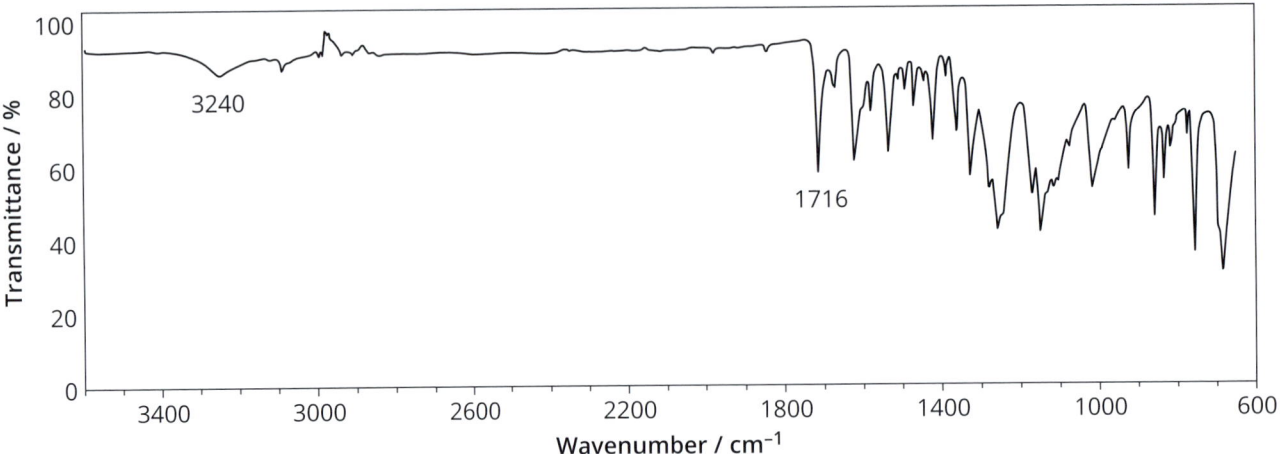

Infrared spectrum

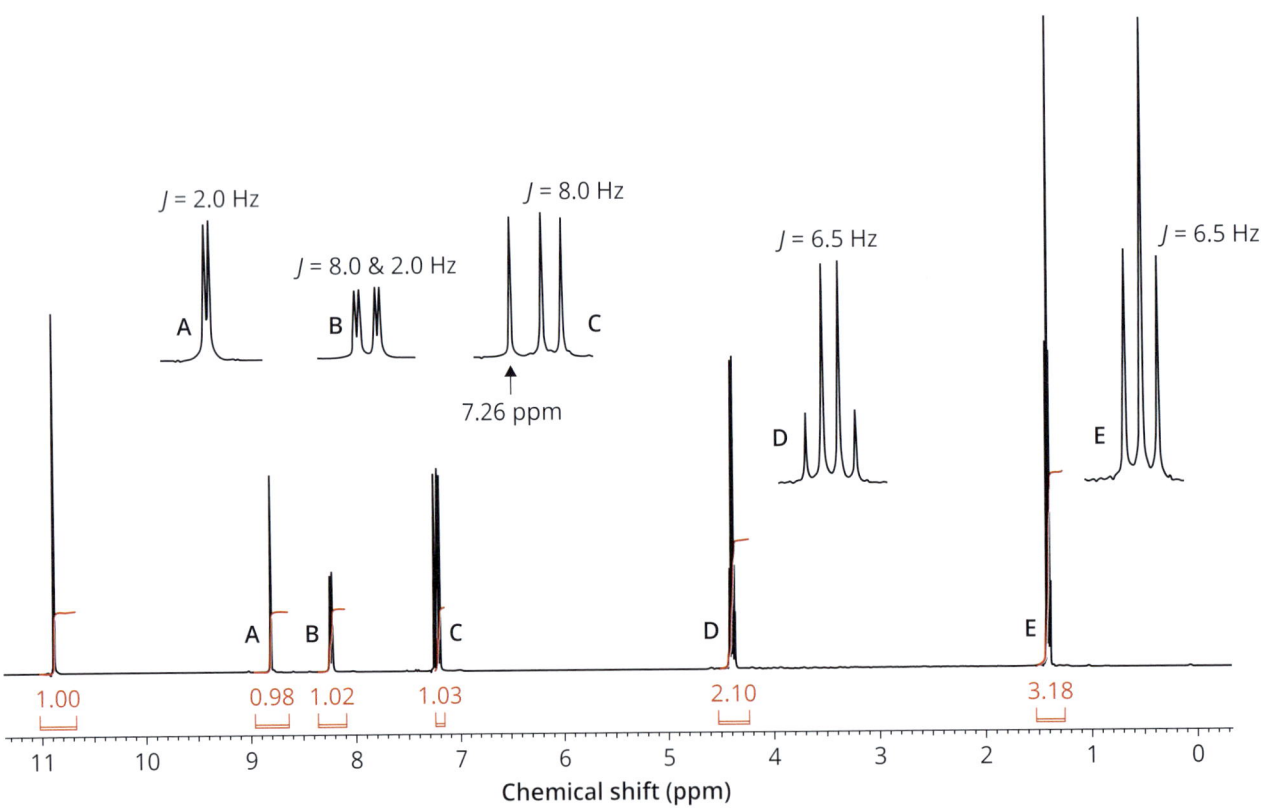

¹H NMR spectrum (400 MHz, CDCl₃)

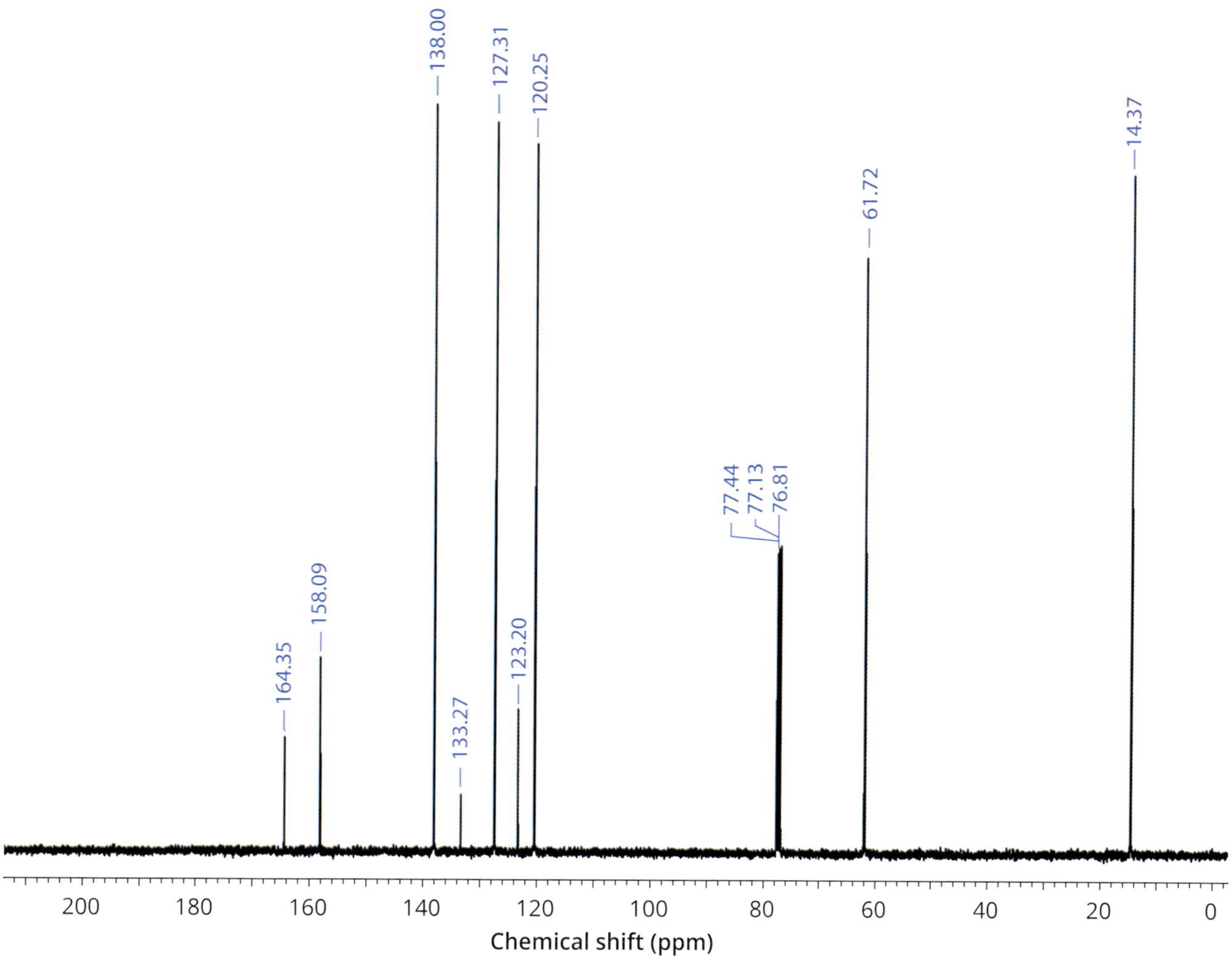

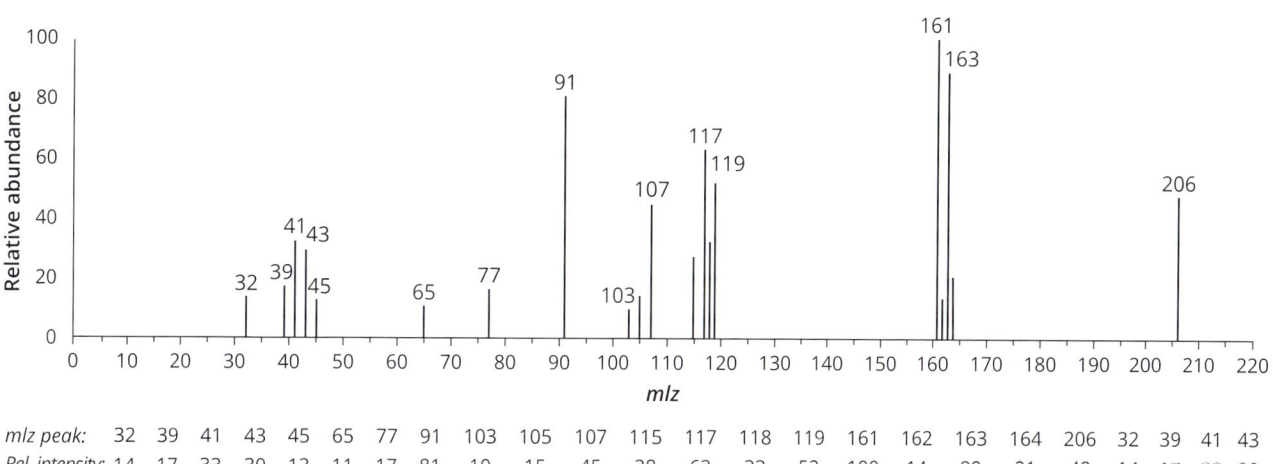

— 138.00
— 127.31
— 120.25
— 14.37
— 61.72
— 77.44
— 77.13
— 76.81
— 164.35
— 158.09
— 133.27
— 123.20

Chemical shift (ppm)

^{13}C NMR spectrum (100 MHz, CDCl$_3$)

9.13 **Problem 13**

Relative abundance

m/z

m/z peak:	32	39	41	43	45	65	77	91	103	105	107	115	117	118	119	161	162	163	164	206	32	39	41	43
Rel. intensity:	14	17	33	30	13	11	17	81	10	15	45	28	63	33	52	100	14	89	21	48	14	17	33	30

Mass spectrum

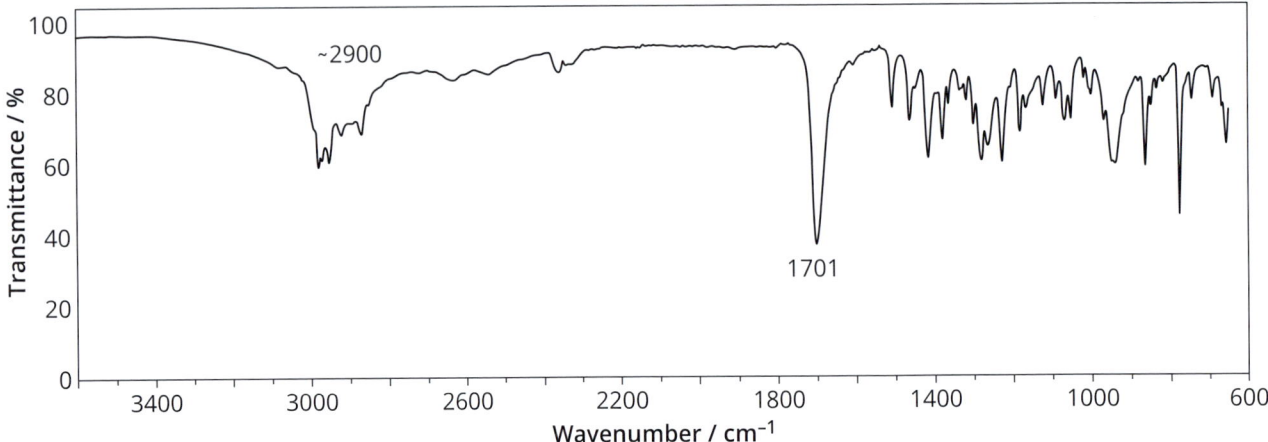

Infrared spectrum

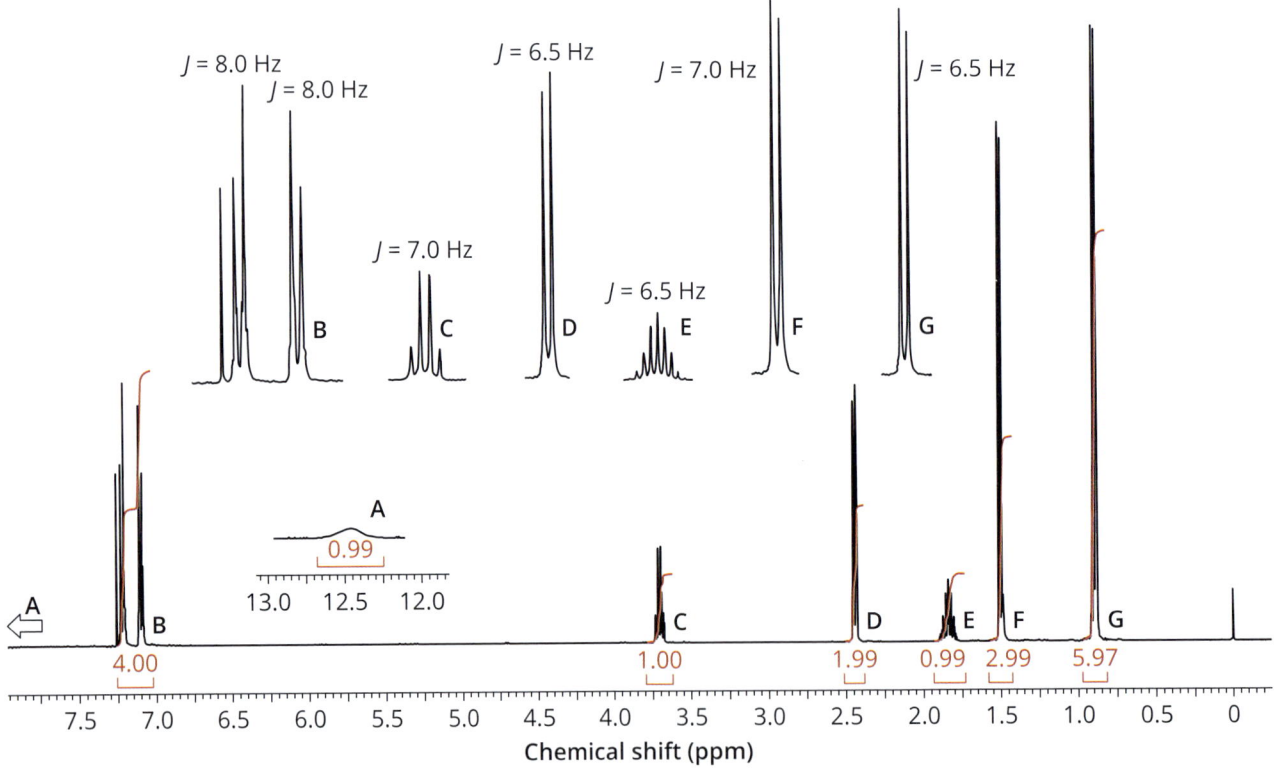

¹H NMR spectrum (400 MHz, CDCl₃)

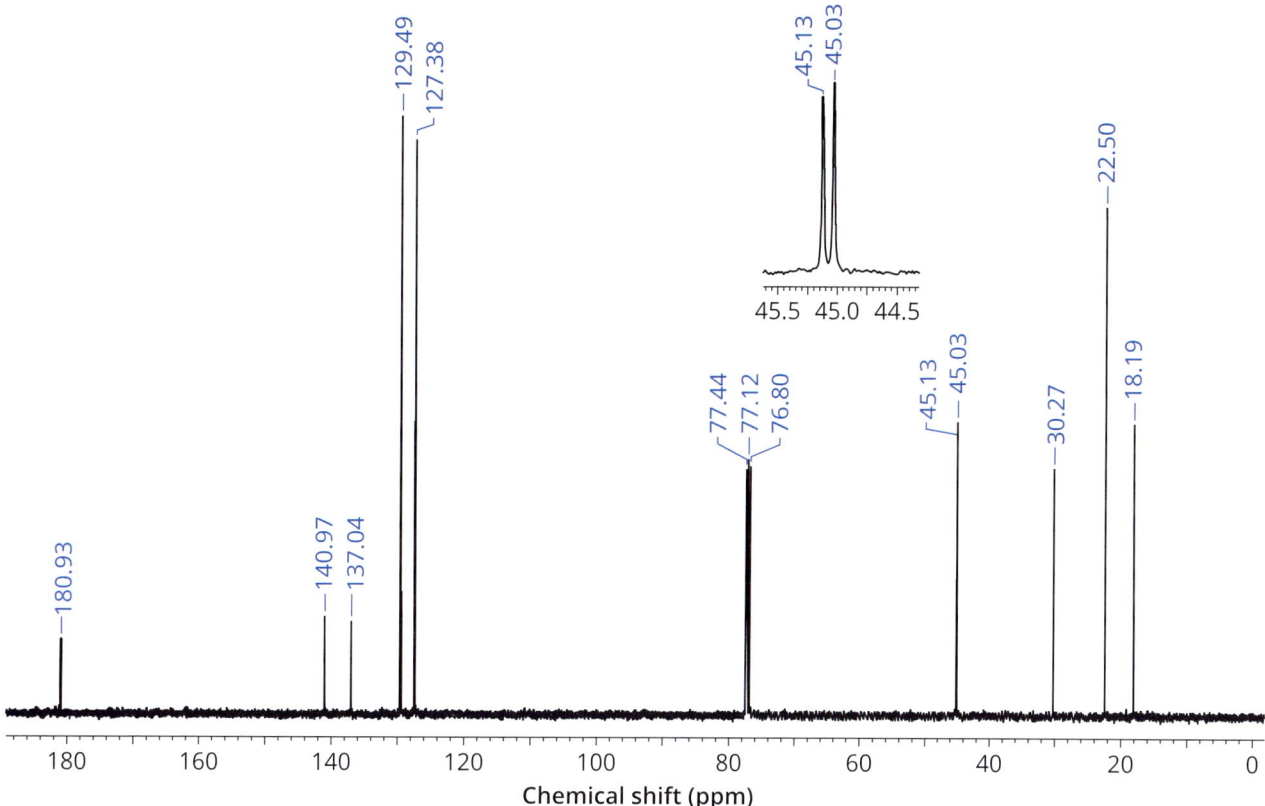

^{13}C NMR spectrum (100 MHz, CDCl$_3$)

9.14 **Problem 14**

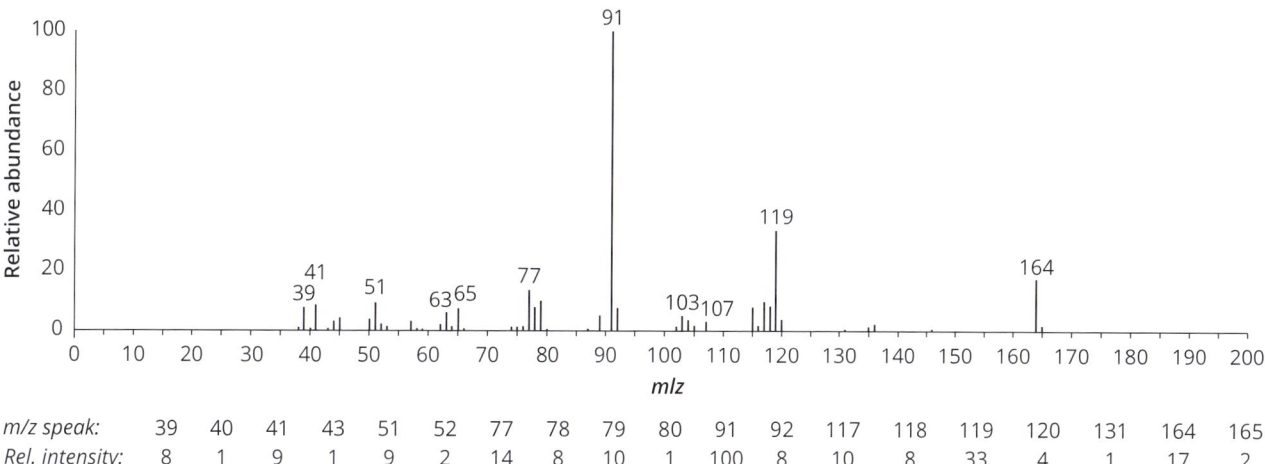

m/z speak:	39	40	41	43	51	52	77	78	79	80	91	92	117	118	119	120	131	164	165
Rel. intensity:	8	1	9	1	9	2	14	8	10	1	100	8	10	8	33	4	1	17	2

Mass spectrum

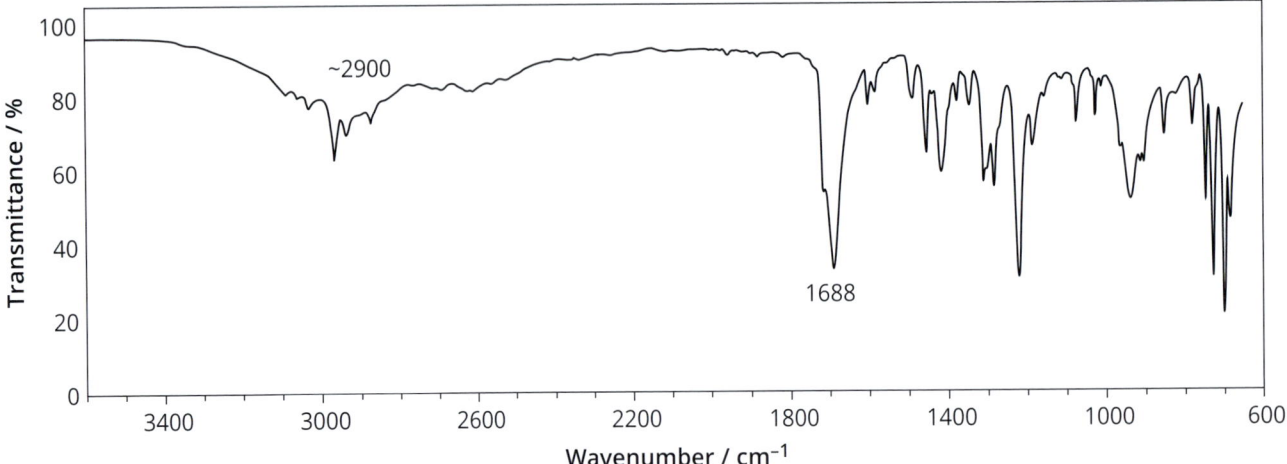

Infrared spectrum

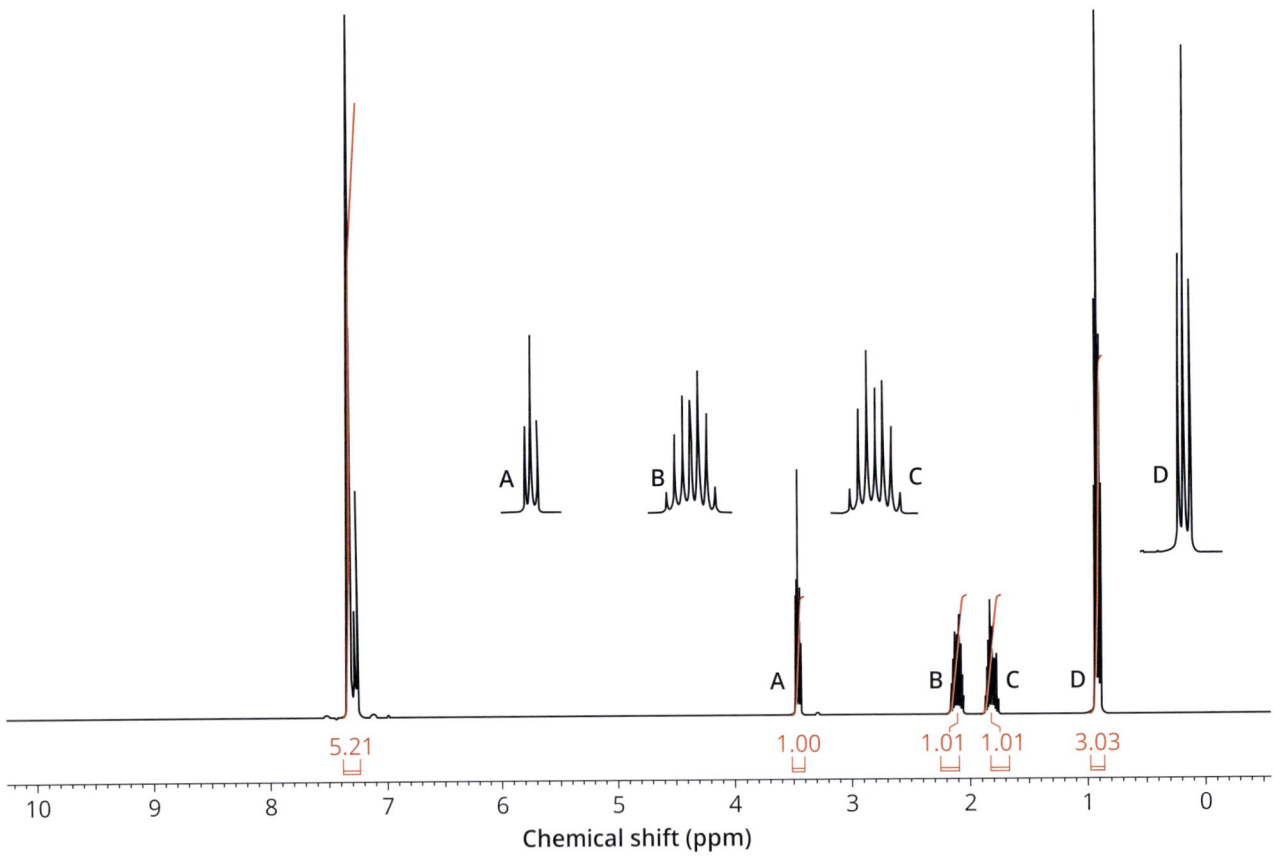

¹H NMR spectrum (400 MHz, CDCl₃)

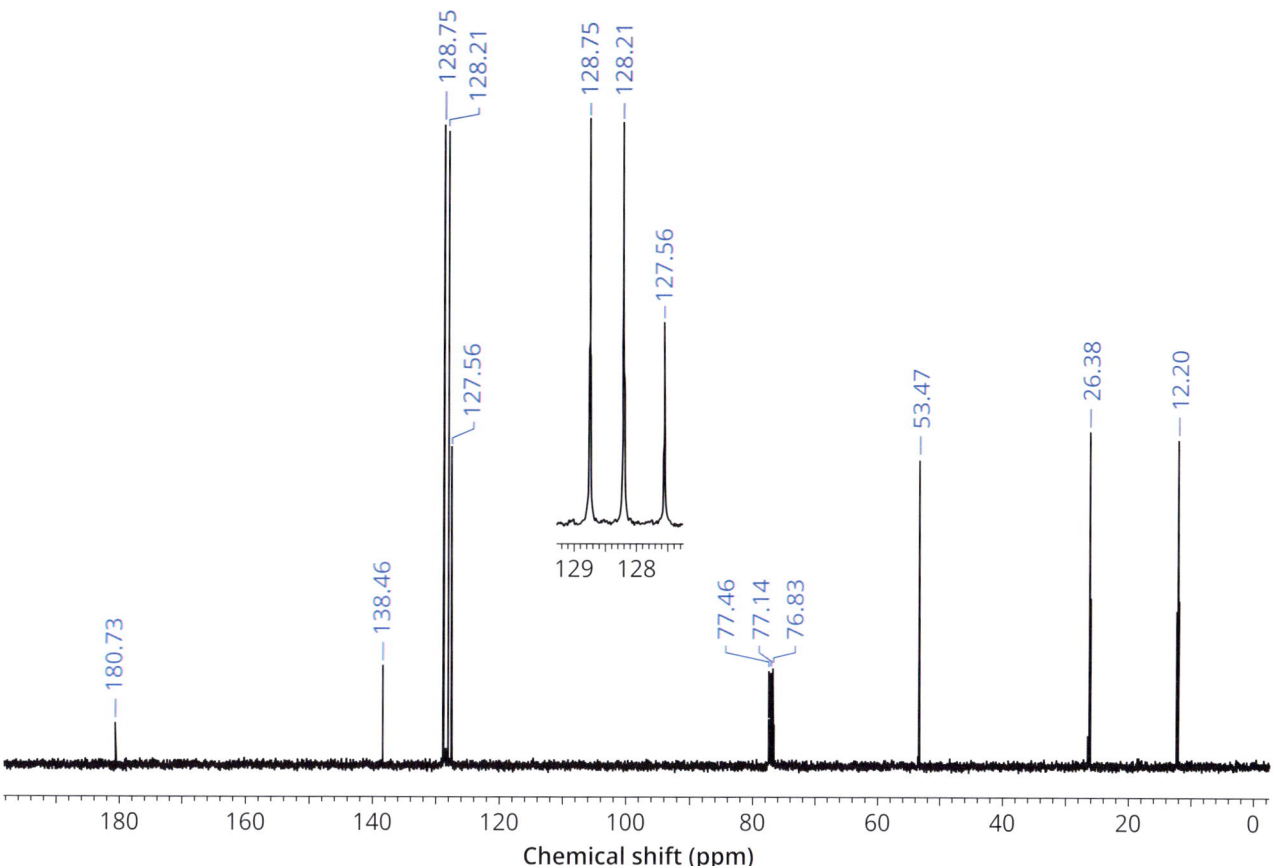

^{13}C NMR spectrum (100 MHz, CDCl$_3$)

9.15 Problem 15

HINT

This compound has a high degree of symmetry.

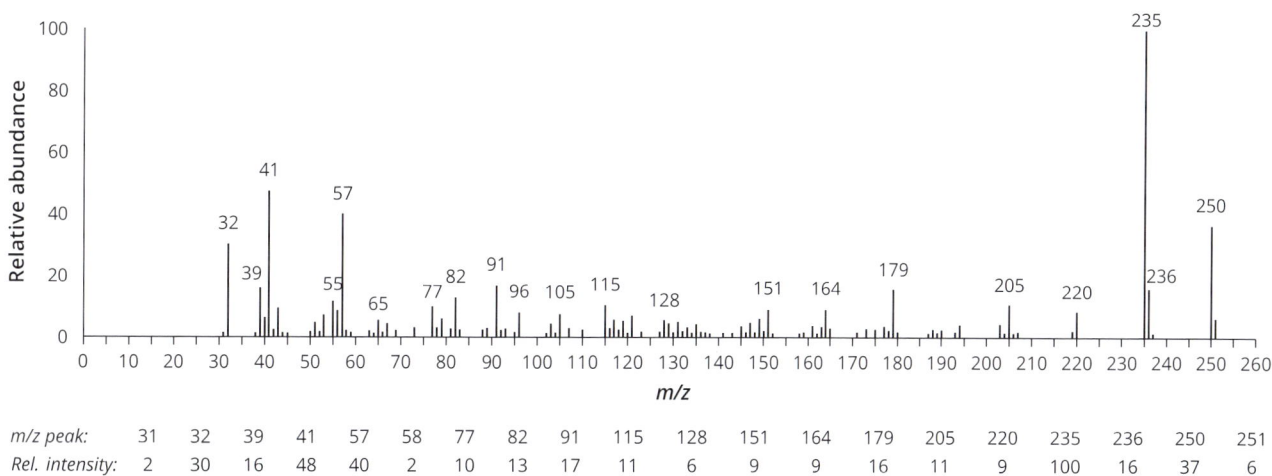

m/z peak:	31	32	39	41	57	58	77	82	91	115	128	151	164	179	205	220	235	236	250	251
Rel. intensity:	2	30	16	48	40	2	10	13	17	11	6	9	9	16	9	11	100	16	37	6

Mass spectrum

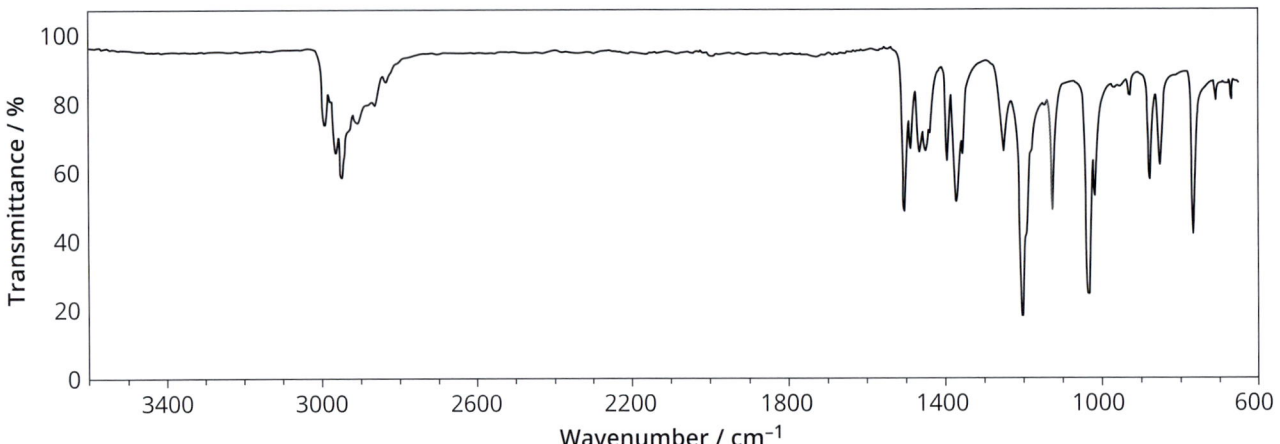

Infrared spectrum

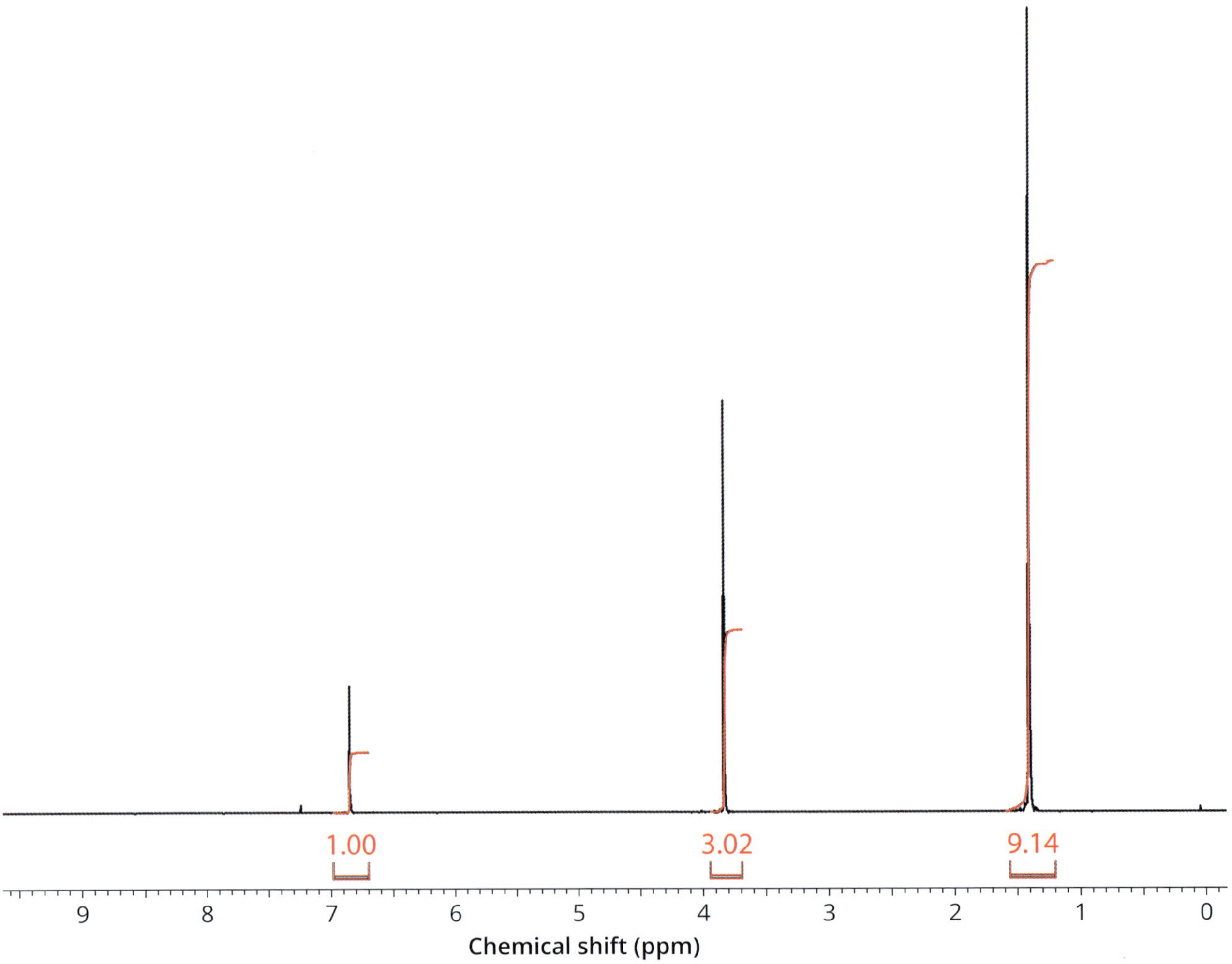

^{1}H NMR spectrum (400 MHz, CDCl$_3$)

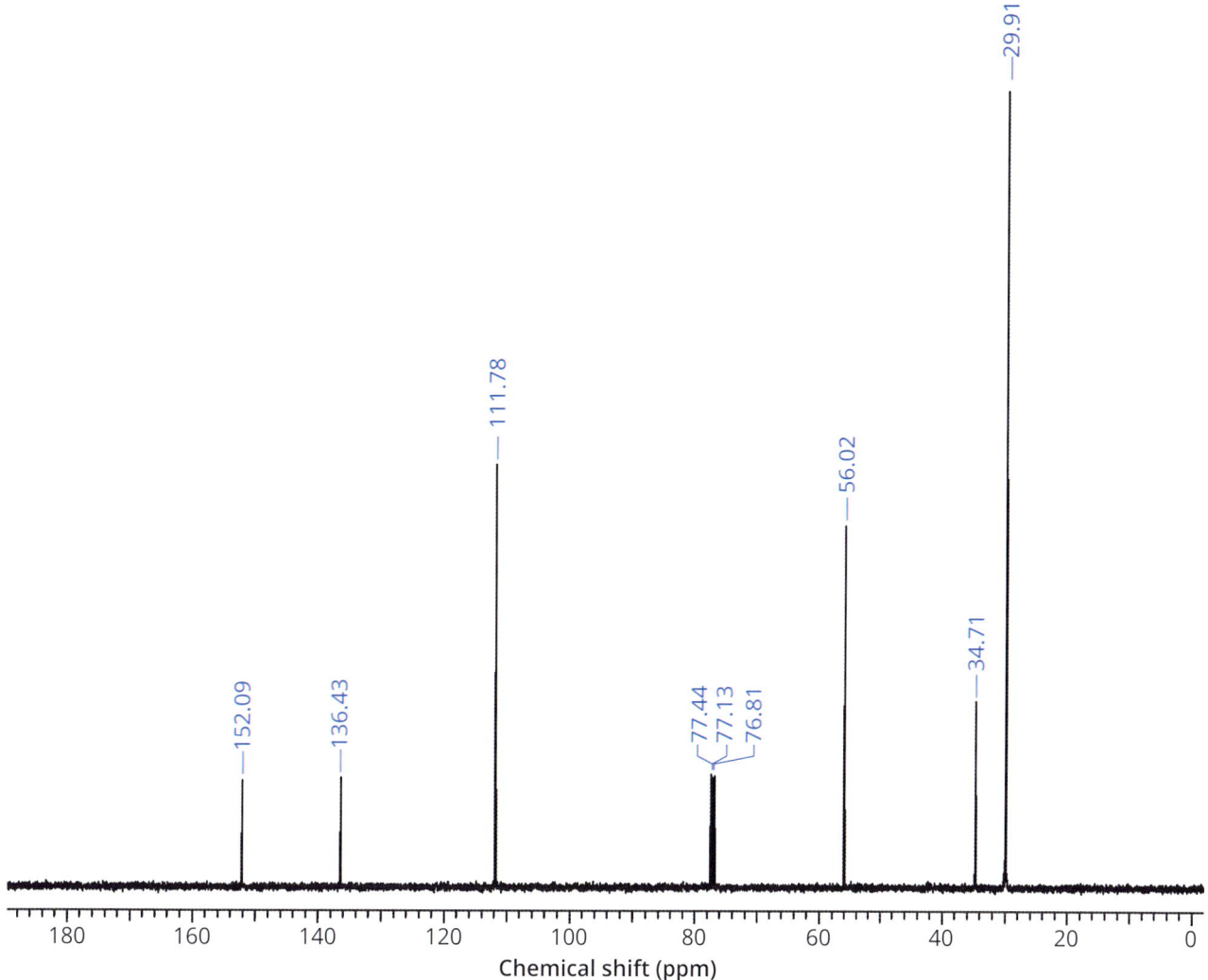

152.09
136.43
111.78
77.44
77.13
76.81
56.02
34.71
29.91

180   160   140   120   100   80   60   40   20   0
Chemical shift (ppm)

^{13}C NMR spectrum (100 MHz, CDCl$_3$)

CHAPTER 10

Advanced NMR methods

The ability to interpret the features of spectroscopic data of organic molecules, and from this deduce their structure, is a critical skill to develop for anyone working with such molecules. The material we have presented so far in this book has been designed to take you along this learning path and provide both strategies and practice for developing this skill. However, in dealing with organic molecules there are often times when direct inspection of their standard spectroscopic data leads to an incomplete or ambiguous definition of their structure, no matter how skilled or experienced you might be. In particular, the high information content found in NMR spectra can make their analysis challenging, especially for molecules that provide highly complex spectra or spectra with heavy overlap. In some cases, it may be possible to deduce the structure of a series of molecular structural features by direct inspection, but there may be multiple solutions for piecing these together. Each of these options may be equally compatible with the spectroscopic data, making a final solution ambiguous or uncertain (Figure 10.1).

In cases such as these, NMR can offer a range of additional techniques that can assist with providing a complete spectrum assignment and defining the molecular structure. We will introduce the most common of these methods in these later chapters and apply them to problems to exemplify their use. However, we recommend that you build your confidence with the approach to direct spectrum assignment modelled in the earlier chapters before considering using these more advanced methods. Indeed, it is possible that you will only come across these advanced techniques in organic spectroscopy lectures presented later in your studies—and the following sections of our book will only then become relevant to you. We will therefore only provide a brief introduction to these methods, which we hope will serve as a pathway towards more advanced spectroscopy textbooks and courses.

10.1 Carbon spectrum editing

In Chapter 3 we mentioned that carbon spectra are routinely collected in the presence of proton decoupling to remove splittings resulting from proton–carbon coupling interactions. The collapse of the C–H multiplet structure into a singlet leads to better signal separation and increased signal intensity, which is of benefit in spectrum analysis. It does also, however, lead to a loss of potentially valuable information contained within the C–H multiplet structures, as we can no longer tell whether each resonance belongs to a CH_3, CH_2, CH, or non-protonated carbon centre. If the $^1J_{CH}$ multiplet structures were

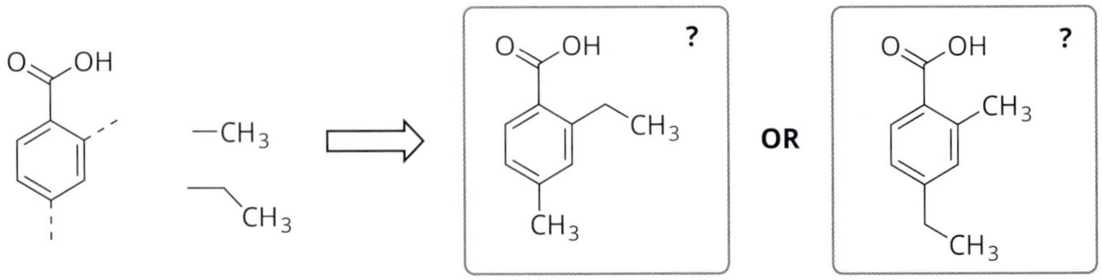

FIGURE 10.1 Structural ambiguity can arise even when one has correctly identified all structural features. More advanced spectroscopic methods can often help resolve such ambiguity.

apparent, each CH group would have a doublet structure due to splitting from the single attached proton. Similarly, each CH_2 would appear as a triplet and each CH_3 as a quartet according to the $n+1$ coupling rule, where n is now the number of directly attached protons (section 2.4.1). Lastly, a non-protonated carbon would produce only a singlet since it has no directly attached protons that could cause splittings. See Figure 3.1 of Chapter 3 for an example of this splitting.

> The loss of splitting information from ^1H–^{13}C J couplings in routine proton-decoupled carbon spectra is problematic. A number of NMR methods have been developed that aim to retain the benefits of complete proton decoupling yet encode the carbon 'multiplicity' information in the carbon spectrum.

The most common of these techniques is known as the DEPT experiment in which the carbon multiplicity information is encoded within the sign and intensity of the carbon resonance. DEPT is an acronym for *Distortionless Enhancement by Polarization Transfer* which describes the editing process used; you do not need to be concerned with the details of this, although it will be explained in more advanced NMR textbooks.

> In the carbon DEPT experiment the resonances of CH_2 groups (even number of attached protons) appear with opposite sign to those of CH and CH_3 groups (odd number of protons) whilst carbon centres without any attached protons are removed from the spectrum altogether (Table 10.1).

The 'editing out' of peaks from non-protonated carbon centres provides a useful means of identifying features such as quaternary centres and carbonyl groups. Thus, DEPT spectra can be recorded in addition to carbon spectra to provide enhanced information content through **spectrum editing**.

The carbon and DEPT spectra of ethyl benzoate illustrate these features (Figure 10.2). The CH_2 methylene peak at 61 ppm is clearly inverted (negative sign) with respect to the methyl CH_3 resonance at 14 ppm and the three methine CH peaks of the benzene ring clustered around 130 ppm.

> It is common practice for DEPT spectra to be presented with CH and CH_3 groups with positive sign and CH_2 groups inverted with negative sign.

The substituted aromatic carbon at 130.5 ppm is not apparent in DEPT as this has no attached proton; the same applies to the carbonyl resonance at 167 ppm. Notice also that the chloroform solvent peaks around 77 ppm are removed from DEPT. This is because the three-line resonance comes from *deuterated* chloroform ($CDCl_3$), meaning strictly (and certainly as far as the DEPT experiment is concerned) that the solvent carbon also bears no *protons* and is therefore also edited out. We can observe this editing out of solvent carbon peaks in most cases since common NMR solvents are deuterated.

TABLE 10.1 Peak signs obtained from DEPT editing of carbon spectra. +ve indicates peaks standing above the spectrum baseline, −ve indicates peaks pointing below the baseline.

Carbon type	DEPT peak
C	none
CH	+ve
CH_2	−ve
CH_3	+ve

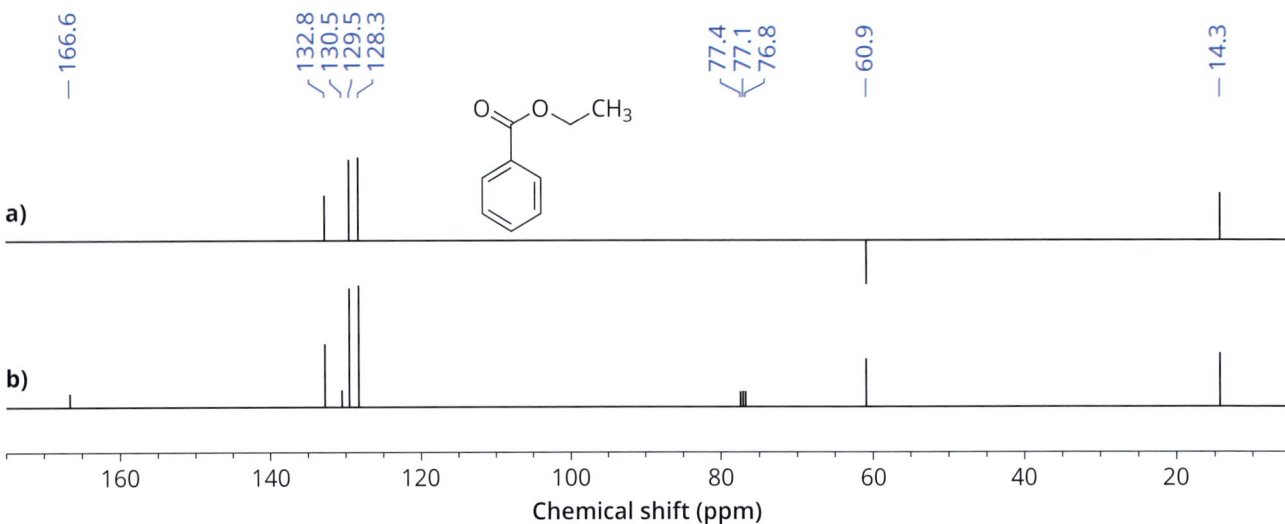

FIGURE 10.2 The a) DEPT-edited carbon and b) conventional carbon spectra of ethyl benzoate in $CDCl_3$.

10.2 Two-dimensional (2D) NMR

In Chapter 1 we likened the process of structure elucidation to that of solving a jigsaw puzzle in which small molecular structural elements are identified and then gradually pieced together to build up the solution. In the NMR data we aim to identify the types of chemical environments present as reflected in the proton and carbon chemical shifts, and then link these together by recognizing proton coupling patterns that provide evidence for these groupings. The correlation of the CH_2 quartet with the CH_3 triplet of the ethyl group is one example of this process, and the previous chapters should have helped you develop your skills at doing this.

> The grouping of molecular fragments can be greatly assisted by NMR methods known as two-dimensional (2D) correlation experiments that can tell us directly which nuclei share coupling interactions.

2D methods are very helpful in building structures. They effectively guide us in how our molecular jigsaw pieces should be linked together. By providing evidence for bonding connectivity between atoms, they ultimately provide us with much more reliable and definitive structural assignments.

> The most commonly employed 2D methods come in two distinct classes:
>
> - 2D **homonuclear correlation** experiments map the *J*-coupling interactions between similar nuclei, most commonly coupled protons.
> - 2D **heteronuclear correlation** experiments map the *J*-coupling interactions between different nuclei, most commonly protons and carbon atoms.

These methods are explained briefly in the sections that follow and their application is illustrated by worked examples in the next chapter, which will be followed by a small number of problems that exploit 2D correlation experiments. In all of this, it is important to remember that whilst these methods can assist us in identifying structures, a sound understanding of the basic principles of spectrum assignment, as illustrated and practised in earlier chapters, remains an essential skill.

10.2.1 What is a 2D spectrum?

The proton and carbon spectra studied in previous chapters are termed one-dimensional spectra in the nomenclature of NMR spectroscopy as they comprise a single chemical shift axis (despite the fact that they have another dimension that represents intensity or peak height, of course).

- Two-dimensional spectra have two (orthogonal) chemical shift axes which may be the same, as in proton–proton correlation spectra, or may be different, as in proton–carbon correlation experiments. Peaks within a 2D spectrum, known as **cross peaks**, serve to correlate a chemical shift on one axis with a chemical shift on another axis and so provide a map of nuclei that couple.

- Since *J*-coupling is transmitted through bonds, a cross peak implies the presence of a bonding pathway between the correlated nuclei.

In proton–proton correlation experiments, the detected *J*-couplings are simply those that give rise to the multiplet structures we see in 1D ^{1}H spectra, and it follows that these 2D experiments detect (or map out) couplings operating across, typically, two or three bonds. Proton–carbon correlation experiments may be set up to detect ^{1}H–^{13}C couplings operating over *either* one-bond (denoted $^1J_{CH}$) *or* over two or three bonds ($^{2/3}J_{CH}$), as exemplified in Figure 10.3 and further illustrated in the following sections.

> The great power of 2D correlation maps is that they provide unambiguous evidence for couplings between nuclei or groups of nuclei. In this way, they provide us with more reliable data on which we can assign spectra and build molecular structures by linking molecular structural elements.

10.2.2 2D ^{1}H–^{1}H Correlation Spectroscopy (COSY)

The basic principles of 2D correlations may be seen in the spectrum presented in Figure 10.4 of the proton–proton correlation map for ethyl benzoate. As this experiment was one of the very early 2D NMR methods to be developed, it is known simply as **CO**rrelation **S**pectroscop**Y**, or more commonly as **COSY**. The dashed grey line in the 2D map shows us where the two chemical shift axes have the same value and is known as the diagonal line.

> We distinguish two different types of peaks in this context:
>
> - Peaks that sit on the diagonal line are known as **diagonal peaks**, and these simply reflect the chemical shifts of the protons in the 1D spectrum. They provide a reference point for identifying our correlations of interest.
>
> - The off-diagonal peaks are the interesting ones! These **cross peaks** correlate different chemical shifts and so indicate coupled protons.

FIGURE 10.3 Examples of coupling interactions in ethyl benzoate that could be identified directly by cross peaks in three different 2D correlation experiments. The correlation in a) directly identifies the ethyl group whilst that in c) links this as part of the ester.

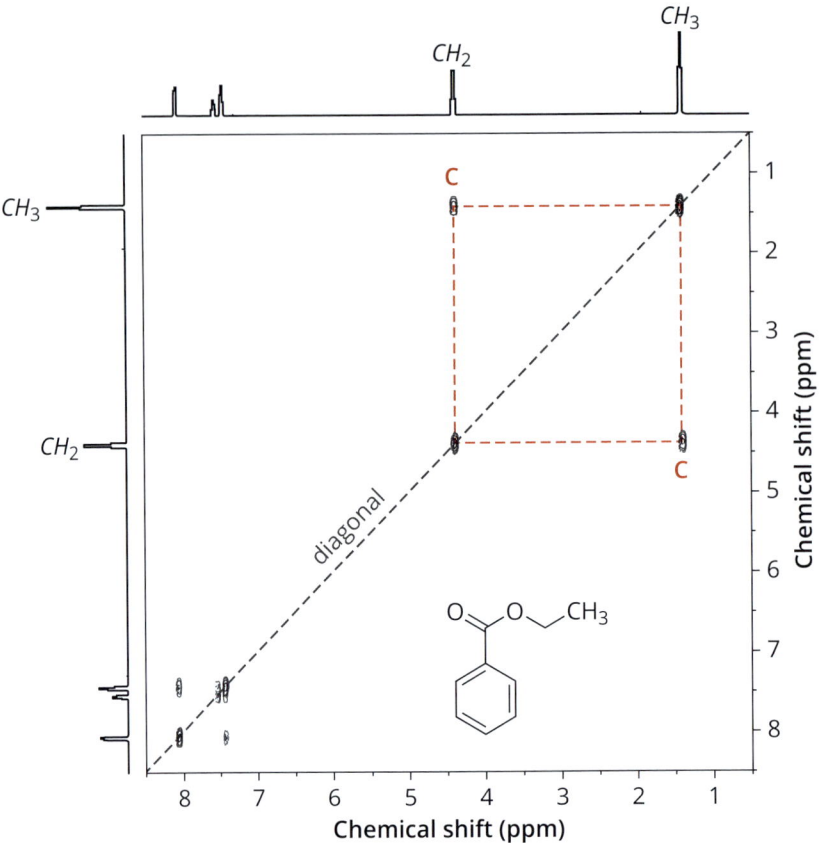

FIGURE 10.4 The ^1H–^1H 2D COSY spectrum of ethyl benzoate. Along each axis the standard ^1H spectrum is shown as a reference. The dashed grey line marks the diagonal (reference) line, and the peaks marked C are the informative cross peaks indicating the ethyl group coupling (connected by the red dashed line).

For the ethyl benzoate sample, cross peaks clearly correlate the CH_2 (4.4 ppm) and CH_3 (1.4 ppm) groups, showing us that these protons are mutually coupled. The fact that we see correlations both above and below the diagonal merely reflect the equivalent CH_2 to CH_3 and CH_3 to CH_2 coupling interactions. Thus, with COSY to hand, the process of identifying mutually coupled protons is simplified to a process of 'joining-the-dots' to map the interactions. The correlations between the aromatic protons of the benzene ring can be mapped in a similar manner in this COSY spectrum.

10.2.3 2D ^1H–^{13}C One-bond Correlation Spectroscopy (HSQC)

The correlation of protons with their directly attached carbon-13 atoms provides the basis for another valuable 2D experiment that is commonly used to assist with spectrum assignment.

> In the one-bond heteronuclear correlation experiment known commonly as HSQC, proton multiplets on one chemical shift axis are mapped on to carbon signals on the other axis, identifying C–H linkages.

HSQC stands for *Heteronuclear Single Quantum Correlation* which, in essence, tells us that the experiment correlates information between different nuclei, in this case ^1H and ^{13}C. Again, you do not need to understand the details of this to make good use of the experiment.

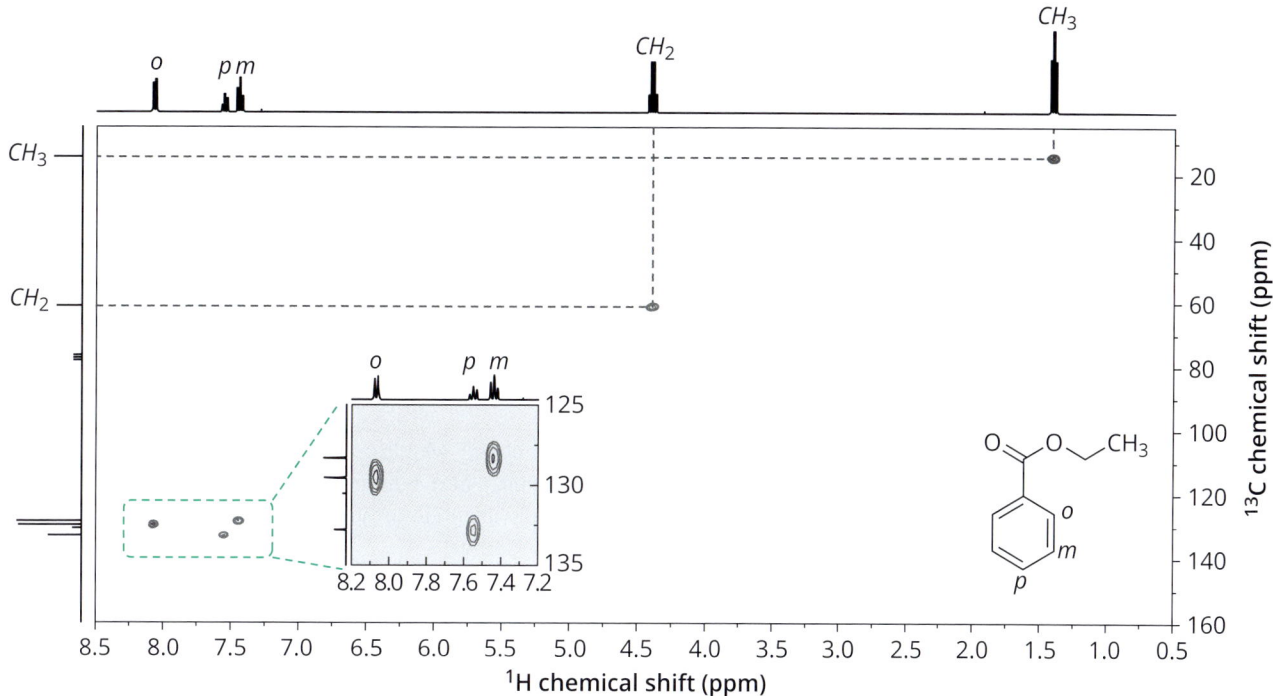

FIGURE 10.5 The ^1H–^{13}C one-bond correlation spectrum (HSQC) for ethyl benzoate. The cross peaks clearly map the one-bond coupling interactions for the CH_2 and CH_3 groups (traced by dashed lines). The expanded insert shows the three correlations expected for the aromatic ring.

In the case of HSQC, there are no diagonal peaks since the two axes reflect different nuclei, so the process of mapping interactions via the cross peaks is quite straightforward. The one-bond correlation experiment for the ethyl group of ethyl benzoate (Figure 10.5) clearly demonstrates the correlated chemical shifts for the CH_2 group (4.4 and 61 ppm for ^1H and ^{13}C respectively), and likewise for the CH_3 group (1.4 and 14 ppm). The ability to correlate two chemical shifts to a group provides us with greater confidence in the group's assignment. For example, a proton multiplet around 3–4 ppm correlating with a carbon around 60–80 ppm would present strong evidence for a HC–O functionality in the molecule.

Another great benefit of the one-bond correlation experiment is that it enables us to separate proton multiplets that may be overlapped (or partially so) in the 1D proton spectrum by distributing them according to the shift of their attached carbon centre.

The dispersion of signals across two dimensions can be of great assistance in the analysis of more complex proton spectra. It is also useful to note that, although we do not introduce this here in detail, a common implementation of HSQC (especially in research laboratories) includes spectrum editing in a manner similar to DEPT that is referred to as *multiplicity-edited HSQC*. In this technique, the CH_2 cross peaks have opposite signs to the CH and CH_3 cross peaks, and a separate DEPT experiment is therefore no longer necessary!

10.2.4 2D ^1H–^{13}C Long-range Correlation Spectroscopy (HMBC)

In addition to the one-bond proton–carbon couplings detected in HSQC, it is also possible for ^1H and ^{13}C nuclei to share couplings across two or three bonds, in a similar manner to ^1H–^1H couplings.

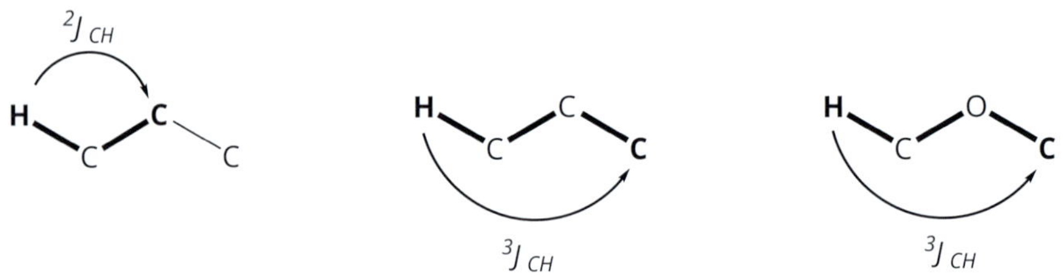

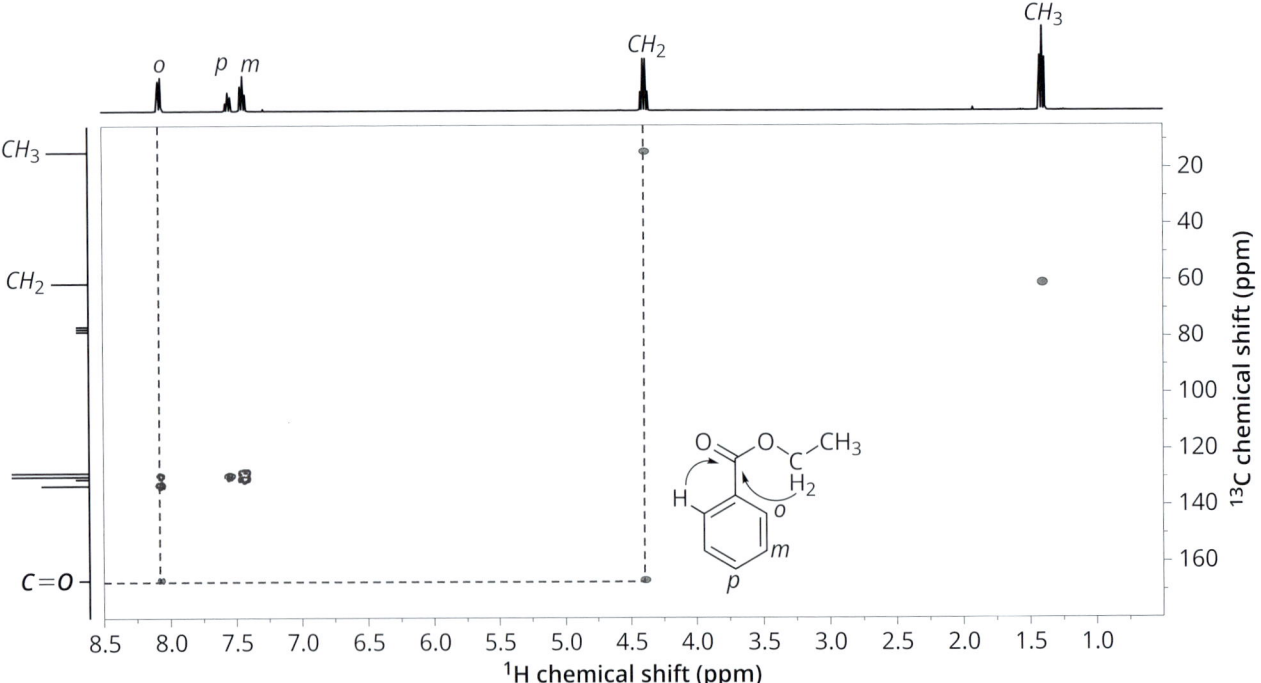

$^2J_{CH}$

$^3J_{CH}$

$^3J_{CH}$

FIGURE 10.6 Example two- and three-bond proton–carbon correlations that may be observed with the HMBC experiment. The arrows start at the proton and point to the correlated carbon centre and are labelled according to the number of bonds on the coupling pathway (shown in bold). See also Figure 10.3c.

> The experiment known as Heteronuclear Multiple Bond Correlation, or HMBC, maps out long-range ^1H–^{13}C couplings and hence identifies bonding pathways within a structure. It can be especially useful in providing evidence for how molecular structural elements are linked together.

HMBC can also provide correlations of protons to non-protonated carbon centres such as carbonyl groups of ketones, acids, esters, and amides, or to quaternary centres in molecules (Figure 10.6), thereby adding to the information we gather from COSY and HSQC spectra. The key features of the HMBC may be seen in the spectrum of ethyl benzoate in Figure 10.7. The protons for the ethyl CH_2 at 4.4 ppm now correlate across two bonds to the adjacent methyl carbon at 14 ppm and similarly the methyl protons at 1.4 ppm link to the neighbouring methylene carbon at 61 ppm. Most significantly, the CH_2 protons also display a cross peak correlating to the ester carbonyl chemical shift at 166 ppm arising from a three-bond coupling pathway. Notice also a correlation from the *ortho*-aromatic protons at 8.1 ppm to the same carbonyl carbon (dashed in lines in Figure 10.7). These correlations provide *direct evidence* for the linkage between the aliphatic ethyl group and the aromatic phenyl via the

FIGURE 10.7 The ^1H–^{13}C long-range correlation spectrum (HMBC) of ethyl benzoate. Dashed lines highlight correlations from the aliphatic and aromatic protons to the carbonyl group, as illustrated for the inset structure.

ester carbonyl functionality. No other experiment presented so far will have provided direct evidence for this connectivity—we might only have surmised its existence from the combined spectroscopic data. HMBC provides us with a greater degree of certainty in many cases of spectrum assignment. In cases of ambiguity, it can help us clarify how molecular structural elements are linked, as we will illustrate in what follows.

> When interpreting HMBC spectra, it is important to realize that, in most cases, any observed correlation may arise from a coupling operating across either two *or* three bonds. But in isolation we are unable to say which of these is responsible for any given cross peak.

Both 2-bond and 3-bond possibilities should be considered when analysing HMBC spectra; the correct assignment for the correlation should become apparent when all spectroscopic data are considered and a structure consistent with all data is proposed. Couplings across four bonds may also be apparent in HMBC spectra in some instances, but as these couplings are often rather small (again, in a similar manner to four-bond ^{1}H–^{1}H couplings) the cross peaks tend to be weak.

As an example of the utility of HMBC, let us reconsider the results of Example 13 in Chapter 8 (see section 8.13). From the spectroscopic data provided, we were able to conclude that the structure had to be one of two possible positional isomers (*regioisomers*) **5** and **6**, which differed only in the relative positioning of the groups attached to the aromatic ring. However, the data could not define which of these was correct. The correlations from a HMBC experiment allow us to arrive at a conclusive result (Figure 10.8). If we were to consider the correlations from the aldehyde proton (which was identified at 9.9 ppm), we would expect different outcomes for the two structures. For **5** we would expect to see three-bond correlations to one carbon around 120 ppm (dashed arrow to carbon shown in red) and another with a significantly greater shift around 150 ppm due to the attached oxygen atom (solid arrow to carbon shown in black). In contrast, for **6** we should see the aldehyde proton correlate to two large carbon shifts around 150 ppm since both neighbouring aromatic carbons bear an attached oxygen (the OH and the OMe). The actual HMBC shows aldehyde correlations to carbons at 125 and 148 ppm (plus a two-bond correlation to the aromatic carbon that carries the aldehyde group at 121 ppm), which confirms that structure **5** is in fact the correct isomer.

Note also that the HMBC spectrum confirms that the 152 ppm carbon is the one bearing the OMe group since the methyl protons at 3.9 ppm correlate to this over three bonds (Figure 10.9). The peak at 148 ppm can therefore confidently be ascribed to the carbon bearing the OH group.

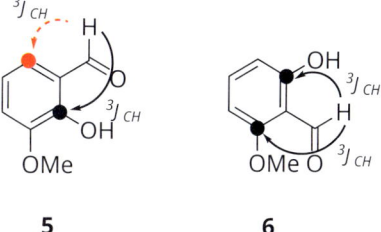

5   **6**

FIGURE 10.8 Distinguishing the regioisomers **5** and **6** of Chapter 8, Example 13 with HMBC correlations. The correlation shown by the dashed arrow (red) is to a protonated carbon centre, whereas those illustrated with solids arrows (black) are to non-protonated centres.

10.2.5 Strategy for interpreting 2D NMR spectra

As for the analysis of 1D NMR spectra, we recommend adopting a systematic approach to employing 2D NMR spectra in structure characterization. An effective strategy to follow that utilizes the methods introduced in this chapter is summarized in Figure 10.10.

The HSQC spectrum provides an overview of proton and carbon environments within a structure by correlating the shifts of these atoms. It is often the clearest and most informative place to start. The assignment of carbon atom type can be derived from chemical shifts by using data from spectrum editing (either from DEPT spectra or directly from the multiplicity-edited form of HSQC itself), as well as information from proton integrals. For instance, a three-proton peak can be correlated directly to identify the methyl carbon chemical shift.

The COSY spectrum can then be used to map out groups of coupled protons within a structure and proves useful for tracing groups of protons in aliphatic chains or rings, or within aromatic structural elements.

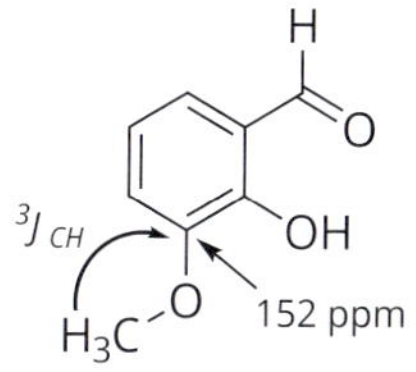

FIGURE 10.9 Assignment of the aromatic carbon bearing the O-methyl group with a three-bond HMBC correlation.

5

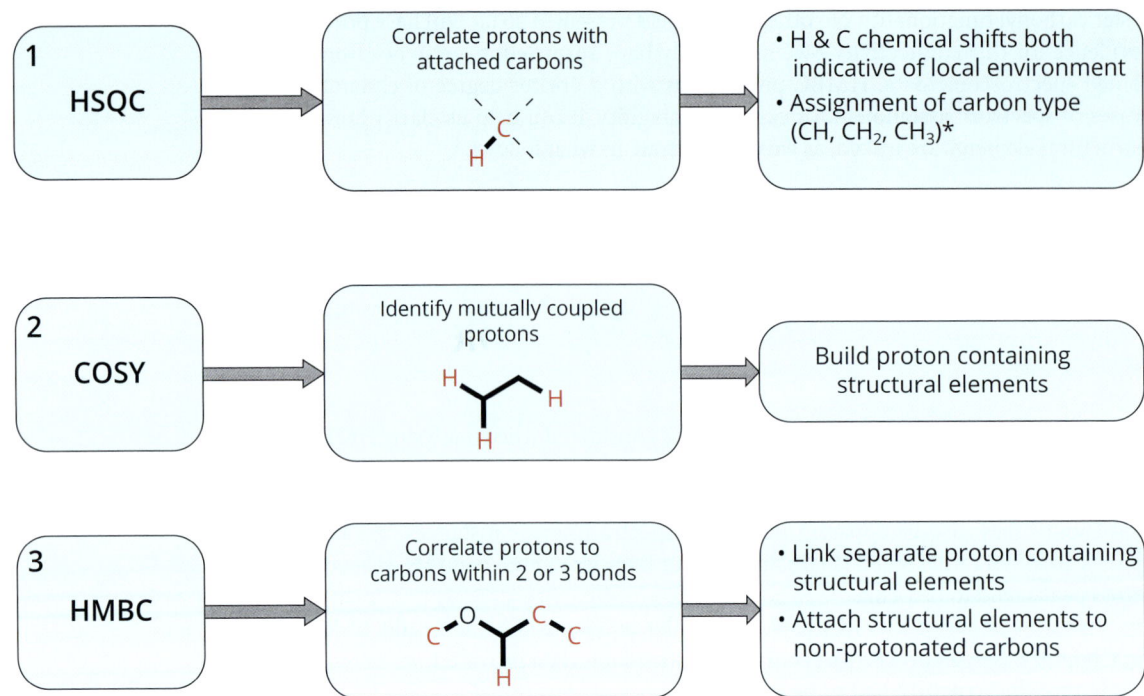

FIGURE 10.10 A strategy for employing 2D NMR spectra in structure characterization. *Assignment of carbon type is often aided by information from spectrum editing (either via DEPT or multiplicity edited HSQC).

Finally, the HMBC spectrum will contain correlations that enable molecular structural elements to be joined together (similar to the linking of jigsaw pieces), commonly by linking these to non-protonated carbon centres. Attaching proton-containing structural elements (e.g. those identified by COSY) to aromatic rings or to carbonyl groups is a common example of this process, as highlighted in the previous section. These principles are illustrated in the worked examples of Chapter 11 and can also be applied in the problems included in that chapter.

CHAPTER 11

2D NMR and spectrum editing

Worked examples and problems

In this chapter we present worked assignment problems that utilize NMR methods alone and in particular emphasize how to interpret the outputs of the methods introduced in Chapter 10. It should always be remembered that these more advanced techniques would not be used by themselves but rather in addition to the analysis of 1D NMR spectra and information from other spectroscopic methods illustrated in earlier chapters. Thus, a firm grasp of basic spectrum interpretation remains a critical skill for the correct application of these techniques. In the worked examples that follow, this basic interpretation will be summarized only briefly, as our primary focus is now on the additional information that can be extracted from the more advanced techniques.

The worked examples are followed by some problems that allow you to work through the assignment process unaided and provide a greater level of challenge. You can find the solutions to the problems in Appendix 2.

11.1 **Example 1**

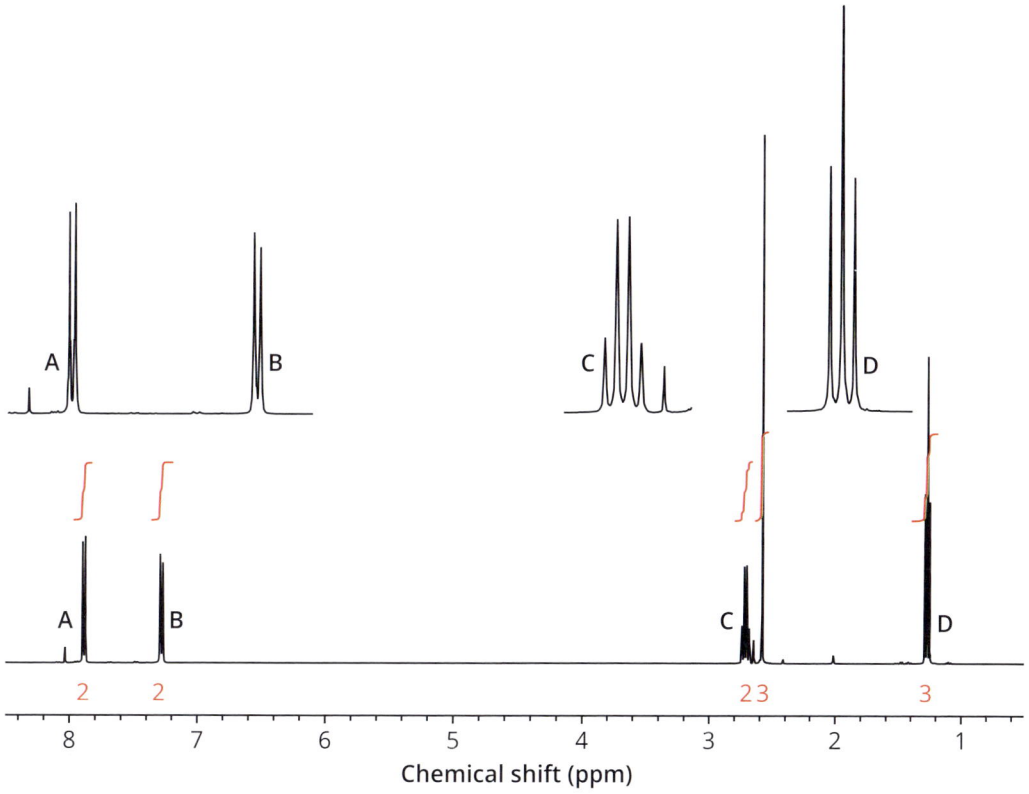

^{1}H NMR spectrum (400 MHz, CDCl$_3$)

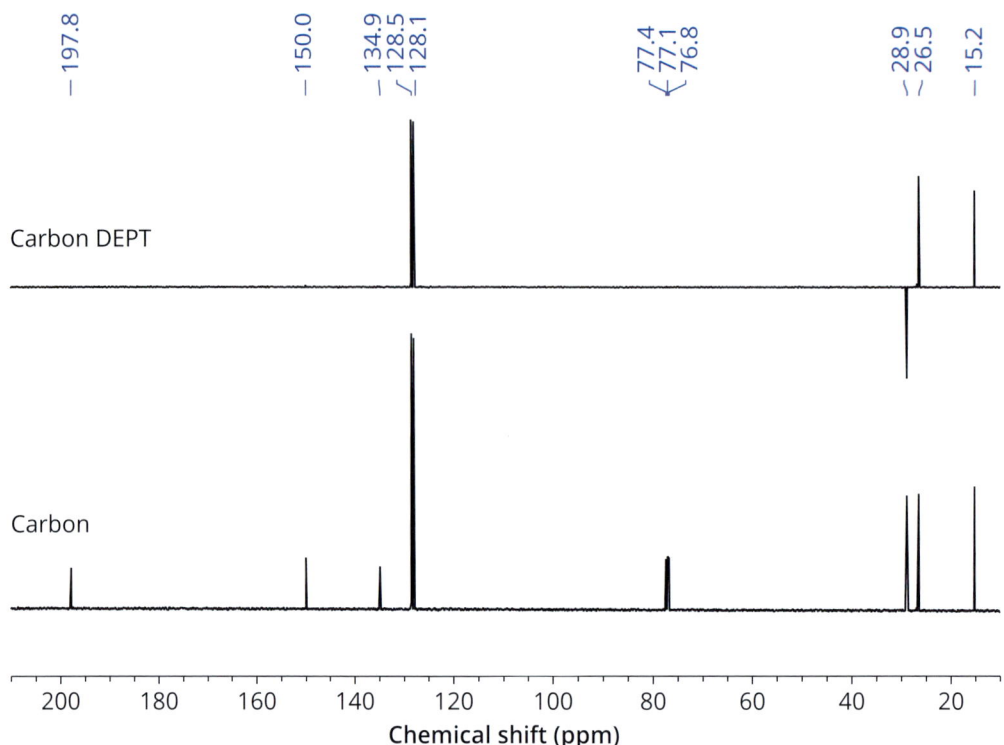

^{13}C and DEPT NMR spectrum (100 MHz, CDCl$_3$)

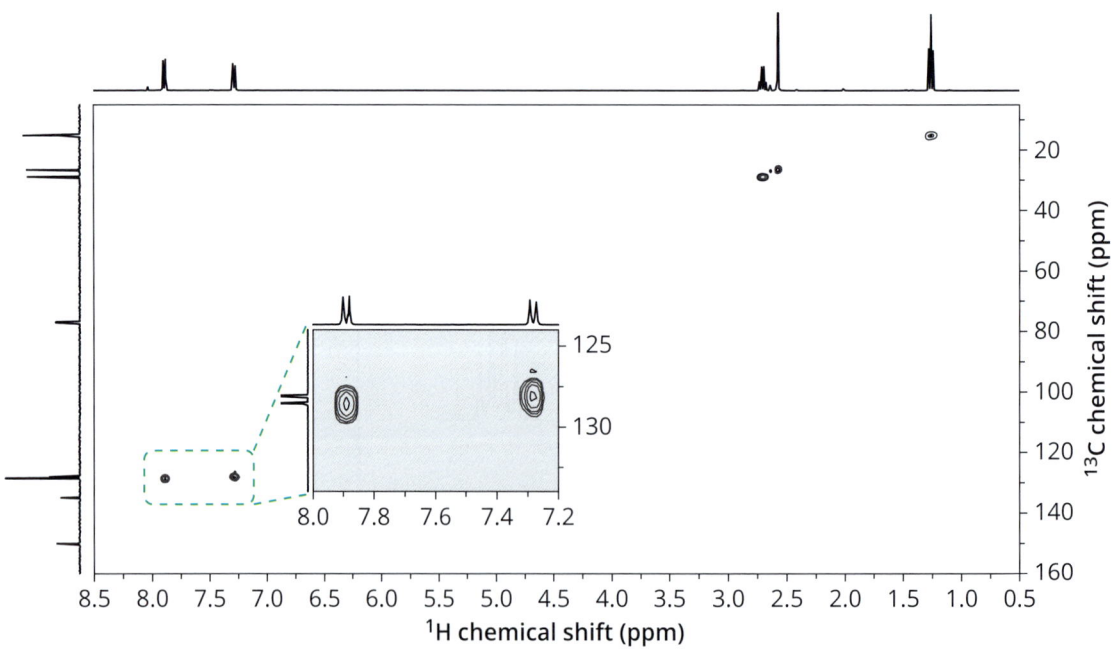

^{1}H–^{13}C HSQC NMR spectrum (400 MHz, CDCl$_3$)

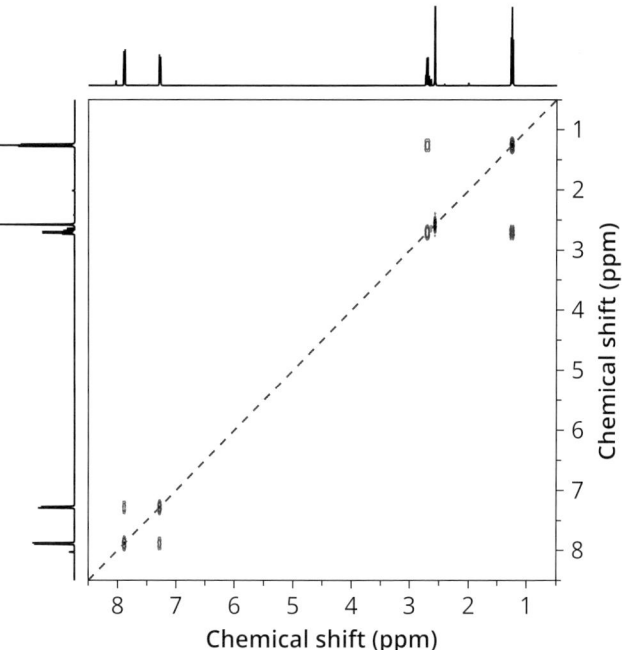

¹H–¹H COSY NMR spectrum (400 MHz, CDCl₃)

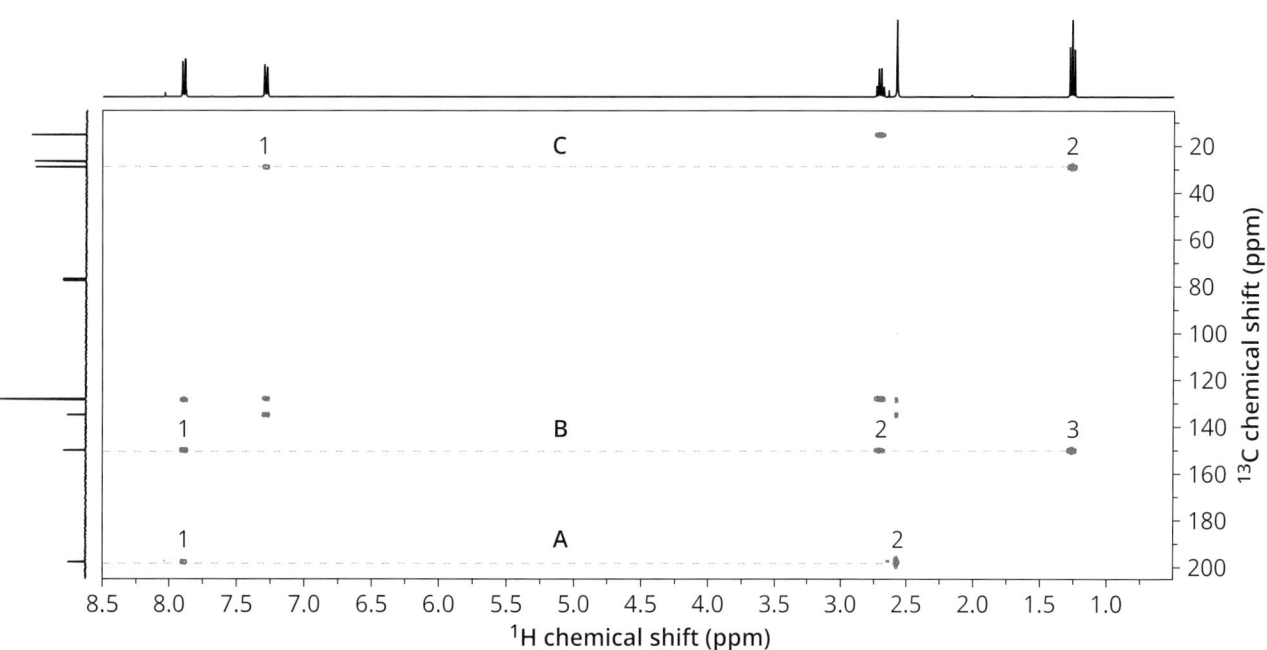

¹H–¹³C HMBC NMR spectrum (400 MHz, CDCl₃)

TABLE 11.1 Chemical shifts for Example 1

$^1H\ \delta$ /ppm	$^{13}C\ \delta$ /ppm
7.90	197.8
7.28	150.0
2.71	134.9
2.58	128.5
1.26	128.1
	28.9
	26.5
	15.2

11.1.1 Analysis of Example 1

The chemical shifts are tabulated in Table 11.1.

a) Following the standard analysis of the ^1H spectrum, we can identify an aromatic benzene ring substituted at the 1 and 4 positions. In the aliphatic region below 3 ppm, the quartet and triplet indicate the presence of an ethyl group and the 3-proton singlet is suggestive of an isolated methyl.

b) Likewise, the carbon spectrum shows four signals in the aromatic region, two of which are of low intensity (135 and 150 ppm), consistent with the 1,4-disubstituted benzene ring. These two weak signals are missing from the DEPT-edited carbon, which confirms that they lack attached protons.

c) At a higher carbon chemical shift we see evidence for a carbonyl group (198 ppm), likely a ketone or aldehyde. The corresponding peak is missing in DEPT, indicating that the carbon lacks an attached proton, which therefore rules out an aldehyde.

d) In the aliphatic carbon region, three signals appear below 30 ppm, and the higher of these (29 ppm) is inverted in DEPT, indicating that this belongs to the CH_2 of the previously suggested ethyl group. The two positive signals are consistent with the assumption that these are methyl resonances.

e) For completeness, note that the solvent $CDCl_3$ peaks clustered at 77 ppm do not appear in the DEPT-edited carbon spectrum.

f) In the HSQC spectrum, cross peaks correlate the protons of the ethyl group with their directly attached carbons; the ethyl CH_2 at 2.7 ppm correlates with the carbon at 29 ppm that was inverted in the carbon DEPT spectrum, which confirms that this belongs to a CH_2 fragment. The two methyl proton resonances correlate to different carbon shifts and allow us to now confirm the methyl carbon resonances as the ethyl CH_3 (1.3, 15.2 ppm) and the isolated methyl CH_3 (2.6, 26.5 ppm).

g) The expansion of the aromatic region of HSQC correlates the proton and carbon centres of the benzene ring at 7.9/128.5 and 7.3/128.1 ppm.

h) Considering the 2D COSY spectrum, a matching pair of cross peaks can be seen to link the CH_2 and CH_3 protons of the ethyl group (2.7 and 1.3 ppm), formally identifying these as being mutually coupled (in this case across three bonds). In the aromatic region, we also see evidence for the mutual coupling of the adjacent protons on the benzene ring (7.9 and 7.3 ppm).

i) The methyl singlet peak at 2.6 ppm displays only a diagonal peak but no cross peaks in COSY; this is simply because it lacks any coupled proton partners with which to correlate.

This analysis so far has identified and determined ^1H and ^{13}C peak assignments for the following structural elements:

It is possible to piece together these structural elements in two ways, yielding positional isomers (regioisomers) that can both be considered compatible with the presented NMR data and would have identical mass (148 Da) and similar IR spectra (C=O at ~ 1680 cm^{-1}):

(1)   (2)

To determine which of these isomers is the correct compound, we consider the remaining data in the HMBC spectrum, which will identify couplings between protons and carbon-13 centres across two or three intervening bonds.

j)   A clear correlation is seen from the aromatic proton at 7.9 ppm to the carbonyl resonance at 198 ppm, together with a corresponding correlation to this carbonyl from the methyl singlet proton at 2.6 ppm (line **A**, peaks **1** and **2**). At the same time, we note a lack of correlations from the ethyl protons to the carbonyl carbon. This proves the attachment of the methyl, and not the ethyl, to the carbonyl group.

k)   The ethyl CH_2 and CH_3 protons correlate to the substituted aromatic carbon at 150.0 ppm (line **B**, peaks **2** and **3**). This proves the attachment of the ethyl group directly to the aromatic ring. This determination is supported further by the correlation of the aromatic proton at 7.3 ppm to the ethyl CH_2 carbon (line **C**, peak **1**).

These key correlations along lines **A–C** can be summarized in the following diagram (the arrows indicate correlations from proton to carbon and the numbers match the cross peaks in the HMBC spectrum) and provide definitive proof for structure **2**, showing unambiguously how the structural elements must be put together. Furthermore, the collective NMR data provide definitive chemical shift assignments for every carbon and proton in the molecule, as summarized below. In particular, assignments for the two methyl carbon resonances are unambiguous from the 2D data.

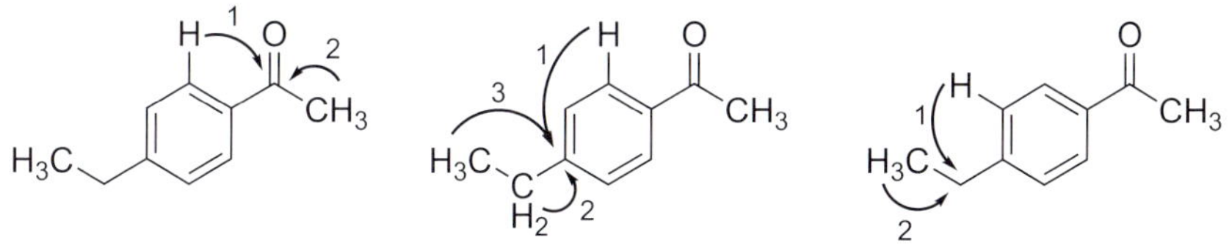

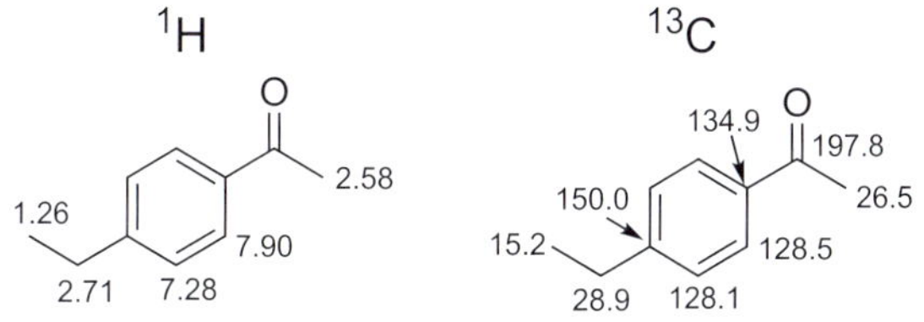

These completed assignments can be presented in tabulated form, as shown in Table 11.2.

TABLE 11.2 Completed assignments for Example 1

$^{13}C\ \delta$ /ppm	Carbon type	$^1H\ \delta$ /ppm
197.8	C	–
150.0	C	–
134.9	C	–
128.5	CH	7.90
128.1	CH	7.28
28.9	CH_2	2.71
26.5	CH_3	2.58
15.2	CH_3	1.26

11.2 **Example 2**

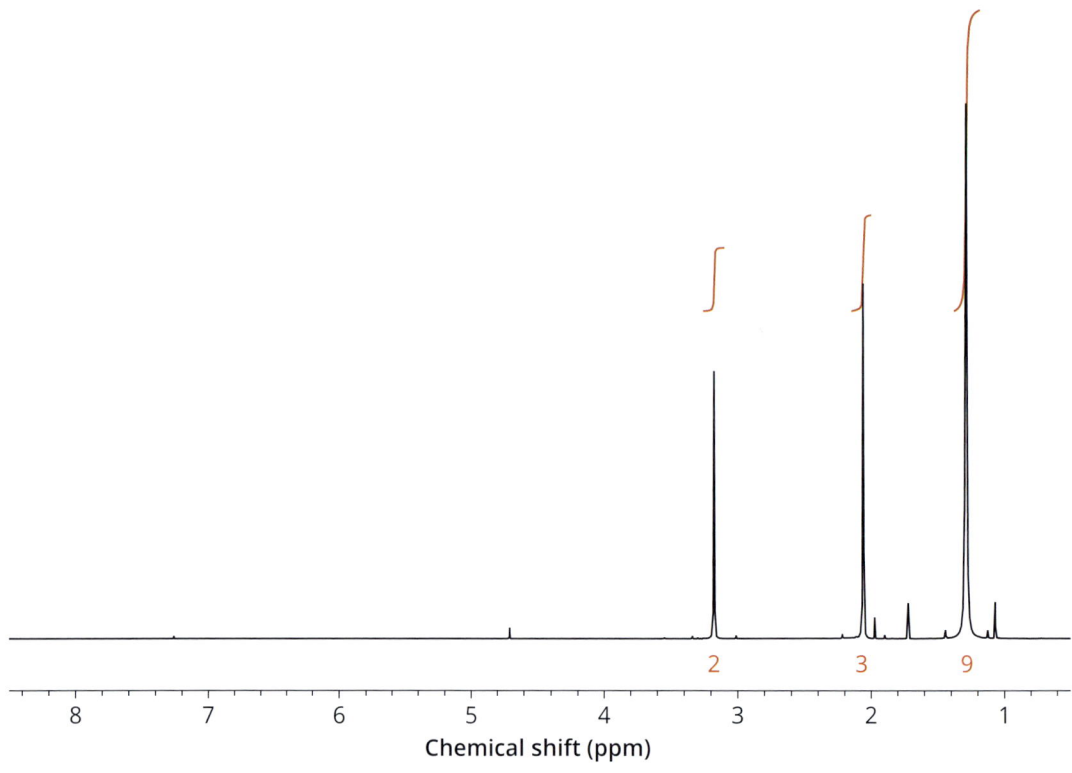

¹H NMR spectrum (400 MHz, CDCl₃)

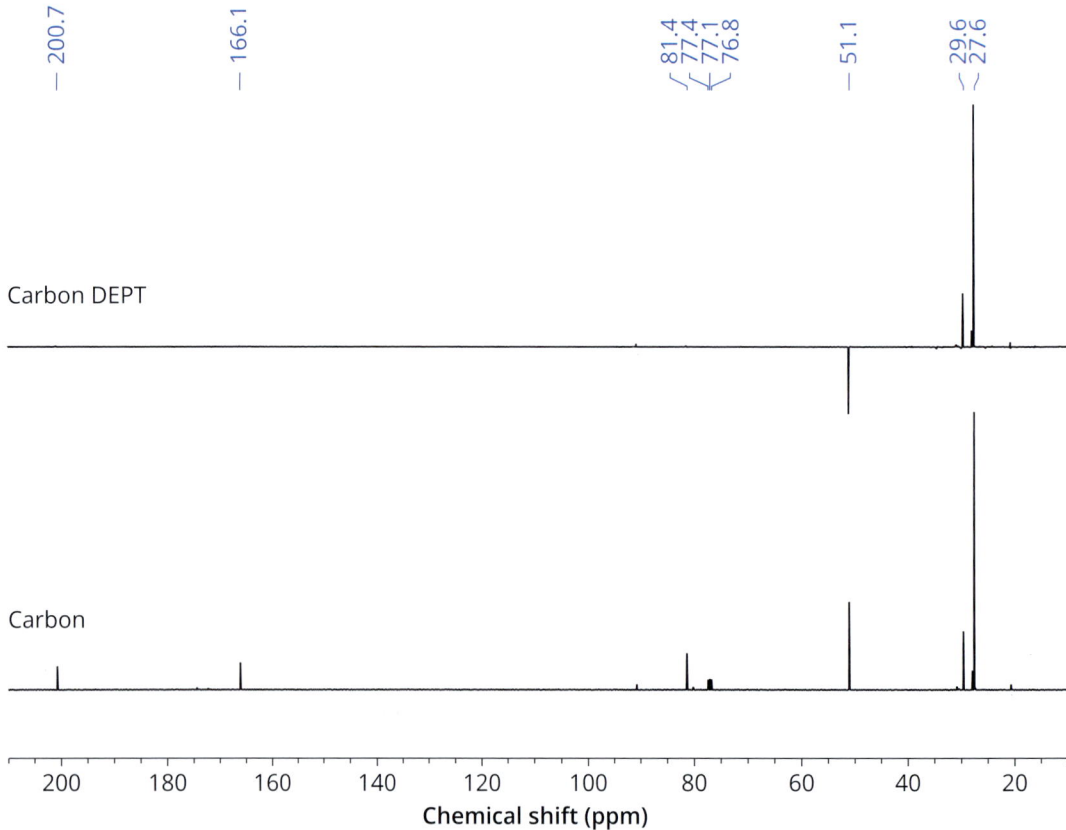

¹³C and DEPT NMR spectrum (100 MHz, CDCl₃)

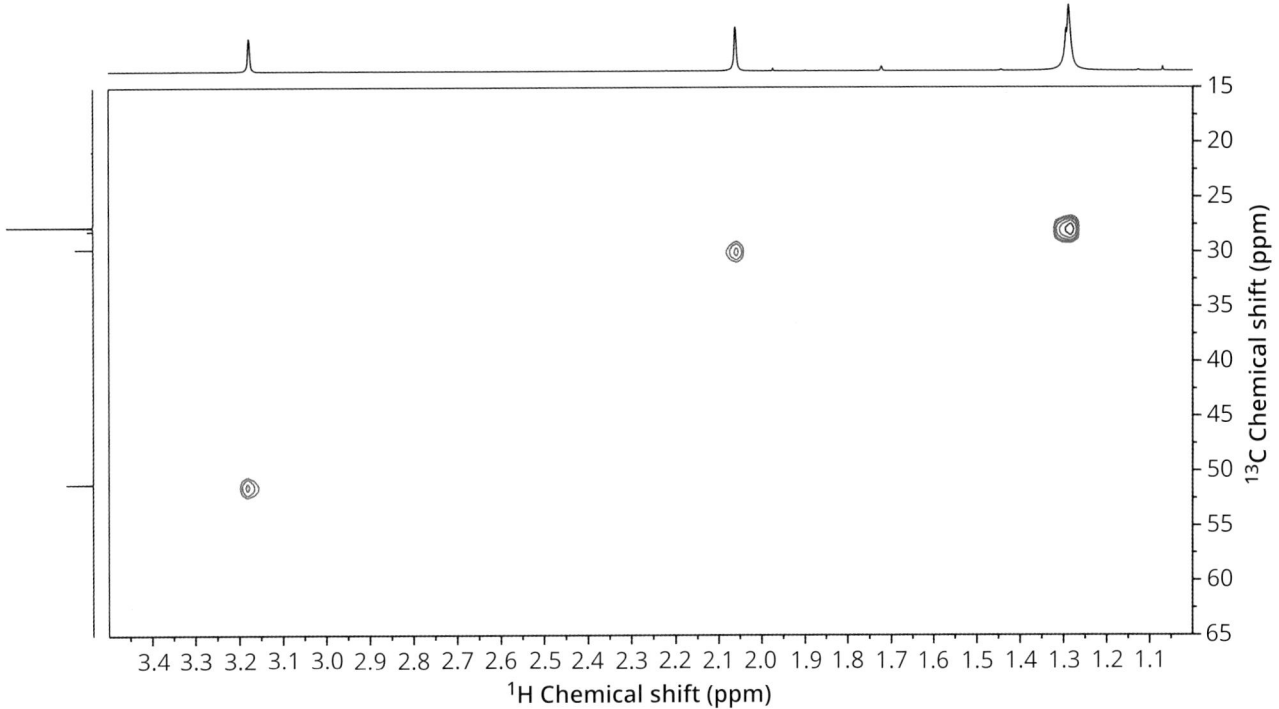

¹H–¹³C HSQC NMR spectrum (400 MHz, CDCl₃)

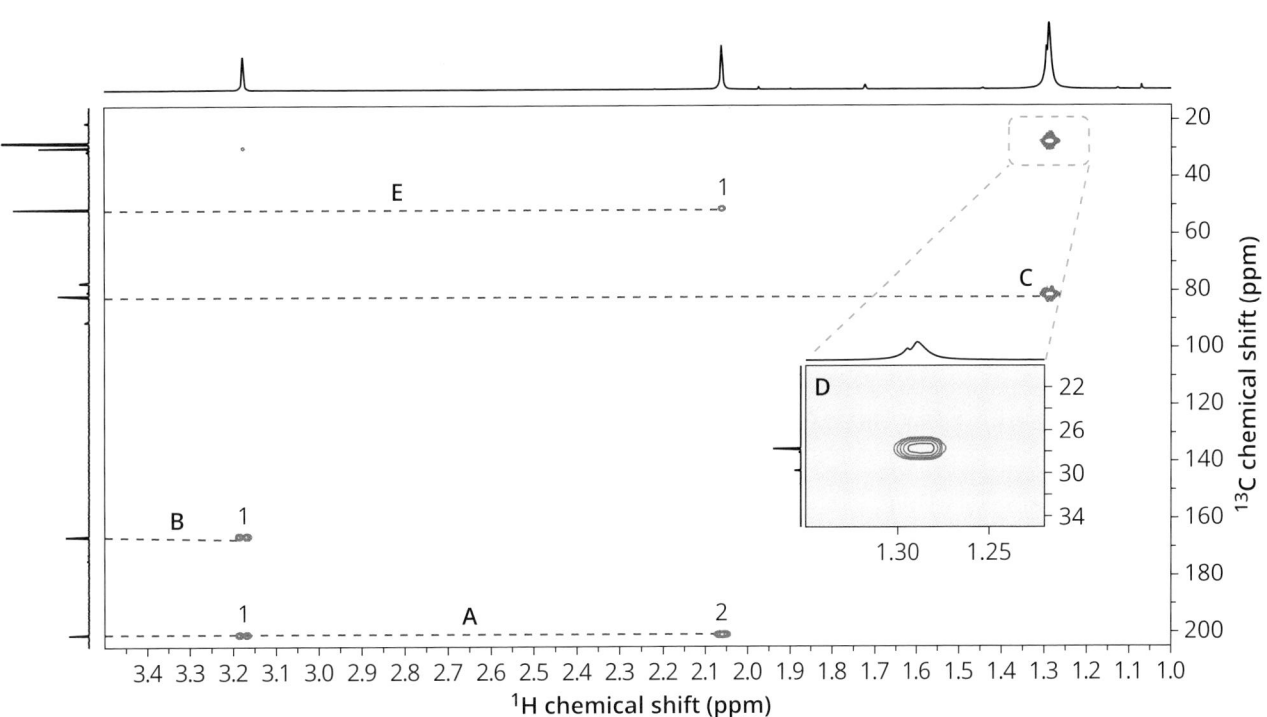

¹H–¹³C HMBC NMR spectrum (400 MHz, CDCl₃)

TABLE 11.3 Chemical shifts for Example 2

$^1H\,\delta$/ppm	$^{13}C\,\delta$/ppm
3.18	200.7
2.05	166.1
1.29	81.4
	51.1
	29.6
	27.6

11.2.1 Analysis of Example 2

The chemical shifts are tabulated in Table 11.3.

a) The proton spectrum shows only three singlets; the lack of coupling indicates that these are all remote from each other in the structure. The 2:3:9 integral ratio suggests that CH_2 and CH_3 groups are present, whilst the 9-proton singlet indicates three equivalent methyl groups, and is thus likely due to the presence of a tertiary butyl (tBu) group $C(CH_3)_3$.

b) The carbon spectrum displays three signals in the aliphatic region, in addition to the chloroform cluster around 77 ppm (we ignore some low-level impurities that have not been peak picked), consistent with the ^1H analysis. This region is dominated by the intense signal at 28 ppm, which would belong to the three equivalent methyls of the tBu group.

c) The carbon signal at 51 ppm is inverted in DEPT, indicating that this belongs to the CH_2 group, leaving the peak at 30 ppm as the remaining methyl carbon.

d) The carbon spectrum shows three further peaks above the solvent, all of which are missing in DEPT, indicating that they lack attached protons. One of these peaks is an aliphatic quaternary centre (81 ppm) whilst the other two are above 160 ppm and are likely to be from carbonyls. The signal at 201 ppm may be a ketone or an aldehyde, but its absence in the DEPT spectrum identifies it as a ketone. The carbon at 166 ppm may originate from an ester, acid, amide, or similar, but its identity cannot be established at this stage.

e) The 2D HSQC spectrum maps the direct ^1H–^{13}C connectivity and confirms the resonance assignments of the three protonated carbon centres.

f) As the proton spectrum lacks any coupled multiplets, there are no ^1H–^1H correlations to trace, so the COSY spectrum has not been recorded in this case.

g) The data so far suggest the presence of the following structural elements:

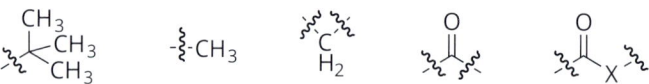

The mass spectrum of the molecule indicates the molecular weight to be 158 Da, and from this it is possible to determine that the missing element X is oxygen (consistent with an ester carbonyl at 166 ppm) and that there are no other components to consider. It is conceivable to piece these structural elements together in a number of ways, but in this case we can use the HMBC data to guide this process.

h) In HMBC, two correlations to the ketone peak at 201 ppm (Line **A**) can be seen, which indicates that the CH_2 (peak **1**) and the CH_3 (peak **2**) protons are within two or three bonds of this carbonyl.

i) The CH_2 protons also correlate with the 166 ppm carbonyl (Line **B**, peak **1**), indicating that these also connect with the ester group.

j) The tBu protons correlate to the lone carbon peak at 81 ppm (Line **C**), confirming this to arise from the quaternary centre of this fragment. The high carbon shift suggests a connection to a heteroatom; in this case, the ester oxygen is the only possibility.

k) We also observe (perhaps surprisingly), an apparent correlation from the tBu methyl proton to its 'own' carbon at 28 ppm (Expansion **D**). In fact, this arises due to the internal symmetry of the tBu group such that each methyl proton correlates to the carbon of an *adjacent* methyl group (across three bonds). Such a correlation is therefore consistent with the $(CH_3)_3C-$ functionality.

l) Finally, we note a correlation of the methyl protons to the CH_2 carbon at 51 ppm (Line **E**, peak 1). As we know that the CH_2 and CH_3 groups are not immediately adjacent (because they do not share ^1H–^1H coupling), yet they both correlate to the ketone in HMBC, these data suggest that these groups sit on either side of this carbonyl.

The collective HMBC data indicates the following connectivity and structure, with the associated ^1H and ^{13}C assignments summarized below.

These completed assignments can be presented in tabulated form, as in Table 11.4.

TABLE 11.4 Completed assignments for Example 2

$^{13}C\,\delta$ /ppm	Carbon type	$^1H\,\delta$ /ppm
200.7	C	–
166.1	C	–
81.4	C	–
51.1	CH_2	3.18
29.6	CH_3	2.05
27.6	CH_3	1.29

11.3 Problem 1

In this first problem, the challenge is not to identify the structure itself from the spectroscopic data, but to show that these data are consistent with a proposed structure and, in doing so, to determine the complete chemical shift assignments for all proton and carbon atoms in the molecule. This process reflects that faced most commonly in organic chemistry when a synthetic procedure has been completed and the chemist is required to prove the identity of the resulting compound (and hence the success, or otherwise, of the synthesis).

Using the spectra provided, determine complete ^1H and ^{13}C assignments for the compound shown (recorded in deuterated methanol, CD_3OD; the corresponding peaks are labelled as 'Solv' in the spectra). Note that the use of fully deuterated methanol means that all exchangeable OH protons have been replaced with deuterons (to become OD) and are thus not visible in the ^1H spectrum. The proton and carbon chemical shifts have been provided in tabulated form (Table 11.5) as an additional resource.

TABLE 11.5 Chemical shifts for Problem 1

$^1H\,\delta$ /ppm	$^{13}C\,\delta$ /ppm
7.54	170.0
4.37	138.8
4.00	130.7
3.67	72.7
2.70	68.4
2.19	67.3
	31.6

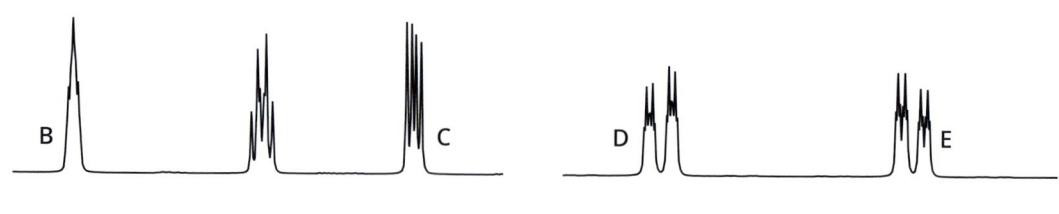

HINT

You should focus on using the 2D spectra to make assignments rather than attempting to interpret all the detail seen in the multiplet structures of the ¹H spectrum.

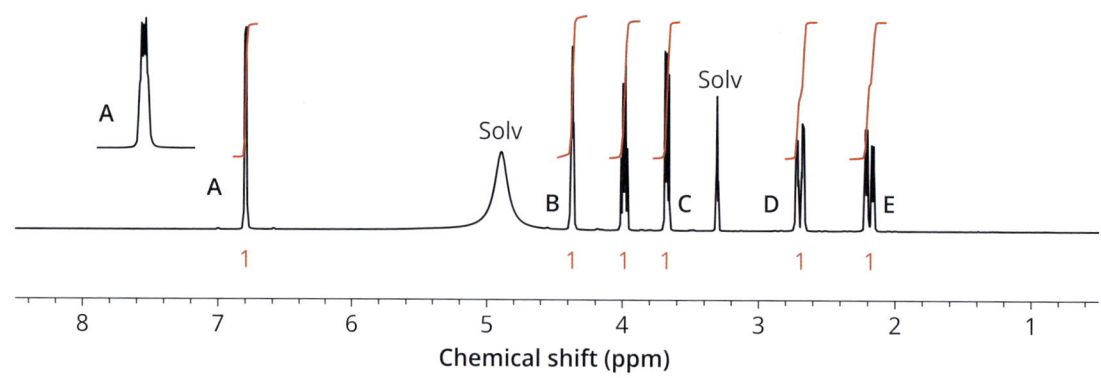

¹H NMR spectrum (400 MHz, CD₃OD)

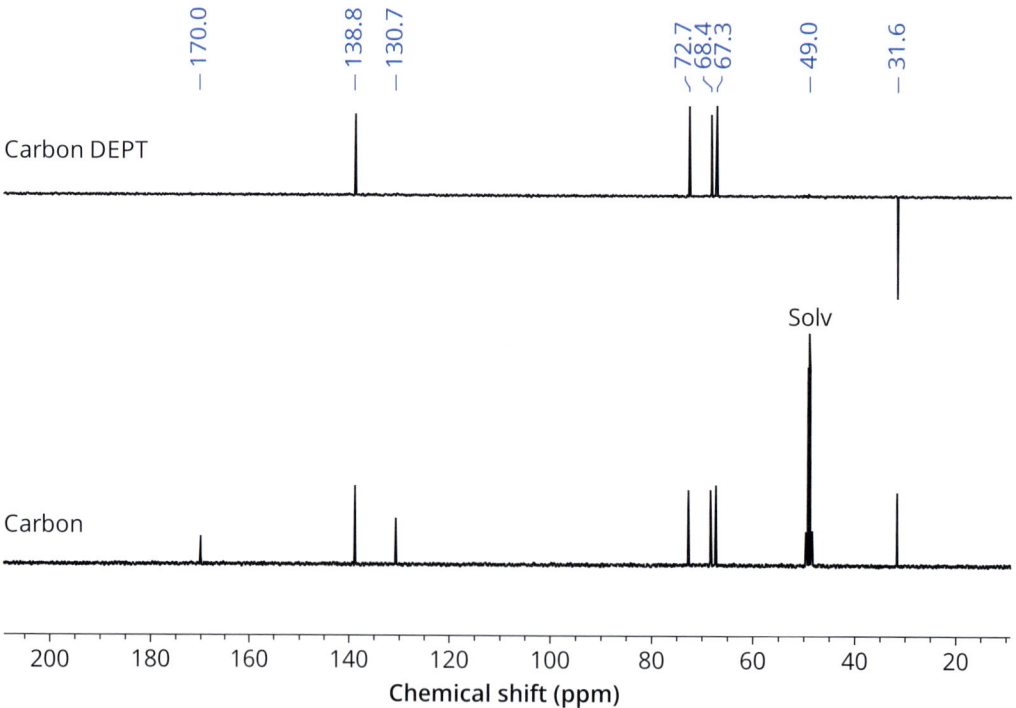

¹³C and DEPT NMR spectrum (100 MHz, CD₃OD)

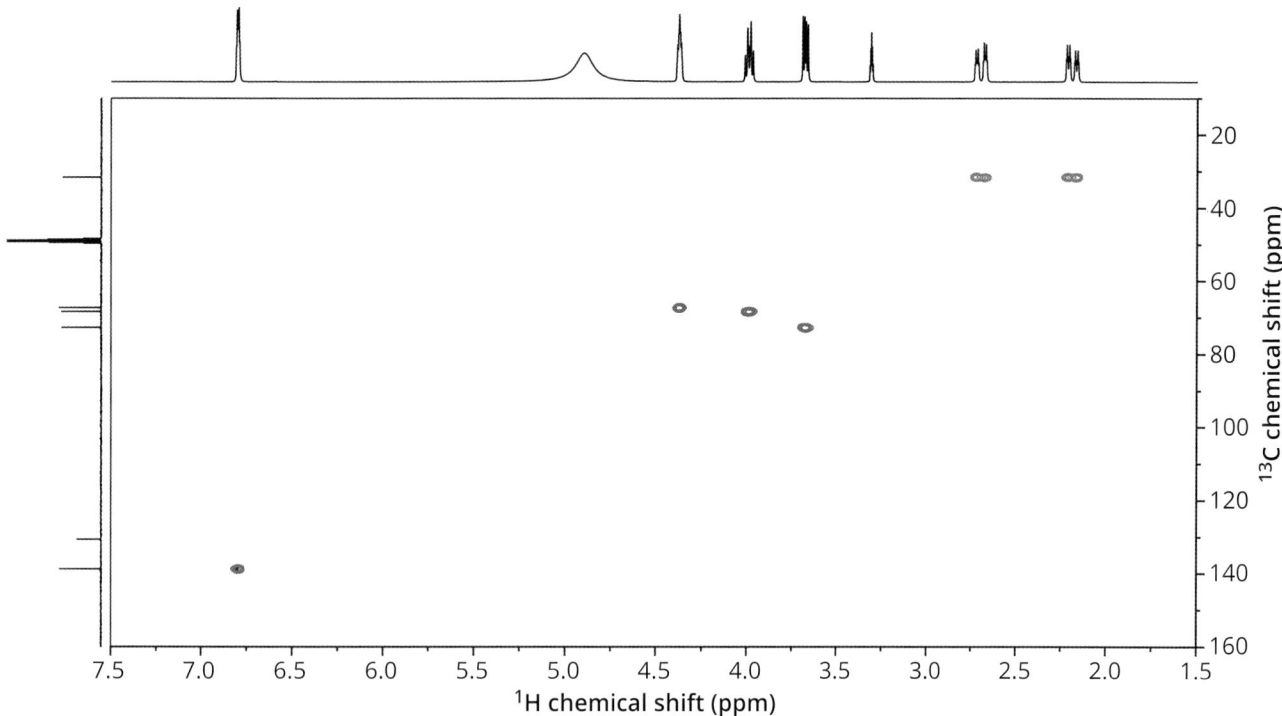

¹H–¹³C HSQC NMR spectrum (400 MHz, CD₃OD)

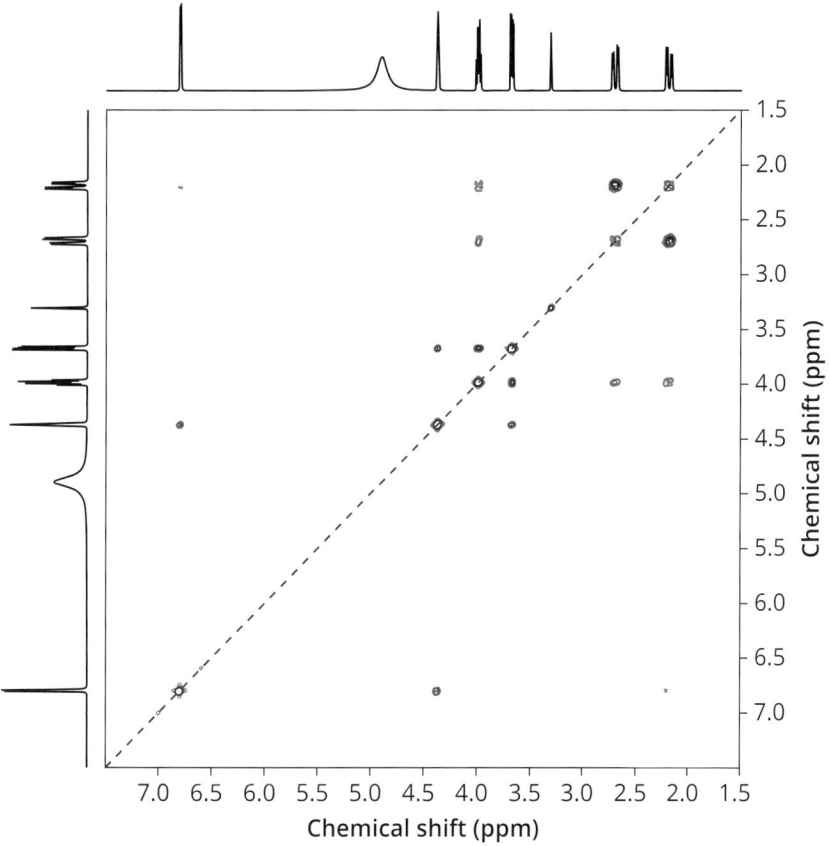

¹H–¹H COSY NMR spectrum (400 MHz, CD₃OD)

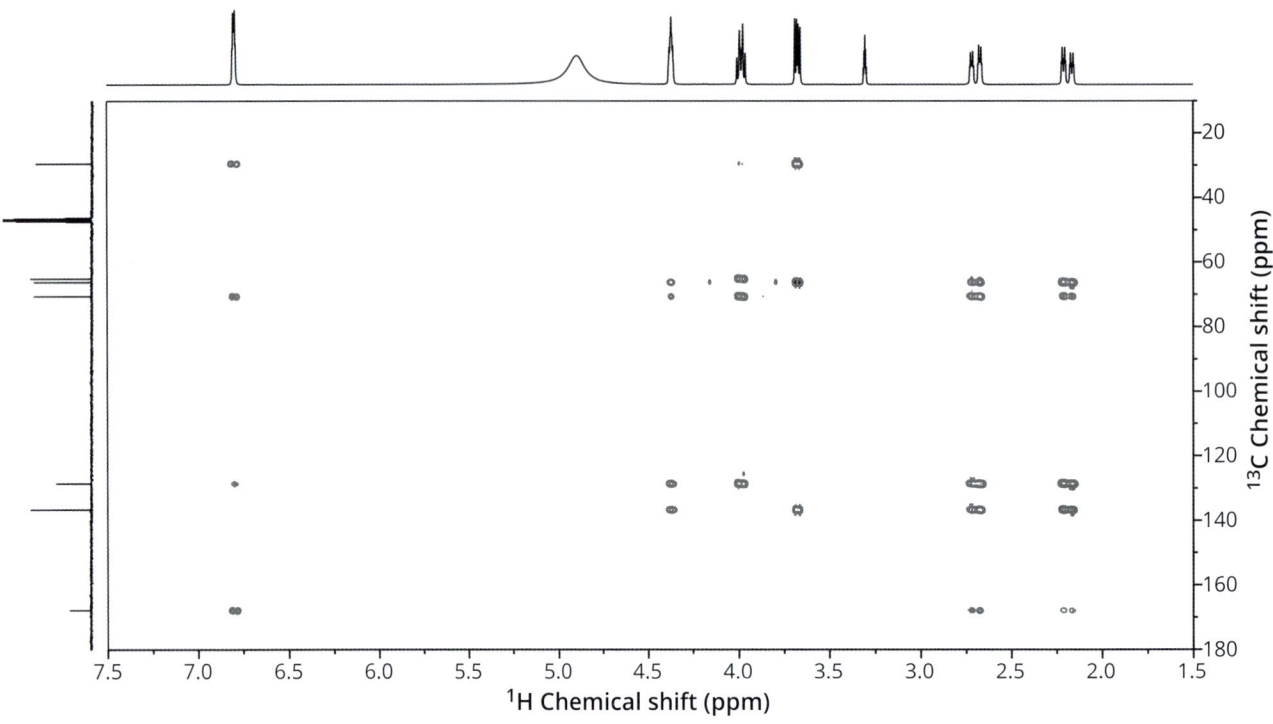

^{1}H–^{13}C HMBC NMR spectrum (400 MHz, CD$_3$OD)

TABLE 11.6 Chemical shifts for Problem 2

^{1}H δ /ppm	^{13}C δ /ppm
7.71	188.7
7.58	155.7
7.51	152.7
7.48	128.3
7.32	127.1
2.62	123.9
	123.3
	113.1
	112.5
	26.5

11.4 Problem 2

In this second problem, the challenge is again to use the 1D and 2D NMR data to show that they are consistent with a proposed structure and, in doing so, to determine the complete chemical shift assignments for all proton and carbon atoms in the molecule.

The proton and carbon chemicals shifts have again been provided in tabulated form (Table 11.6).

HINT

Note that when interpreting the HMBC data, it is important to remember that this type of spectrum exhibits correlations between protons and carbon centres across two or three bonds. In aromatic ring systems it is often the case that $^2J_{CH}$ coupling constants are vanishingly small, giving rise to rather weak two-bond cross peaks. In contrast, $^3J_{CH}$ couplings are rather large and often give rise to more intense cross peaks.

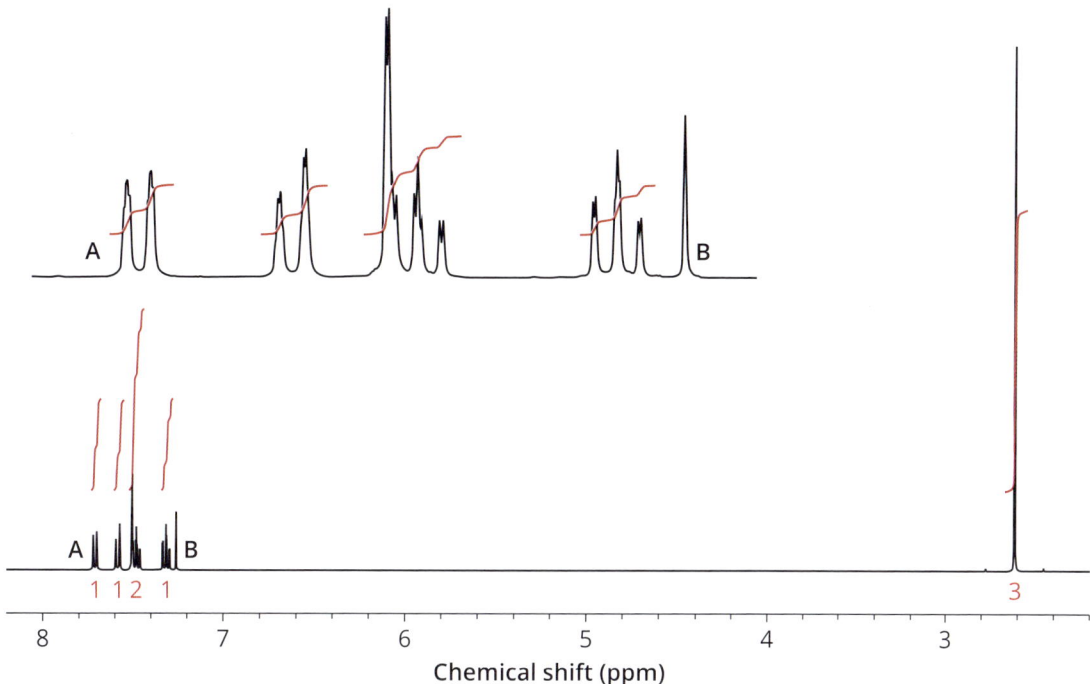

¹H NMR spectrum (400 MHz, CDCl₃)

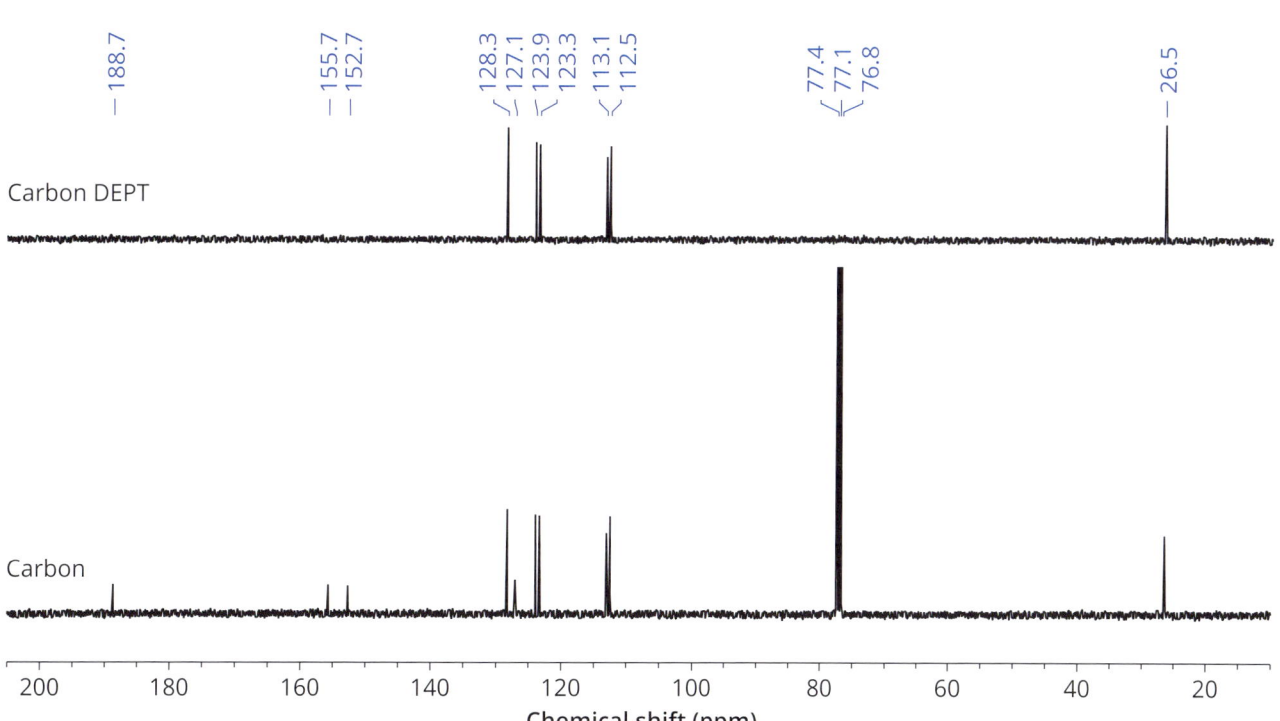

¹³C and DEPT NMR spectrum (100 MHz, CDCl₃)

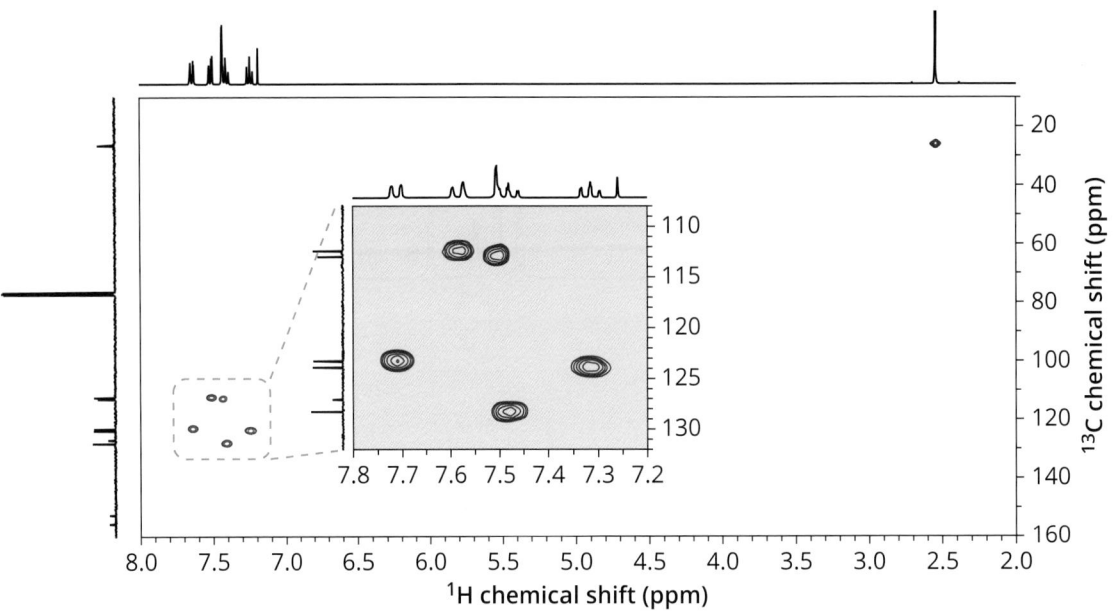

¹H–¹³C HSQC NMR spectrum (400 MHz, CDCl₃)

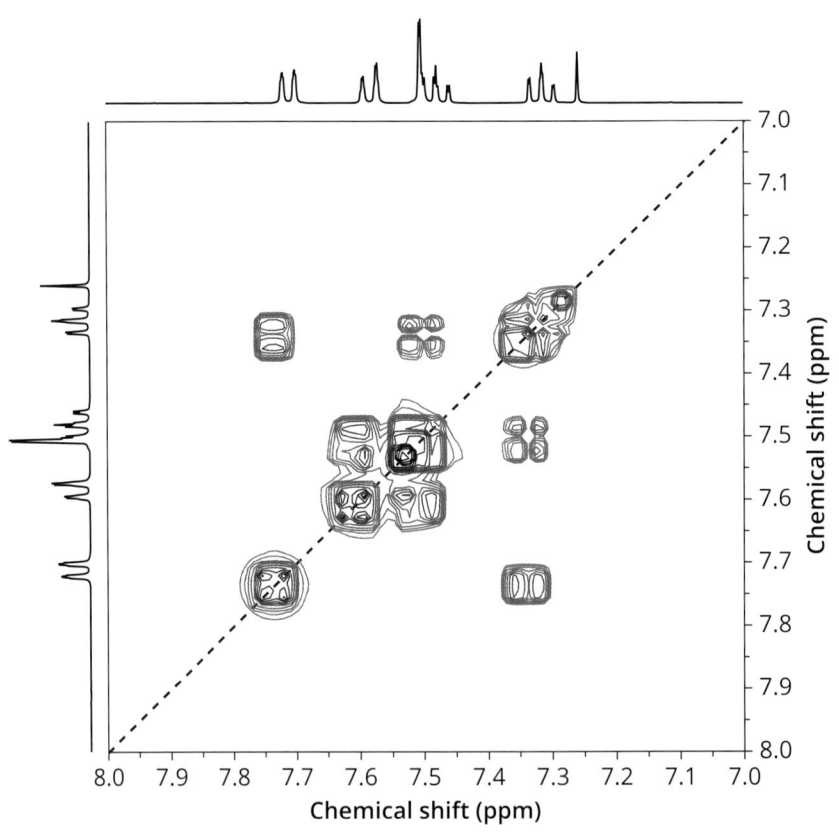

¹H–¹H COSY NMR spectrum (400 MHz, CDCl₃)

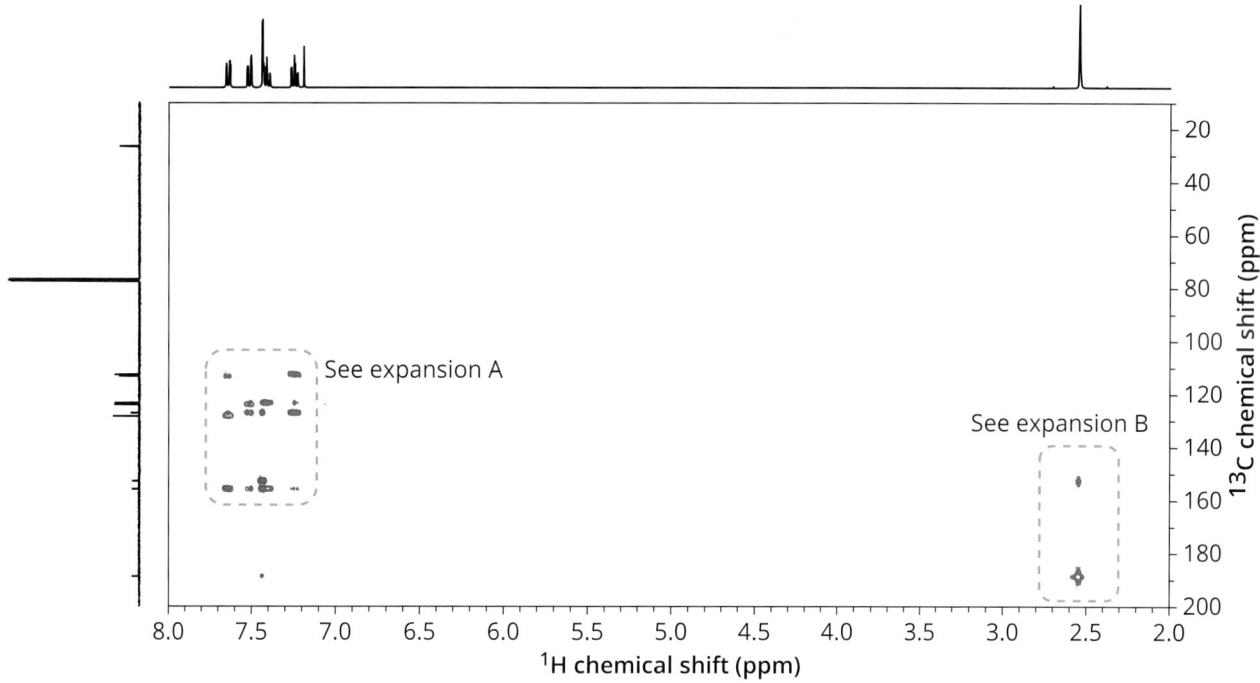

¹H–¹³C HMBC NMR spectrum (400 MHz, CDCl₃)

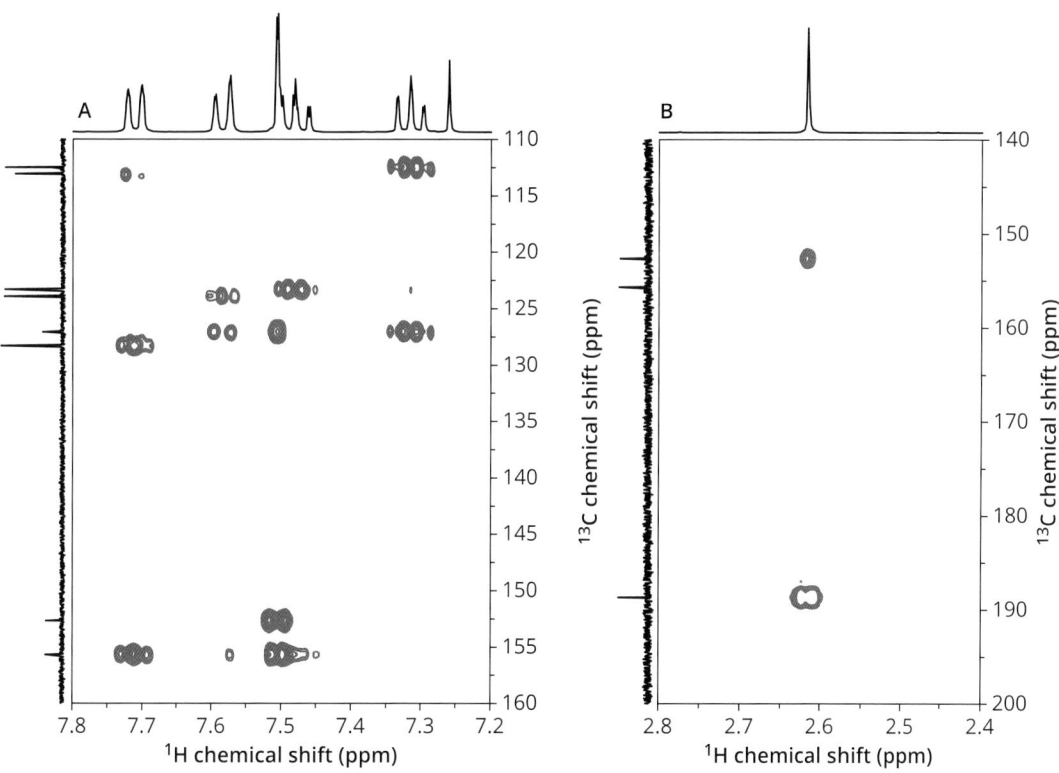

¹H–¹³C HMBC NMR spectrum expansions (400 MHz, CDCl₃)

CHAPTER 12

Advanced mass spectrometry methods

In Chapter 5 we focused exclusively on the interpretation of electron ionization (EI) mass spectra. However, despite the historical and continued importance of EI in mass spectrometry, the technique has a number of limitations for chemical identification and structural elucidation. The most significant limitation is that EI mass spectrometry is restricted to the analysis of volatile, thermally stable compounds. In addition, there is also little control over analyte fragmentation, which limits EI applications largely to the analysis of purified compounds unless it is coupled to gas chromatography. Although not an inherent limitation, we also only dealt with the measurement of nominal masses in Chapter 5. Higher mass measurement accuracy can also play an important role in the interpretation of chemical identity and structural elucidation.

In recent decades dramatic and transformational technical developments in mass spectrometry have led to new capabilities that address many of the limitations of EI-MS. In particular, new ionization methods, types of mass analyser, and separation methods have provided hyphenated mass spectrometry techniques capable of highly sensitive, high-resolution analyses of individual molecular structures in complex mixtures. These 'high-resolution' mass spectrometry systems have extended capabilities well beyond those available using EI-MS. In this chapter we introduce and discuss some of these more advanced capabilities, focusing on how they can be used for the elucidation of chemical identity and chemical structure using practical examples.

12.1 Soft ionization methods and their mass spectra

EI is referred to as a 'hard' ionization technique and is only compatible with analytes that are thermally stable and sufficiently volatile to be transferred into the gas phase during the ionization process. The energy that results from the interaction of accelerated electrons with neutral gas-phase molecules is relatively high, leading to molecular fragmentation. Although useful for structural interpretation, this process is only suitable for small, volatile organic molecules and therefore can only be used successfully with a minority of organic compounds. A number of alternative ionization techniques were subsequently developed, such as chemical ionization (CI), which effectively reduces the amount of energy transferred to the analyte but still requires analytes to be volatile. However, it wasn't until the development of atmospheric pressure ionization (API) approaches, namely electrospray ionization (ESI) and matrix-assisted laser desorption ionization (MALDI), in the 1980s that a much broader range of organic molecules could be analysed. These API techniques are not dependent on analyte volatility and involve very low energy transfer. They are now widely used in mass spectrometry.

API techniques function in fundamentally different ways to EI. They are called 'soft' ionization techniques because they involve a much lower energy transfer to the analyte during the 'ionization process'. This leads to a very different mass spectrum, characterized by very few, if any, fragmentation peaks. Figure 12.1 shows the mass spectrum of *sec*-butylamine ($C_4H_{11}N$) produced using electrospray ionization. There are no fragment ions present and, as a consequence, the signal for the analyte is the base peak. ESI is generally a higher-sensitivity ionization method compared to EI, in part because the signal that represents the analyte has not been split into multiple parts via fragmentation. It is worth noting that the ionization process taking place in soft ionization methods does not involve high-energy electrons, or indeed true molecular ion formation (where an electron is removed during the ionization process), in contrast to EI. Instead, both ESI and MALDI (and other API methods in general) produce *charged molecules* from the neutral analyte, via mechanisms involving protonation or deprotonation or other cation or anion adduct formation.

> Ions produced through adduct formation are referred to as *charged molecules* rather than *molecular ions*. They can also be referred to as *protonated* or *deprotonated* molecules (when the adduct is due to the addition or loss of a proton, which is very common), or more generally *cationized* or *anionized* molecules. Note the terms *quasi-molecular* ions and *pseudo-molecular* ions were in common use previously, but these are now discouraged, in part due to their descriptive ambiguity.

One of the main advantages of API methods is that they do not require analytes to be in the gas phase. In fact, ESI forms charged molecules from analytes in solution and MALDI ionizes crystallized solids. API methods, in particular ESI and MALDI, are now extremely common in mass spectrometry but are generally used for different applications. For the analysis of organic small molecules, ESI and atmospheric pressure chemical ionization are most common. Table 12.1 provides an overview of a selection of common ionization methods used in small molecule analysis.

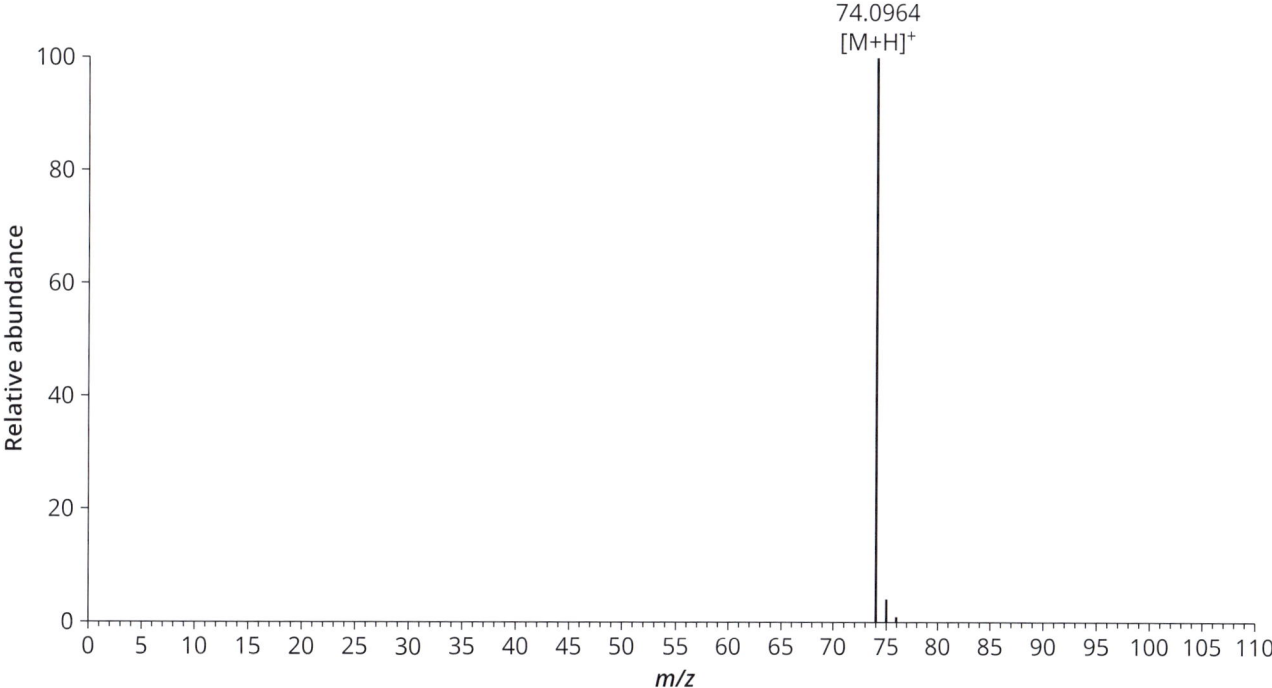

FIGURE 12.1 ESI mass spectrum of *sec*-butylamine ($C_4H_{11}N$) showing the peak representing the protonated $[M+H]^+$ adduct.

TABLE 12.1 An overview of common ionization methods used in mass spectrometry.

Ionization method	Acronym	Mechanism of ionization	Sensitivity	Applications	Type
Electron ionization	EI	Electron removal	Medium	GC-MS	Hard
Chemical ionization	CI	Electron removal	Medium	GC-MS	Soft
Electrospray ionization	ESI	Protonation/deprotonation	High	LC-MS	Soft (API)
Atmospheric pressure chemical ionization	APCi	Protonation/deprotonation	High (for selected compounds)	LC-MS	Soft (API)
Matrix-assisted laser desorption	MALDI	Protonation/deprotonation	High	Direct studies/MALDI imaging	Soft (API)
Inductively coupled plasma	ICP	Plasma ionization	High	Trace metal analysis	Hard
Desorption electrospray ionization	DESI	Protonation/deprotonation	Medium	Surface analysis	Soft (API)
Secondary ionization MS	SIMS	Electron removal	High	Surface analysis	Hard

The following are some of the key differences between 'hard' (EI) and 'soft' (API) ionization methods:

- 'Soft' ionization techniques transfer less energy to the analyte, leading to very little fragmentation in contrast to EI.
- The formation of ions using API methods takes place at atmospheric pressure and does not require analytes to be volatile.
- Soft ionization methods are amenable to a much broader range of small organic molecules compared to EI.
- Soft ionization techniques generally provide higher sensitivity than EI.
- Due to the lack of fragments and a different ion formation process, the mass spectra of API methods look very different from EI mass spectra.

12.2 The formation of charged molecules

There are now a wide range of soft ionization techniques available and, although these may differ in terms of the analytes to which they are amenable or the type of samples, the fundamental mechanisms of forming charged analyte molecules are similar. A charged molecule is formed when a proton forms an adduct with a neutral molecule (positive ion mode), or a proton is removed (negative ion mode). Figure 12.2 illustrates the process of protonation that occurs during electrospray ionization. This process simultaneously removes solvent molecules through evaporation (exemplified by water in the example, but this can also be a mixture of organic solvent and water molecules), which leads to the formation of a desolvated (water removed) charged adduct in the gas phase.

Protonated molecules (positively charged) are referred to using the shorthand $[M+H]^+$ and deprotonated molecules (negatively charged) are referred to using the shorthand $[M-H]^-$. In some cases, other cation or anion adducts can also form (e.g. Na^+, K^+, NH_4^+ in positive ion mode and Cl^- and CH_3COO^- in negative ion mode), and it is quite common to have more than one adduct-type present in the spectrum for an analyte at the same time. Only a single proton is required when a singly charged adduct is formed, which means that during electrospray ionization the autolysis of the water in an aqueous solvent (sample diluent) provides sufficient protons to facilitate the process of charged molecule formation. In practice a small amount of weak acid, such as 0.1% formic or acetic acid, is often

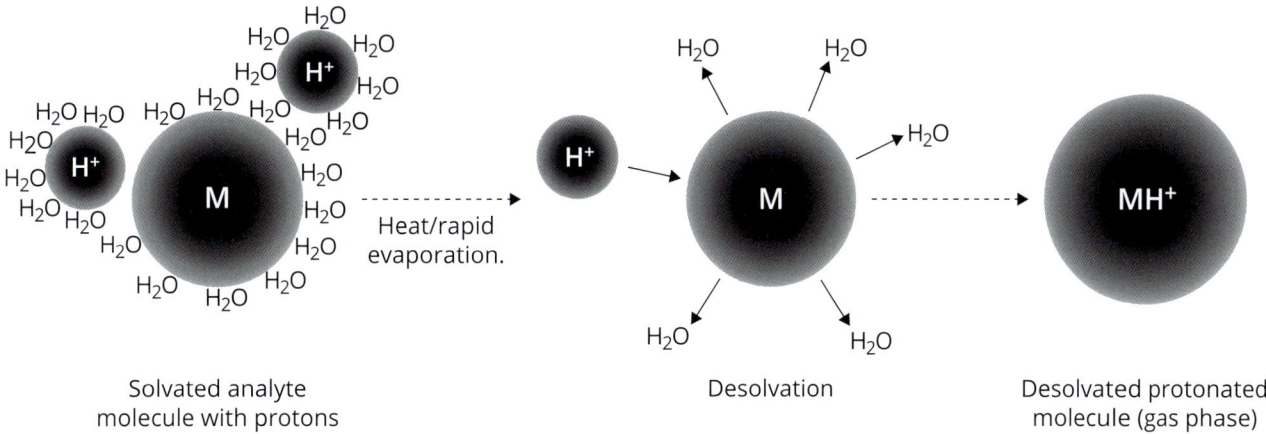

FIGURE 12.2 A schematic representation of charged molecule formation during electrospray ionization. A desolvated protonated adduct is formed where the analyte molecule becomes charged whilst moving from the aqueous to the gas phase. Water molecules evaporate rapidly during this process, leading to the desolvated protonated adduct $[M+H]^+$. The relative size of molecular species has been modified to emphasize the process.

added to the analyte sample to enhance proton adduct formation. Using MALDI, where analytes are in the solid phase, co-crystallization with a weak acid facilitates proton transfer from the acid to the analyte during the ionization process.

- Charged species formed through soft ionization processes are not true molecular ions and so are often referred to as 'charged molecules'.
- The protonation and deprotonation process in ESI follows the principles of acid–base chemistry with basic functional groups accepting a proton in positive ion mode and acidic functional groups losing a proton in negative ion mode.
- Many analyte molecules are ionized together during the ionization process and, when transferred to the detector, generate a total ion signal that determines signal abundance in the mass spectrum.

Although the vast majority of ions formed via electrospray ionization and other soft ionization methods are created through adduct formation, as described in this section, it is worth noting that it is also possible to detect M^+ species. These are not formed during ionization but can be detected if the molecule already has a fixed charge in solution prior to ionization.

12.2.1 Multiply charged ions

Although many small organic molecules form singly charged adducts with electrospray ionization, some molecules can be multiply charged. Indeed, as molecular weight increases, so, in general, do the number of protonatable or deprotonatable sites, and one of the benefits of electrospray ionization is its ability to form multiply charged ions. Figure 12.3 illustrates the electrospray mass spectrum of arginylisoleucine, a dipeptide formed via the condensation of isoleucine and arginine. Note the difference in m/z between the singly and doubly charged forms of the molecule and that the distance between isotope peaks is halved for the doubly charged species. For triply charged molecules, the M and M+1 isotope peaks are approximately 0.33 daltons apart in the spectrum, and with four charges they are 0.25 daltons apart. Really large molecules such as proteins and oligonucleotides can have 10s or even 100s of charges, and this is where ultra-high-resolution mass spectrometers become extremely useful as they provide enhanced capabilities for resolving isotope peaks and therefore the means to more accurately identify the monoisotopic mass.

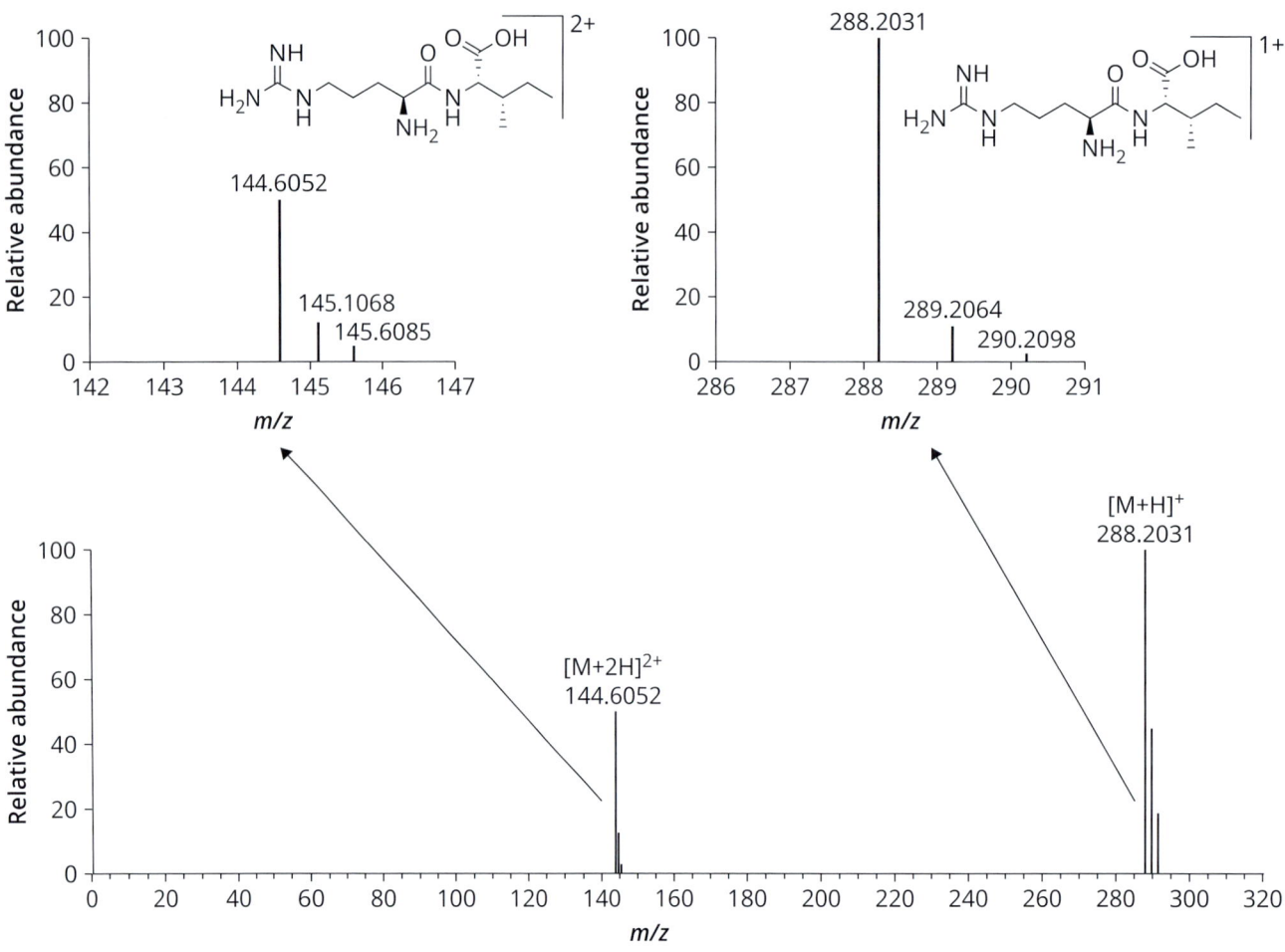

FIGURE 12.3 A mass spectrum showing singly and doubly charged molecules of the dipeptide arginylisoleucine. Note that the doubly charged molecule provides an m/z value that is approximately half of that of the singly charged ion. In addition, the isotope peaks are spaced 0.5 Da apart for the doubly charged ion and 1 Da apart for the singly charged ion.

Although soft ionization mass spectra tend not to have many fragment peaks, they can nevertheless provide multiple peaks representing different charge states and adduct forms of the analyte. It is also possible for analytes to form adducts with themselves, creating dimers or other multimers, and all these forms may be present in various combinations in the soft ionization mass spectrum. These different forms have predictable m/z values, so it is relatively straightforward to identify them, which can help confirm compound identifications. Which specific adducts are formed will depend on the analyte itself (how amenable it is to the process of adduct formation), other components in the sample (e.g. sodium or potassium ions), and the types of solvent being used, as well as analyte concentrations.

- The soft ionization mass spectrum may show multiple non-fragment peaks representing the same analyte.

- The adduct profile for a particular analyte often alters using different soft ionization techniques; for example, MALDI is less likely to provide multiply charged analytes than ESI, but sodium adducts are still common.

12.2.2 Ionization of larger molecules

Due to the low-energy process of forming charged molecules using soft ionization techniques, very little redistribution of internal energy occurs, which in turn results in little or no fragmentation. This means that the charged analyte molecule is almost always the base peak in the mass spectrum. The low energy transfer also means that soft ionization techniques can effectively ionize larger and less stable molecules; indeed, ESI and MALDI are used for the analysis of a very wide range of molecular masses, from small molecules all the way to macromolecular protein complexes with molecular masses in the millions of daltons in the case of ESI.

12.2.3 Ionization efficiency

Due to their ability to ionize a wide range of compound classes, structures, and molecular weights, soft ionization methods can be susceptible to interferences from other sample constituents. At the start of the ionization process, analytes are usually in solution (ESI) or solid mixtures (MALDI), meaning that a phase transition is needed in both cases when forming gas phase analyte ions. The presence of other compounds in the sample, such as salts and other analytes, can have a suppressive effect (or an enhancing effect, but this is less common) on ion formation and may alter the abundance of analyte ions formed. This contrasts with EI, where ionization is highly reproducible. The variability in signal abundance for soft ionization techniques can lead to difficulties in comparing relative amounts of an analyte between samples. To mitigate this, liquid chromatography systems are often coupled directly to the mass spectrometer as prior chromatographic separation can help remove or reduce interferences from other compounds by decreasing susceptibility to ion suppression effects (see section 12.6).

12.3 High-resolution spectra and accurate mass analysis

In addition to the development of soft ionization methods, another important development has been the ability to measure m/z values with high accuracy and resolve masses well beyond unit resolution. We were concerned only with unit resolution (1 dalton accuracy) in Chapter 5, but there are many compounds that have different chemical formulas but whose m/z values are well within 1 Da of each other. These would not be resolved at unit mass resolution, but developments in mass analyser technologies have led to 'high-resolution' instruments that are able to provide m/z measurements at very high mass accuracy due to their narrower mass spectral peak distributions. Time of Flight (TOF), Orbitrap, and Fourier Transform Ion Cyclotron Resonance (FT-ICR) mass analysers (Table 12.2) are some of the most common high-resolution mass analysers. Some types of Orbitrap-MS and FT-ICR-MS are currently able to provide m/z accuracies down to 0.1 part per million (ppm). At this level the error in the m/z determination is between the 4th and 5th decimal place of a single mass unit on the dalton scale. Such instruments can therefore be used to differentiate m/z values that differ by less than the mass of a single electron (~0.0005 Da).

TABLE 12.2 Selected figures of merit for different analysers found in mass spectrometry including their indicative relative performance in terms of resolution and mass accuracy.

Ionization method	Acronym	m/z limits	Resolution	Scan times/s	Applications
Quadrupole	Q	2,000	4,000	0.1–1	Targeted studies
Time of Flight	TOF	>1 million	40,000	100 µs	Untargeted and untargeted studies
Orbitrap	Orbitrap	~20,000	>500,000	0.1–1	Untargeted and untargeted studies
Ion Cyclotron Resonance	FT-ICR	~20,000	>1 million	0.5–10	Targeted and untargeted studies

The IUPAC unit of mass at the atomic scale is the *unified atomic mass unit* (u) and is defined as 1/12 of the mass of ^{12}C (since 1961). In mass spectrometry the term u and dalton (Da) are used synonymously—that is, 1 Da = 1 u. The accuracy of an *m/z* measurement is commonly referred to in parts per million and for high accuracy instruments, this is commonly in the range 0.1–5 ppm.

Soft ionization techniques can be used with both low-resolution and high-resolution mass analysers, but high-resolution instruments are particularly useful for determining the chemical composition of analyte ions. This is because the high decimal place accuracy they provide enables elemental compositions to be calculated directly from the *m/z* measurement itself in the case of many small molecules. Although accurate mass measurements are not always necessary for structural elucidation using EI-MS (due to the presence of structurally relevant fragment peaks), it can greatly help in the interpretation of soft ionization mass spectra where fragments are not typically present.

12.3.1 Calculating elemental compositions

The elemental composition of analytes associated with an *m/z* value in the mass spectrum can often be predicted directly with a high degree of confidence using high-resolution monoisotopic accurate mass measurements. This is achieved by calculating all the elemental combinations that are theoretically possible within the constraints imposed by the accuracy of the *m/z* measurement and then evaluating the practical possibilities. It helps to narrow down the options if chemical knowledge about an analyte is available, which is often the case. This could, for instance, be knowledge about the possible elements present or absent from a reaction mixture or experimental information from the mass spectrum itself, such as the presence of certain elements based on the isotopic composition, such as chlorine, bromine, or sulfur, for example. Most instrument vendor software has inbuilt elemental formula prediction capabilities, and similar tools can also be found in open-source programs online.

12.4 Tandem mass spectrometry

One of the significant advantages of soft ionization techniques over EI is that very few fragment ions are generated, meaning that there is no ambiguity as to whether the intact charged analyte is present. On the other hand, the absence of fragment peaks means that only chemical composition information, and no structural information, is found in the mass spectrum. This represents a problem for studies focused on structural elucidation. *Tandem mass spectrometry* is a solution that introduces an analyte fragmentation step that takes place in a controlled manner inside the mass spectrometer after ionization. Also referred to as MS/MS or MS^2, tandem mass spectrometry usually requires a mass spectrometer to have multiple mass analysers present in the same instrument. A major benefit of tandem mass spectrometry is that it provides significant control over the fragmentation process, which makes it particularly useful for investigating chemical structure.

Tandem mass spectrometry incorporates two or more mass analysers in the same instrument from which its name derives. Tandem mass spectrometry experiments provide a more controlled way to fragment analytes compared to fragmentation during EI.

There are two basic approaches to tandem mass spectrometry, sometimes referred to as 'tandem mass spectrometry in space' and 'tandem mass spectrometry in time'. Each requires different instrumental configurations. Tandem mass spectrometry in space utilizes two sequential mass analysers. The first mass analyser is used to select a *precursor ion*, which is then fragmented (this can be via various mechanisms). The first mass analyser

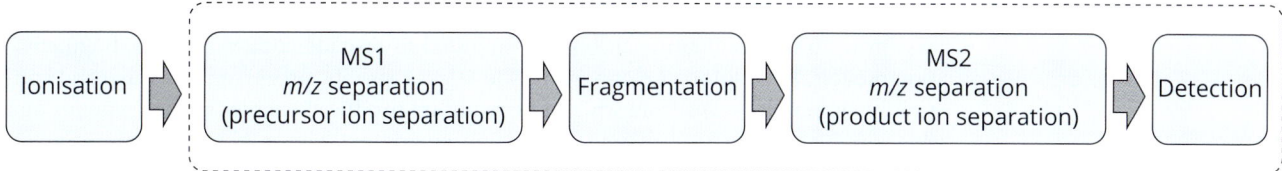

FIGURE 12.4 A schematic of the tandem mass analysis process leading to fragmentation. The steps inside the dotted line take place in various evacuated compartments of the mass spectrometer.

provides a high degree of selectivity; charged molecules with a specific *m/z* value can be selected for fragmentation even when a complex mixture is present. The *product ions* produced are the fragments, and these are subsequently detected using the second mass analyser. Figure 12.4 provides a schematic of the multiple analyser configuration used for tandem mass spectrometry in space experiments.

There are a number of different techniques that can be used to physically fragment charged analytes in a mass spectrometer. One of the most common is known as *collision-induced dissociation* (CID). CID involves precursor ions being accelerated to increase their kinetic energy followed by entry into a collision cell containing a low pressure of inert gas molecules (such as nitrogen, argon, or sometimes helium). This induces gas-phase molecular collisions which raise the internal energy of the analyte, leading to fragmentation. Figure 12.5 illustrates the process of CID fragmentation.

- *Collision-induced dissociation* (CID) can fragment charged analytes by inducing collisions with inert gas molecules within a collision cell inside the mass spectrometer.

- Tandem mass spectrometry can provide structural information for specific analytes even when analysing complex mixtures.

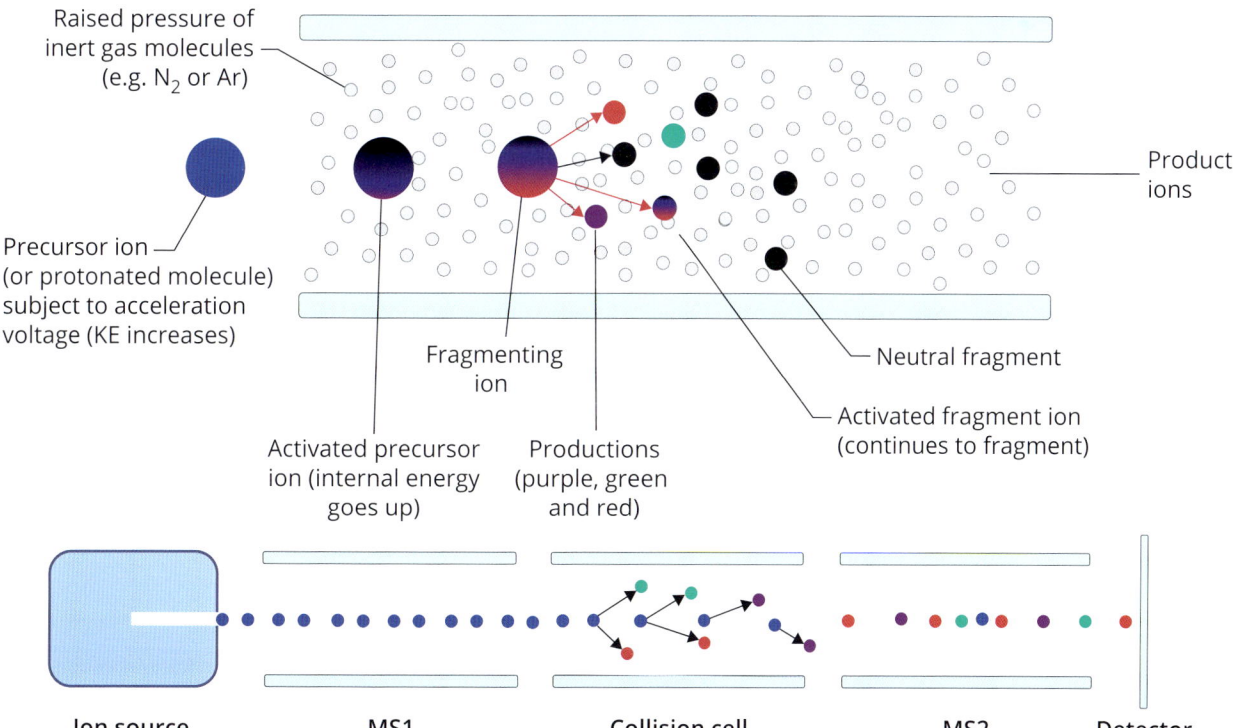

FIGURE 12.5 The process of precursor ion fragmentation in a CID fragmentation cell to form product ions and neutral fragments.

The inert gas pressure in the CID cell and the amount of kinetic energy analyte ions possess prior to fragmentation can both be adjusted to provide control over the fragmentation process. In addition, the opportunity to select a specific precursor m/z value and control the number of scans from which fragments are produced enables both precursors and their product ions to be measured in the same experiment. Figure 12.6 illustrates examples of common tandem mass spectrometry in space experiments.

> The charged analyte selected for fragmentation is referred to as a *precursor ion* and the fragments produced are called *product ions*. You may also encounter the terms 'mother' and 'daughter' ions, which have been used in the past to refer to precursor and product ions, but we do not recommend using this now outdated terminology.

'Tandem mass spectrometry in time' experiments can be performed on instruments which contain an 'ion-trap' mass analyser. Quadrupole-ion traps, FT-ICR-MS systems, and hybrid orbitraps are the most common examples of these types of instruments. The ion trap can store 'packets' of ions and fragment these sequentially over time using a process of resonant excitation. This enables multiple rounds of precursor selection and fragmentation to be performed within the ion-trap because ions of different m/z value can be selected, stored, and fragmented sequentially with a high degree of control. When multiple rounds of fragmentation are performed using the same analyte, this is called an MS^n experiment. Repeated fragmentation creates a fragmentation cascade or fragmentation tree (Figure 12.7). From a single precursor, the first generation of product ions is formed via fragmentation; one of these fragments is selected to be the next precursor ion and subsequently

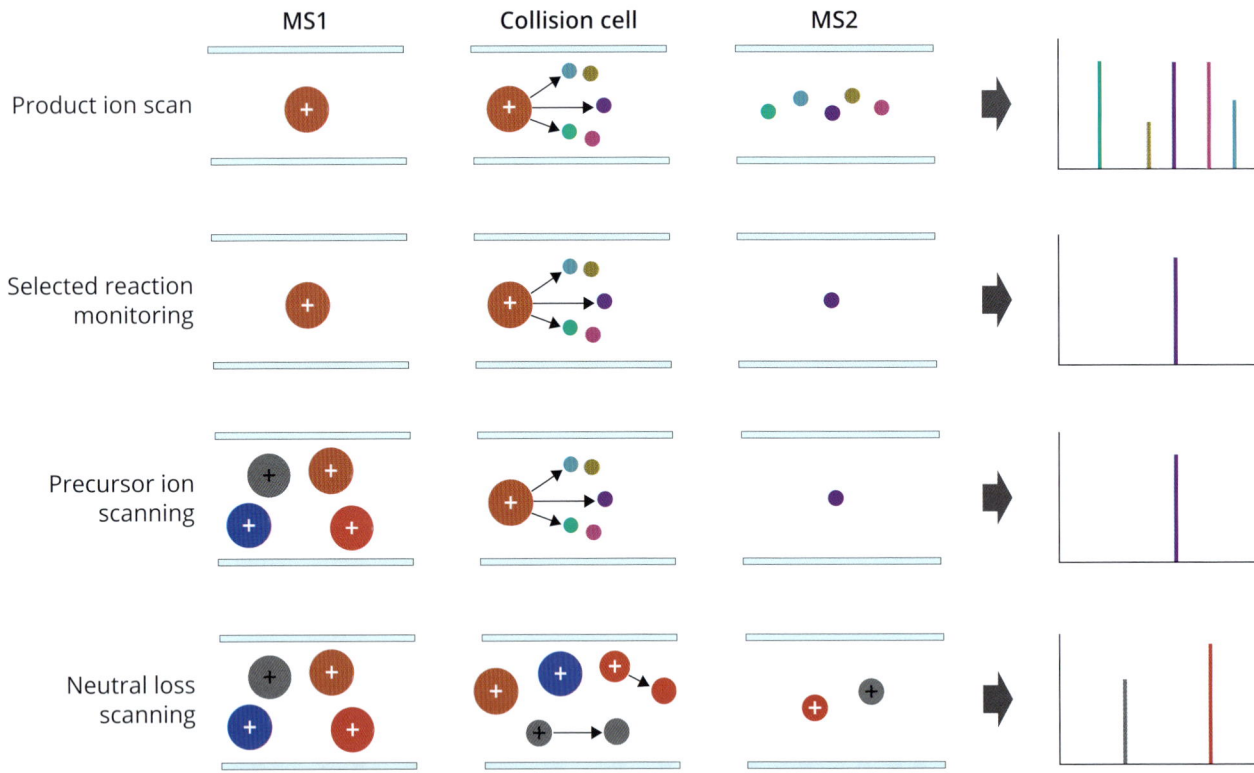

FIGURE 12.6 The combination of tandem mass analysers and fragmentation options has led to four common fragmentation experiments: *Product ion scanning* enables the product ions of a precursor to be measured even in complex mixtures. *Selected reaction monitoring* provides high specificity and sensitivity analysis by monitoring only a single fragment ion of interest. *Precursor ion scanning* provides an experiment for identifying whether any precursor ions present produce a specific product ion. *Neutral loss scanning* enables precursors, which lead to specific neutral losses, to be identified.

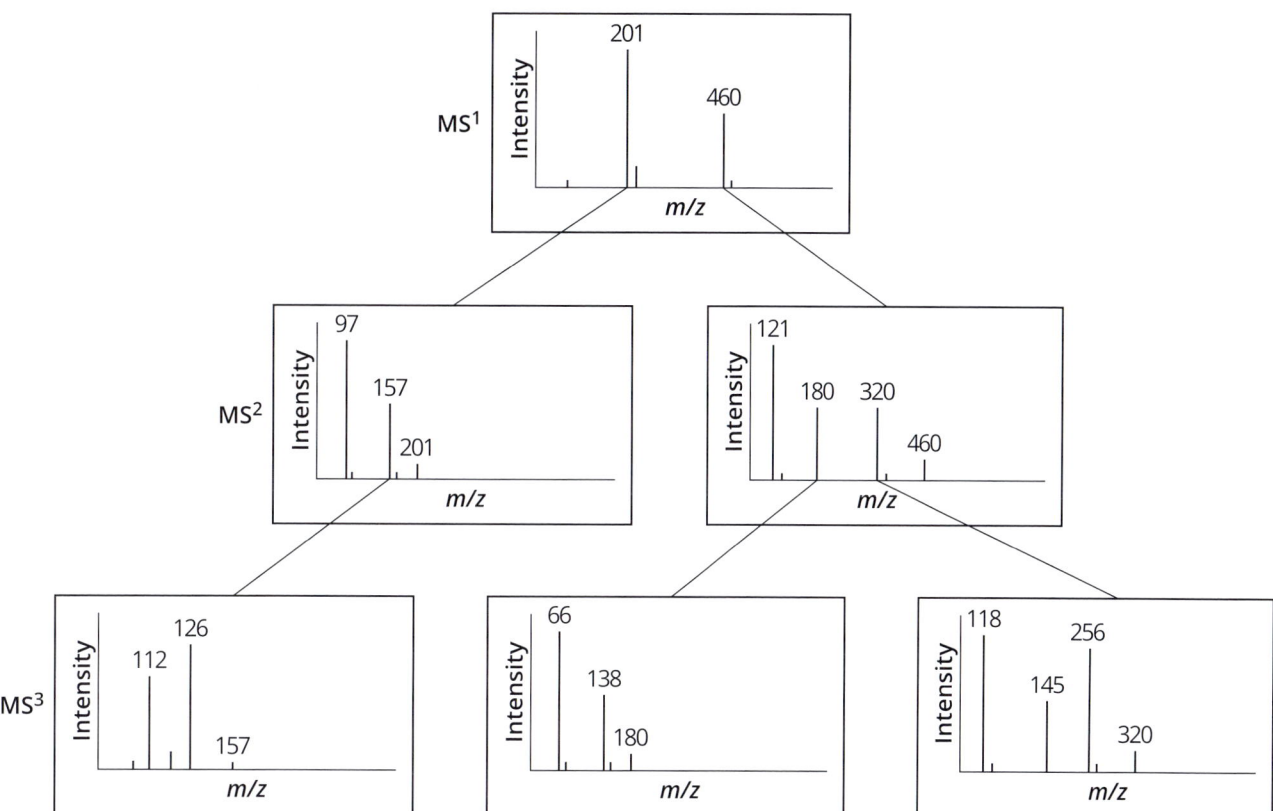

FIGURE 12.7 Resonant excitation CID using ion traps enables MS^n tandem mass spectrometry that provides a fragmentation tree from a single precursor ion.

fragmented. The process can then be repeated through several generations of precursors and products. The letter 'n' in MS^n refers to the number of sequential precursor/product ion fragmentations performed from the initial m/z precursor. In theory, this is not limited but in practice usually 4–5 rounds of fragmentation are possible before the precursors become too low in abundance or small in m/z to successfully fragment further, it is therefore dependent on the size of the initial precursor and its signal intensity. MS^n therefore provides a high degree of control over the investigation of molecular structure and can be highly effective in both structural elucidation and structural confirmation studies.

Both tandem mass spectrometry in space and time provide a range of analytical capabilities dependent on the type of mass analyser combination and the analyser's performance characteristics. Table 12.3 summarizes the more common tandem mass spectrometry configurations. For example, triple quadrupoles provide a highly versatile configuration that provides a range of different tandem mass spectrometry experimental capabilities. This is largely due to the mass filtering ability of the quadrupole analyser. Triple quadrupoles have high analyte selectivity, e.g. the ability to transmit only selected precursor and product ions, and potential for very high sensitivity (signal-to-noise ratio) due to the low background generated by product ion selection prior to detection. However, triple quadrupoles have relatively low resolution and do not provide accurate mass analysis, unlike Q-TOF and Q-Orbitrap configurations which generate high-resolution, high mass accuracy tandem mass spectra.

Although 'triple quads' only use two quadrupole mass analysers to filter and detect ions, their name is derived from the fact that the collision cell itself is also usually a quadrupole—even if this is only used for the fragmentation step and not to select or detect ions specifically. The abbreviation 'QqQ' is used to denote this.

TABLE 12.3 Common mass analyser combinations currently used in tandem mass spectrometry.

Type	Analyser configuration	Abbrev.	Fragmentation resolution	Tandem MS experiment type
MS/MS in space	Triple quadrupole	QqQ	Low	MS/MS
	Quadrupole–Time of Flight	Q-TOF	High	MS/MS
	Q-Orbitrap	Q-Orbitrap	High	MS/MS
	Time of Flight–Time of Flight	TOF-TOF	High	MS/MS
	Quadrupole–Ion cyclotron resonance MS	Q-ICR	High	MS/MS
MS/MS in time	Quadrupole ion traps	QIT	Low	MS/MS & MS^n
In space & time	Linear ion trap–Ion cyclotron resonance	LTQ-ICR	High	MS/MS & MS^n
In space & time	Quadrupole ion trap-Orbitrap	QIT-Orbitrap	High	MS/MS & MS^n

12.5 Mechanisms of fragmentation

Mechanisms of fragmentation in tandem mass spectrometry of protonated and deprotonated molecules differ from those found in EI, mainly due to the lack of an unpaired electron on the charged analyte molecule. Fragmentation of even-electron ions, such as those formed by soft ionization techniques, are mainly driven by the presence of the analyte charge. This can either be retained on the original atom or migrate to a new atom in the molecule during the fragmentation process. Rearrangements are common although, as in EI, this is a unimolecular process. The most common mechanisms in positive ion mode include inductive cleavages, displacement reactions, and rearrangements. In negative ion mode elimination and displacement reactions are common.

Here we show some selected examples of common fragmentation mechanisms by way of introduction. As in the case of EI, if we know where the charge resides on a precursor ion, this can help predict fragmentation mechanisms. The charged adduct will usually be associated with the most basic atom (positive ion mode) or acidic site (negative ion mode) in the non-ionized molecule. When the internal energy of the charged analyte increases above a threshold level (via molecular collisions in the CID fragmentation cell), bonds break, which commonly leads to the formation of a neutral molecule and a charged fragment. Note this is different from EI fragmentation where a charged fragment and a neutral radical are usually formed. Figure 12.8 illustrates the CID fragmentation

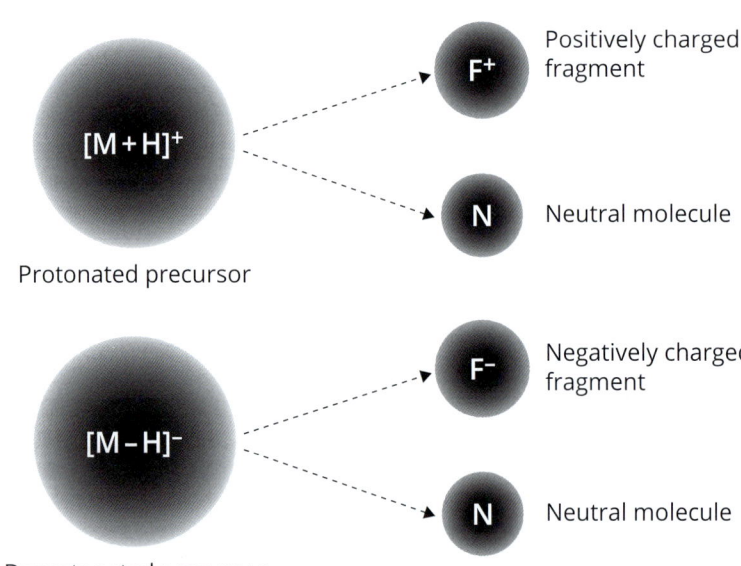

FIGURE 12.8 Protonated and deprotonated precursor ions fragment inside a collision cell to form a neutral molecule and a charged fragment. The proton may be incorporated into the neutral fragment or the charged analyte via migration or rearrangement.

process in both ion modes. Note that the original charged proton that formed the adduct in positive ion mode can be incorporated into the neutral fragment or the charged fragment depending on the mechanism.

There are two basic mechanisms of fragmentation process that occur during CID:

- Charge retention fragmentation (CRF), where the charge remains in the same location.
- Charge migration fragmentation (CMF), where the charge migrates to another atom during fragmentation.

The neutral molecule that is lost in the fragmentation process is often a small molecule such as H_2O, CO, CO_2, SO_2, or NH_3. Specific structures or functional groups can lead to characteristic neutral losses. There is a tendency to produce smaller neutrals, which leaves larger charged fragments better able to stabilize their charge. It is common for a range of different fragmentation pathways to occur for the same precursor, leading to a range of neutrals lost and different fragment ions being present in the same mass spectrum of an analyte.

In general, even-electron ions usually produce fewer but more stable fragments compared to EI fragmentation. Cleavages usually occur at the thermochemically least stable bonds, with longer bonds tending to fragment more easily than shorter bonds. Sometimes, when the internal energy of the charged analyte increases during CID, the proton can become mobile and may migrate to another less basic site in the structure from which fragmentation may be initiated. This 'mobile' proton-initiated fragmentation occurs more often in larger molecules. As in EI the fragmentation process in tandem mass spectrometry is unimolecular, takes place very fast, and does not involve reactions between fragments. This means that fragments only contain atoms that were originally present in the precursor and do not come from reactions with non-precursor fragments via collisions.

- The initial location of the charge is often the most acidic or basic site in the molecule (in positive or negative ion mode respectively).
- Fragmentation mechanisms can be divided into two types: those where the charge is retained on the same atom as the precursor ion (CRF) and those where the charge migrates to another atom prior to fragmentation (CMF).
- Activation can initiate fragmentation at the initial protonation/deprotonation site or another location after proton migration. The actual site of fragmentation can sometimes be difficult to predict.

12.5.1 Charge site-driven cleavage, β-H-rearrangements, and negative ion elimination reactions

Four common fragmentation mechanisms (found in CID and HCD tandem mass spectrometry) are: (a) charge site-driven cleavage (also known as inductive cleavage or '*i*'); (b) heteroatom-assisted inductive cleavage; (c) β-H-rearrangements in positive ion mode; and (d) elimination reactions (*e*) in negative ion mode. The generic mechanisms are illustrated in Figure 12.9.

Figure 12.10 illustrates the charge site-driven cleavage of the beta-blocker Bisoprolol for ESI protonation followed by CID fragmentation.

- In general, protonated and deprotonated analytes, generated via soft ionization, tend to undergo heterolytic bond cleavage during fragmentation.
- β-H-rearrangements commonly occur during fragmentation of protonated and deprotonated molecules.
- Fragmentation patterns tend to be specific less to functional groups (in contrast to EI) and more to structural elements, with the degree of stability of the product ions most often determining the mechanism of fragmentation.

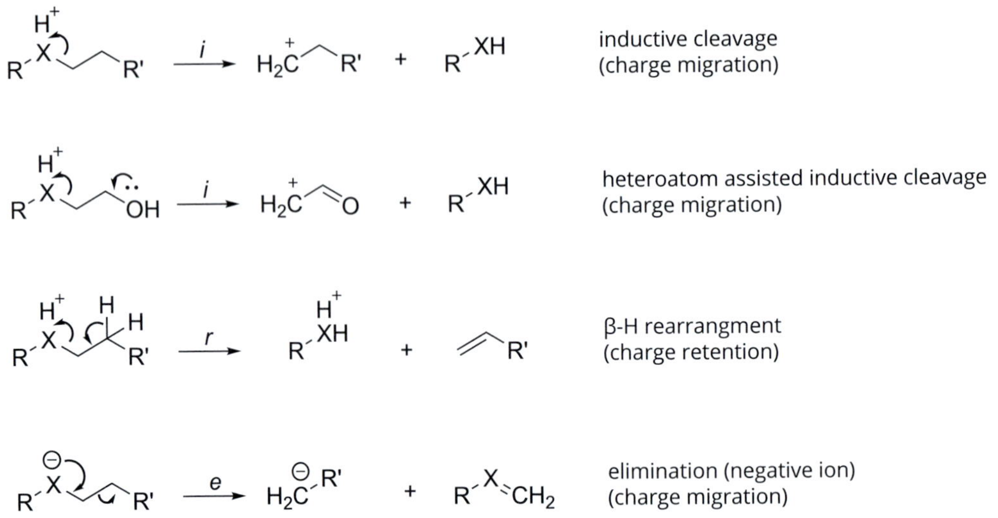

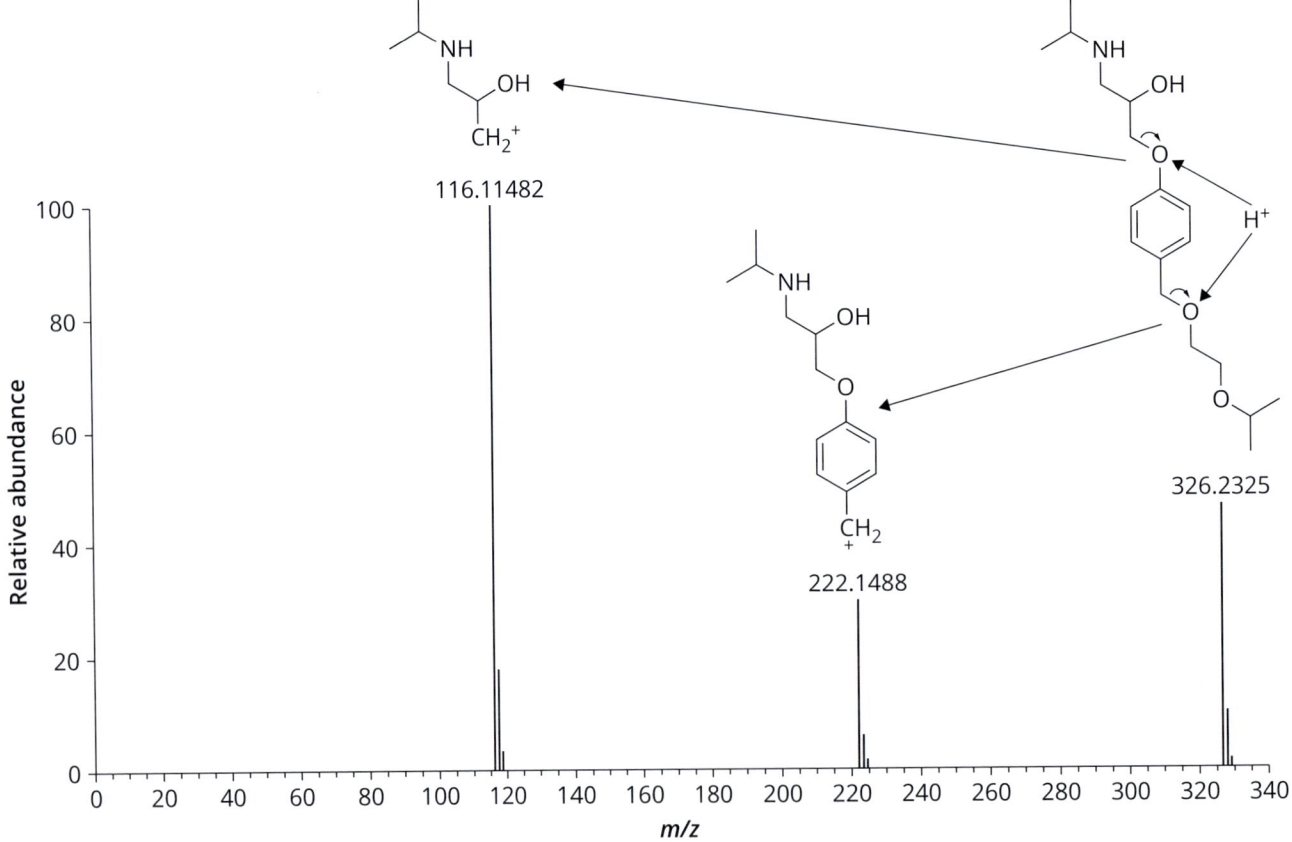

FIGURE 12.9 Heterolytic fragmentation mechanisms for even-electron protonated and deprotonated small molecules. Inductive cleavages (*i*) are very common and can take place in a variety of ways. One of the most common types of rearrangement involves β-H-migration (*r*) and in negative ion mode elimination reactions (*e*) are also often found. A wider range of mechanisms can occur but those illustrated the ones that are most often occur in small molecule CID fragmentation.

FIGURE 12.10 Bisoprolol is a cardio-selective β1-adrenergic blocker used to treat high blood pressure and prevent myocardial infarction. Fragmentation of its [M+H]⁺ adduct by CID leads to two fragmentation products via a simple charge-site driven heterolytic cleavage mechanism based on two equivalent protonation sites.

The loss of neutral molecules in the CID of positively charged protonated adducts is illustrated by cysteine in Figure 12.11. Accurate mass analysis provides the chemical formula for each fragment and supports structural assignment.

In general, there is more complexity in tandem mass spectrometry fragmentation mechanisms compared to EI methods, even though fewer fragments tend to be produced in the mass spectrum. As a result, product ion spectra can be more difficult to predict and interpret. In practice, it is extremely difficult to elucidate full chemical structures from tandem MS spectra alone for anything beyond relatively simple small molecules. Nevertheless, tandem mass spectra can provide crucial structural information in structural elucidation studies and are often most powerful when combined with additional analytical information. Tandem mass spectrometry is typically used for the following purposes:

- For structural confirmation where fragments from an unknown compound in an experimental sample are matched to the fragmentation spectrum of an authentic standard using tandem mass spectrometry.

- As part of a range of tools and approaches to gather structural information; commonly including NMR and other spectroscopic techniques, whose data can, in combination with tandem mass spectrometry, provide a compelling case for structural assignments.

- As a standalone structural elucidation tool when the amount of sample is insufficient for, or amenable to, NMR analysis.

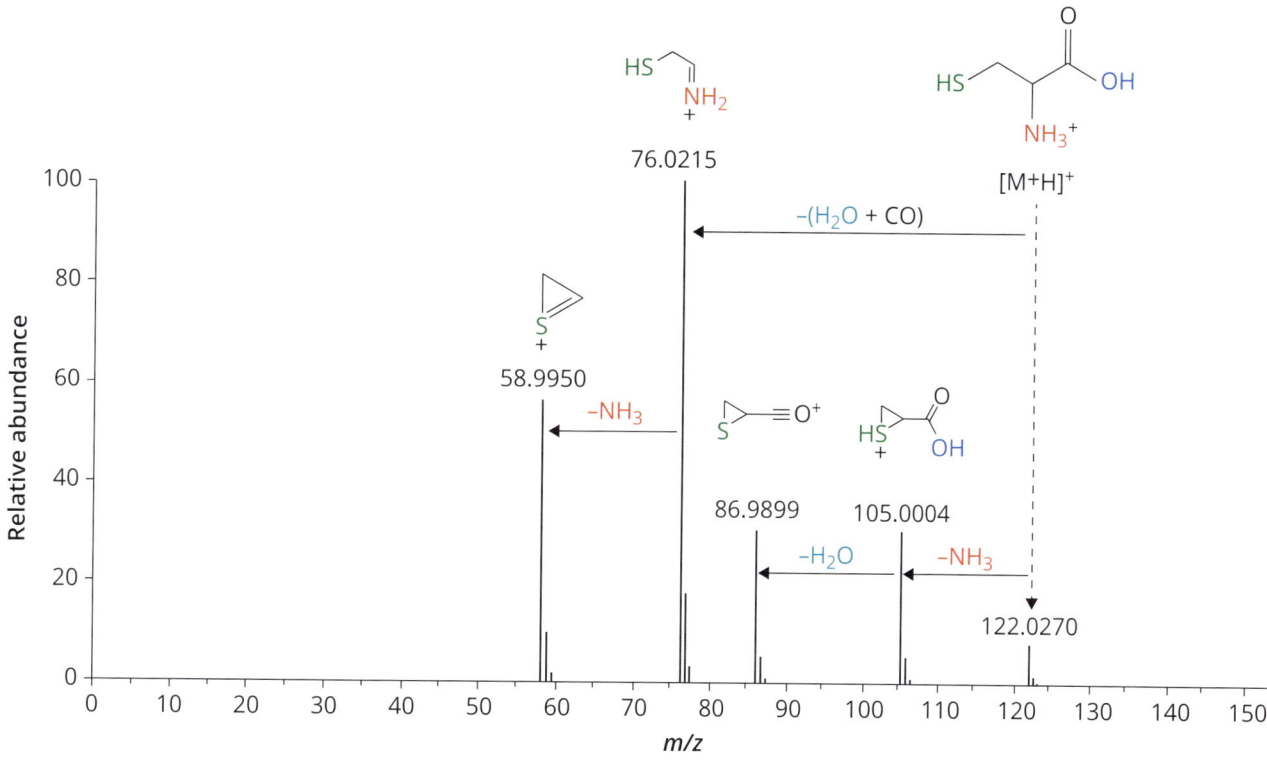

FIGURE 12.11 The CID fragmentation spectrum of the amino acid cysteine shows the different neutral losses and structural rearrangements that can occur via two competing fragmentation pathways. The first is driven by inductive cleavage and rearrangement leading to loss of NH_3 with subsequent loss of H_2O to form the cation shown at m/z 86.9899. The second is driven by loss of H_2O and CO with heterolytic cleavage of the C–C bond adjacent to the carboxyl group.

12.6 Chromatography and other separation techniques in structural elucidation

The development of ESI has enabled analysis of a much wider range of compounds than was previously possible. Another important outcome has been the ability to couple liquid chromatography directly with mass spectrometry (LC-MS). LC-MS has been a very important development in mass spectrometry as it has not only facilitated the analysis of complex mixtures but brings a range of other benefits, including:

- The ability to perform relative and absolute quantification.
- Reduces matrix and ion suppression effects.
- Enhances sensitivity by increasing the signal-to-noise ratio of analyte signals.
- Provides a sample inlet that uses relatively small amounts of sample.
- The ability to separate structural isomers and stereoisomers.
- Is compatible with orthogonal mass spectrometry techniques, including tandem mass spectrometry.

A number of these benefits can be particularly useful in structural confirmation and elucidation studies. Probably the most important of these is providing evidence of structural identity through the matching of retention times for experimental analytes and authentic standards. This is because the chromatographic retention time of an analyte is highly sensitive to chemical structure, and so exact retention time matching can be used to confirm correct structural assignments. In the absence of an authentic standard, the retention characteristics of an analyte can also indicate its relative polarity and therefore reveal clues about its chemical structure.

Compatibility with tandem mass spectrometry experiments (LC-MS/MS) enables structural fragments to be generated from multiple different compounds that are chromatographically separated, for example from a reaction mixture, which can help differentiate reactants from products. Ultraviolet and visible absorbance of the electromagnetic spectrum measurements (e.g. as in UV-VIS spectroscopy) can also be made inline after chromatographic separation, but prior to ionization. These non-destructive measurements provide absorption spectra that can, for example, reveal complementary information about the conjugation state of π-electrons, which can then be correlated with mass spectral fragmentation information to predict the degree of unsaturation of the analyte molecule, as well as potentially where multiple bonds are located. This is complementary to the saturation index which can be calculated for the analyte from the predicted chemical formula (see Chapter 7 Checklist, point 5).

> Fragmentation of multiple analytes in a reaction mixture can be performed using LC-MS/MS.

Finally, as part of the processes of structural elucidation from complex mixtures, LC-MS can be used to physically separate, detect, and isolate analytes of interest. A small fraction of the eluent from the LC system can be collected using the analyte's retention time as a guide. This provides a sample containing the isolated analyte of interest. Automated repeat collection of the eluent can be used to purify sufficient quantity of sample for use in NMR and IR experiments.

12.7 Ion mobility-mass spectrometry (IM-MS)

Ion mobility spectrometry (IMS) is a technique that provides gas-phase separation of ions (or charged molecules) based on a compound's size, shape, and charge rather than on *m/z* alone. IMS separation systems are now built into some tandem mass spectrometers and can generate *collisional cross-section* values for charged analytes (IM-MS).

- IM-MS is a fast molecular separation technique with *m/z* detection.
- *Collisional cross-section* (CCS) values from ion-mobility experiments provide a universal value that can help differentiate structural isomers.

There are a number of different combinations of IMS and MS technologies (IM-MS) available which can provide different capabilities. IM-MS is not yet routinely used in small molecule structural elucidation studies but can be particularly useful where different structural forms of the same compound may be present. CCS values can be calculated from the drift time of a specific chemical structure and therefore represent a universal property for a particular compound's 3D structure. CCS has great potential for identifying chemical structures based on a library search approach. IM-MS can therefore play a useful role in supporting structural confirmation and elucidation studies. Importantly, IM-MS is compatible with the analysis of complex mixtures using chromatographic separations combined with mass spectrometry (e.g. LC-ESI-IM-MS/MS).

12.8 Databases approaches and machine-learning in compound identification studies

Computational approaches have played a major role in the interpretation of mass spectra ever since the development of computing. When it comes to the interpretation of chemical identity and structure, there are three major areas where computational approaches can be particularly useful.

The first is calculating the chemical formula, which we have discussed earlier in section 12.3.1. This is a relatively straightforward prediction of the chemical formula constrained by the accurate mass measurement of precursor and product ions.

The second area is the automated searching of compound databases and mass spectral libraries to help identify structural possibilities for candidate chemical formula predictions. For example, one approach is to search large compound databases such as *PubChem* (https://pubchem.ncbi.nlm.nih.gov/) and *ChemSpider* (www.chemspider.com), which contain millions of structures, to look for matches to chemical formula predictions. This approach will not lead to a single candidate structure but can generate a list of possible chemical structures, narrowing down the structural options considerably. This list can then be reduced further by incorporating orthogonal information gathered, for example from prior knowledge about the chemistry or additional analytical information. In contrast to most compound libraries, *mass spectral libraries* also contain experimental mass spectra and an increasing number of databases also include tandem mass spectra. Such databases currently include *METLIN*, *MassBank* (https://massbank.eu/MassBank/), *m/zCloud* (https://www.mzcloud.org/), and *NIST* (https://chemdata.nist.gov/). These spectral libraries allow you to search for structurally-specific matches using tandem mass spectra.

It is, of course, only possible to find matches if the compound has previously been characterized and uploaded to the database. These database approaches are therefore not suitable for characterizing novel structures or known structures not in the database being used. Whilst there are many different databases available, none of them are fully comprehensive. In addition, although EI fragmentation mass spectra are highly reproducible, irrespective of the specific instrument type, CID fragmentation of protonated and deprotonated species can vary significantly in the abundance of fragment ions and sometimes in which are present, for example when different instrument types and settings are used. The intensity profile of tandem mass spectra are, therefore, not easily matched even when the correct compound is present in a database. An additional confounding factor is that small molecules, representing different chemical structures that share common functional groups, can often provide very similar fragmentation spectra. So, while databases can be extremely helpful for narrowing down chemical structures, you should consider their results with care and always verify potential identifications using authentic standards where possible.

> Databases can be a useful tool to support structural identification, but care must be taken when using them in isolation to assign structural identity.

A third and final area of computational interest in relation to structural studies is the development of *in silico* fragmentation and computational approaches, including machine-learning. These are applicable even when experimental spectra are not available in databases or authentic standards cannot be obtained. *In silico* fragmentation of candidate structures can be used to narrow down potential structural matches. There are two main approaches in this respect:

- Rule-based bond-breaking computational methods which tend to be specific to certain types of fragmentation approach (examples of current software packages include *Mass Frontier* and *ACD/MS Fragmenter*).
- Systematic bond-breaking approaches which essentially break all possible bonds without discrimination (e.g. *MetFrag*).

Systematic bond-breaking methods create lots of redundancy but don't suffer from rule-based bias.

> *In silico* predicted fragmentation patterns can be compared to experimental product ion spectra to determine the closest matches to proposed structures.

Finally, it should be noted that generating algorithms which work equally well for all types and sizes of molecule is technically still not possible, and accurately predicting the relative ion intensities of product ion spectra remains a major challenge. Orthogonal information such as ion-mobility-derived CCS values, product ion spectral database-matching, and generation of *in silico* fragmentation patterns, based on putative structural formulae, are all useful tools. Whilst none of them tends to be sufficient when used in isolation, they can provide convincing evidence for chemical structure when combined. With the development of powerful machine-learning algorithms, *in silico* fragmentation tools are likely to become considerably more effective in the future. It may not be too long before more accurate fragmentation prediction capabilities are developed that can routinely be used with large compound databases.

12.9 Worked example: Analysis of indomethacin

In this section we provide a step-by-step approach to analysing soft ionization, high-resolution mass spectra. As an example, we use the pharmaceutical drug indomethacin, the MS spectrum of which is shown in Figure 12.12. The first step is to determine what type of adducts are present and their chemical composition. In common with EI spectra, the highest *m/z* value will usually represent the unfragmented charged analyte, but it may be in multiple adduct forms. For example, look out for dimers or

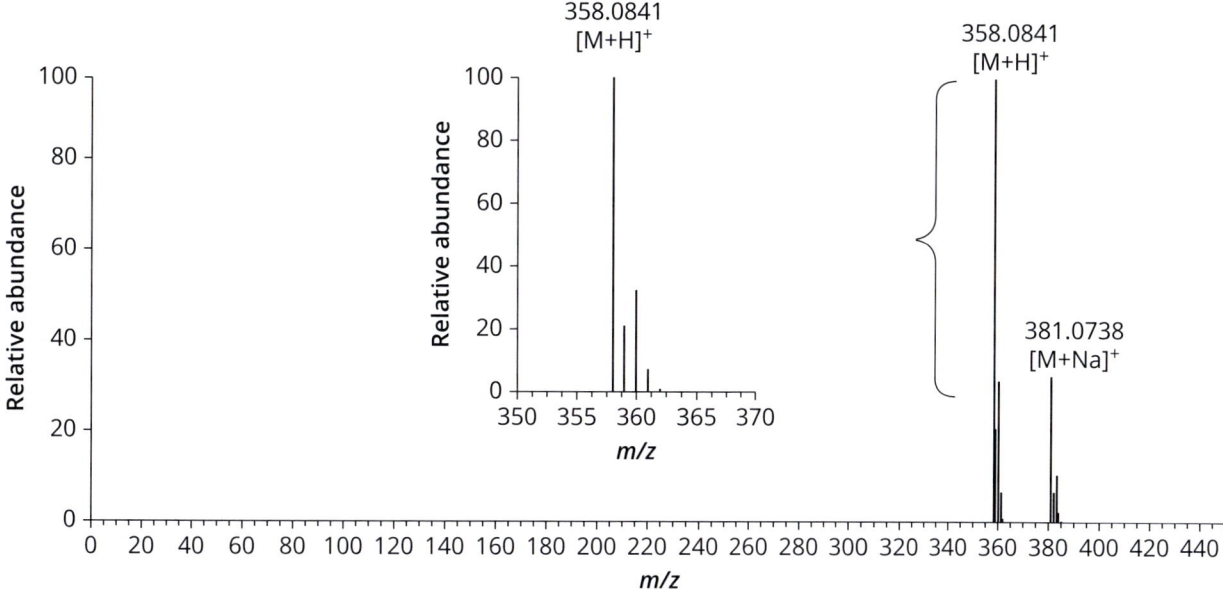

FIGURE 12.12 The ESI mass spectrum of the pharmaceutical drug indomethacin ($C_{19}H_{16}CINO_4$) showing both protonated and sodiated adducts. The inset shows the isotope pattern for the [M+H]$^+$ adduct in more detail, revealing a distinctive A+2 pattern which is commensurate with a chlorine atom being present.

even multimers of the analyte at the highest *m/z* value and/or cationized or anionized adducts other than [M+H]$^+$ or [M–H]$^-$ (depending on the ion mode). The spectrum in Figure 12.12 illustrates the [M+H]$^+$ and [M+Na]$^+$ adduct peaks produced by electrospray ionization with 4 decimal place mass accuracy.

> As different adducts have logical mass differences, and a relatively small number of types are common, careful examination of the mass spectrum can usually identify those most likely to be present. This allows us to deduce the mass of the analyte with confidence.

The 4 decimal place mass accuracy of the *m/z* values shown in Figure 12.12 makes it straightforward to confirm the presence of both the protonated and sodiated adducts by calculation of the mass difference. The easiest way to do this is to use a *chemical formula calculator* which uses a library of the accurate monoisotopic masses for all chemical elements and their isotopes to 5 or more decimal place accuracy. As discussed in section 12.3.1, these calculators simply match all combinations of the elemental masses present and report those combinations that are within the stated mass accuracy tolerance of the *m/z* measurement. A tolerance of 5 parts per million (ppm) error or lower in the mass measurement is a generally accepted threshold for chemical formula prediction. When doing this, remember to apply any information you may have that will constrain the elemental possibilities; for example, only make available those elements found in reactants used in a chemical synthesis. The ppm error can be calculated using the following equation:

$$\left(\left(\text{Theoretical mass} - \text{Measured mass}\right)/\text{Theoretical mass}\right)\times\left(1\times 10^6\right)$$

Note that the 5 ppm error threshold does not guarantee that a single chemical formula will be predicted. Theoretically the number of possible formulae for a specific ppm error threshold increases with mass, which means that, as mass increases, it also becomes progressively more difficult to generate a unique chemical formula unless the ppm error of the measurement is reduced.

Once the analyte masses and adducts have been identified, the next step is to calculate possible chemical formulae based on the accurate mass measurement and the

TABLE 12.4 Chemical formula predictions at different mass accuracies for the *m/z* value 358.0841. The possible chemical formulae can be further narrowed down by recognition of unusual isotope patterns and predicting the number of carbon atoms from the isotope pattern.

	5 ppm mass accuracy		3 ppm mass accuracy		2 ppm mass accuracy		1 ppm mass accuracy	
	Predicted formula	*m/z* value (4dp)	Predicted formula	*m/z* value (4dp)	Predicted formula	*m/z* value (4dp)	Predicted formula	*m/z* value (4dp)
1	$C_8H_{16}O_{11}N_5$	358.08463	$C_8H16O_{11}N_5$	358.08463	$C_8H_{16}O_{11}N_5$	358.08463	$C_8H_{16}O_{11}N_5$	358.08463
2	$C_{19}H_{17}O_4NCl$	358.08461	$C_{19}H_{17}O_4NCl$	358.08461	$C_{19}H_{17}O_4NCl$	358.08461	$C_{19}H_{17}O_4NCl$	358.08461
3	$C_{21}H_8N_7$	358.08411	$C_{21}H_8N_7$	358.08411	$C_{21}H_8N_7$	358.08411		
4	$C_3H_{17}O_9N_9Cl$	358.08377	$C_3H_{17}O_9N_9Cl$	358.08377				
5	$C_{14}H_{18}O_2N_5Cl_2$	358.08375	$C_{14}H_{18}O_2N_5Cl_2$	358.08375				
6	$C_{13}H_{23}O_2N_3Cl_3$	358.08558	$C_{13}H_{23}O_2N_3Cl_3$	358.08558				
7	$C_{20}H_{13}N_5Cl$	358.08594						
8	$C_9H_{12}O_7N_9$	358.08596						
9	$C_9H_{19}N_9Cl_3$	358.08289						

calibration tolerance of the instrument (note that the latter may vary by analyser type). We can demonstrate the impact of different mass accuracies by setting different ppm errors for the *m/z* measurement. Table 12.4 illustrates this by providing chemical formula predictions for the [M+H]⁺ value of 358.0841 at four different ppm error tolerances—5 ppm, 3 ppm, 2 ppm, and 1 ppm, with the underlined values showing where the uncertainty is located in the mass measurement. For 5 ppm and 3 ppm mass accuracy, the last three out of five decimal places are uncertain whilst it is the last two decimal places or the last decimal place out of five for 2 ppm and 1 ppm mass accuracy respectively. These demonstrate that there is a significant decrease in the number of possible chemical formulae as the mass measurement error decreases. Even at 1 ppm there are still two possible formulae in this example.

To differentiate between these two options, we can apply some of the approaches used for the interpretation of EI mass spectra discussed in Chapter 5. For example, we can estimate the number of carbon atoms in the molecule based on the ratio of the '¹³C' isotope peak relative to the '¹²C' peak (see section 5.4.3). Note the algorithms used for chemical formula prediction are usually based on calculation from the *m/z* value alone and do not take into consideration additional chemical information inherent in the mass spectrum such as relative isotope peak abundances. Similarly, we can look for characteristic A+2 isotope peaks that show the presence of elements such as Cl, Br, and S (see section 5.2.1). The inset in Figure 12.12 shows the isotope pattern for an [M+H]⁺ adduct in more detail, revealing the relative height of the M+1 peak (~21%, which suggests that there are approximately 19 carbon atoms in the compound) and a distinctive A+2 pattern commensurate with the presence of a single chlorine atom. When constraining the possible formulae to include 19 carbon atoms and the presence of a single chlorine atom in the formula, the correct chemical formula $C_{19}H_{17}O_4NCl$ is predicted uniquely at 3 ppm accuracy.

- When there is more than one possible formula predicted from an accurate mass measurement, it is usually possible to narrow down the options by employing simple spectral analysis approaches, such as predicting the number of carbon atoms, identifying A+2 elements, and applying the nitrogen rule (see section 5.4.2).

- The number of matching formulae calculated for a fixed mass accuracy increases exponentially with increasing *m/z* value.

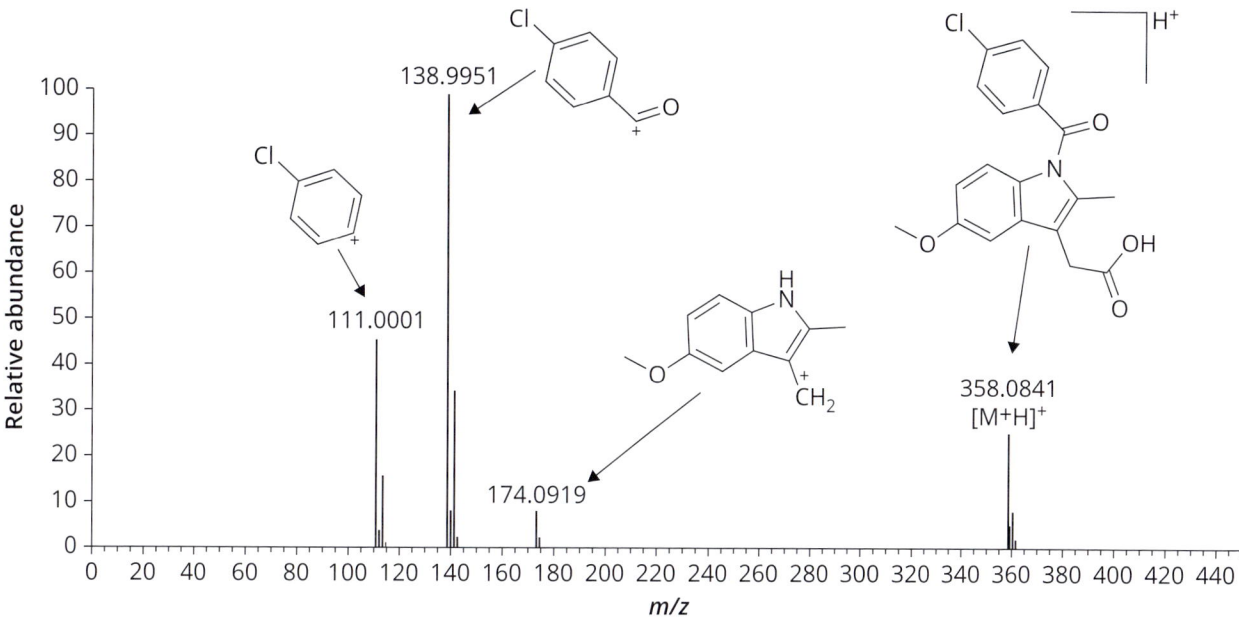

FIGURE 12.13 The tandem mass spectrometry CID fragmentation spectrum of indomethacin. The fragmentation pattern can be explained by the positive charge residing on the nitrogen atom in the cyclopentane ring after ESI. The base peak at m/z 138.9951 can be explained by heterolytic cleavage of the N–C bond adjacent to the carbonyl during activation, via a charge site-driven cleavage mechanism (inductive cleavage) and further inductive cleavage to give m/z 111.0001. The least abundant peak at m/z 174.0919 results from a multi-stage fragmentation with migration of the charge.

At this stage we have a predicted chemical formula, but the structural possibilities are still very diverse without further analytical information. The tandem mass spectrometry CID spectrum for this compound is shown in Figure 12.13. The protonated indomethacin produces a number of fragments via charge site-driven cleavage which convey some useful structural information. For example, the base peak at m/z 138.0841 and the lowest mass peak at m/z 111.0001 clearly still contain a single chlorine atom based on the recognizable A+2 isotope pattern for Cl. It is also clear that the predominant fragments are not due to loss of a simple neutral, such as H_2O, CO_2, and CO, and that inductive cleavage around the chlorine atom is also not a dominant mechanism. This should give us some pause to consider the implications from a structural perspective. For example, we might conclude the presence of a highly stable structure linked directly to the chlorine atom to explain the presence of chlorine in multiple low-mass fragments. This is indeed the case due to the highly stable aromatic ring on which Cl is a direct substituent in indomethacin, making it energetically unfavourable to fragment around the chorine atom. This is further supported by calculating the degree of unsaturation (DU) using the predicted formula (DU calculations are introduced in Chapter 7 Checklist, point 5). This example illustrates the way in which tandem mass spectra can provide information about chemical structure. Nevertheless, very rarely are we able to identify the full chemical structure, other than for very simple molecules, without additional information.

12.10 **Worked example for a 'unknown' compound**

The following worked example illustrates the interpretation of positive ion electrospray mass spectra produced using a high-resolution mass spectrometer calibrated to within 5 ppm mass accuracy. We will treat the compound as an 'unknown' but to make it easier to link structural information with mass spectral information as we work through the problem, we provide both the mass spectra and chemical structure of the analyte. The steps taken for the analysis of the spectra are similar to that for EI (Chapter 5, section 5.5): we aim to piece together the various spectral information, or 'clues', in order to eliminate elemental and structural impossibilities, and ideally arrive at a single structural interpretation. In practice, how close we can get to proposing a unique chemical structure often depends on the analyte's structural complexity, its chemical composition, and what additional physiochemical information is available.

Analyte information	
Chemical formula	$C_5H_{10}N_2O_3$
Structural formula	

MS Peak list	Rel. Intensity	MS/MS Peak list	Rel. Intensity
m/z 147.0758	100	m/z 56.0496	2.7
m/z 148.0799	5.41	m/z 84.0433	80.9
m/z 149.0876	0.62	m/z 101.0707	7.0
		m/z 130.0498	100
		m/z 147.0758	6.5

12.10.1 Analysis of the data

a)   The base peak in the mass spectrum in positive ion mode (Figure 12.14) is at m/z 147.0758, and there are no peaks that suggest additional adducts. We can therefore speculate that the signal at 147.0758 is likely to represent the protonated analyte. The isotope peaks are approximately 1 Da apart, so we are looking at the spectrum of a singly charged analyte. There are no apparent 'A+2' elements present based on visual inspection of the isotope composition. Chemical formula prediction at 5 ppm (using the elements C, H, O, N, P, and F) suggests the following possible formulae:

- $C_5H_{10}O_3N_2$ (−4.2 ppm)
- $C_3H_8O_2N_5$ (+4.9 ppm)

Note that it is worth restricting searches to only the elements that are possible based on isotope patterns and prior knowledge of the chemical system. In this case, our inspection of the isotopic pattern represented by the mass spectral peaks suggests that S, B, Cl, Br, and other A+2/heavy metals are not present, so we exclude these from the formula prediction tool calculation.

b)   To help differentiate the predicted chemical formulae possibilities, we can estimate the number of carbon atoms (abundance of the M+1 isotope peak relative to the base peak (M)) and arrive at a prediction of between 4 and 6 carbon atoms (5.4/1.1 = 4.9), suggesting that the formula $C_5H_{10}O_3N_2$ is the most likely to be correct. Two additional pieces of information can help us confirm this prediction. The first is to compare the full isotope structure from the experimental sample with that of the predicted isotope patterns for the two possible formulae. Figure 12.16 shows that in this case it is possible to confidently match one of the predicted isotope patterns to the experimental isotope pattern by visual inspection, with $C_5H_{10}O_3N_2$ providing the closest match.

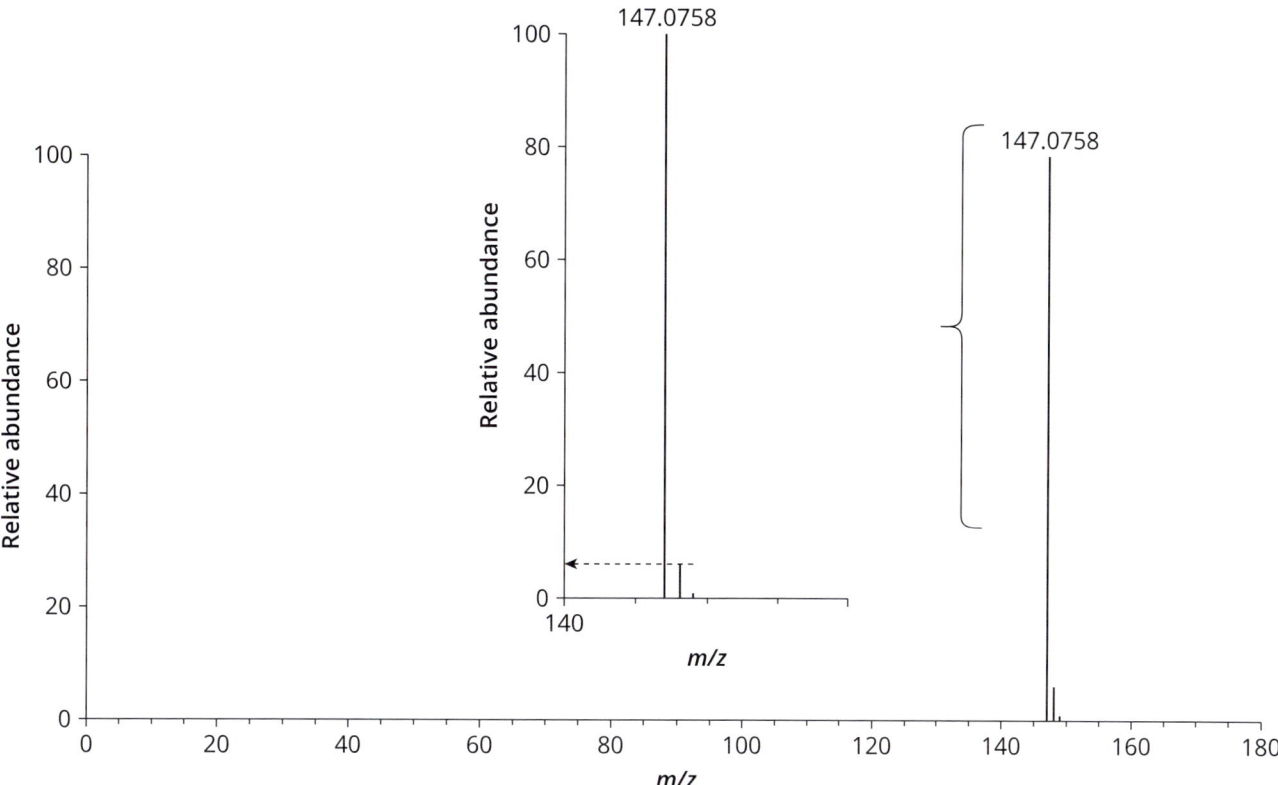

FIGURE 12.14 MS spectrum of the example analyte.

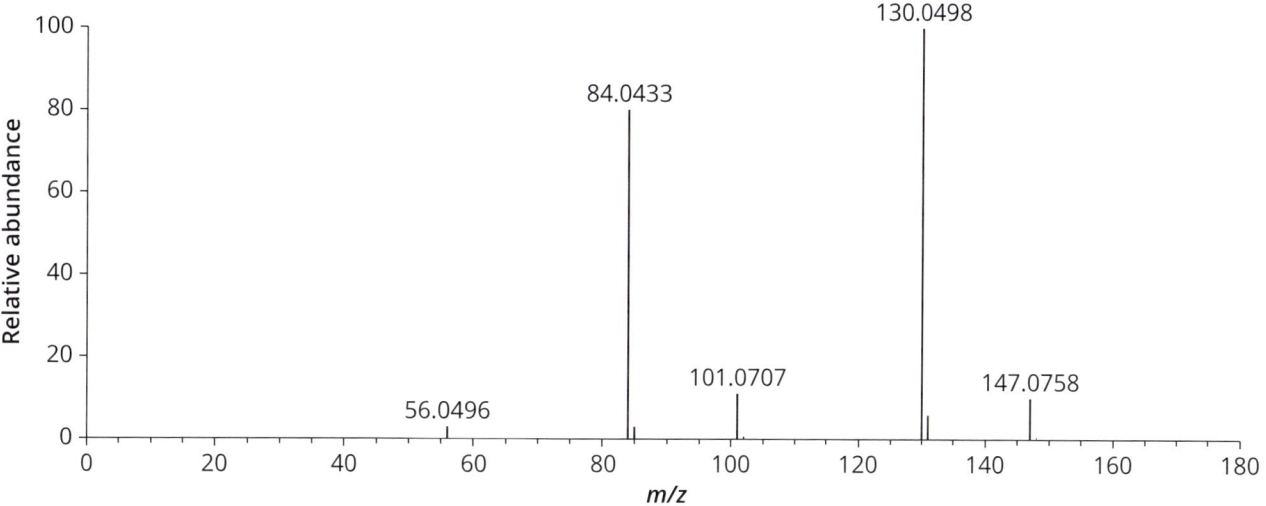

FIGURE 12.15 MS/MS spectrum derived from tandem mass spectral analysis of the precursor ion at *m/z* 147.0758.

For a more accurate confirmation the simulated peak lists (*m/z* values, intensities, and predicted isotopic composition) can also be compared directly in tabulated form. Their relative intensities provide a highly selective match to a chemical formula (note that if high-resolution instruments are used, additional isotope fine structure may be accessible, including the natural abundance ^{15}N and ^{2}H patterns, which provide an extra degree of specificity). Finally, another way to determine the correct formula is to apply the nitrogen rule. In this example, the rule is applied to even-electron ions (recall: all protonated or deprotonated molecules are even-electron ions) and as the

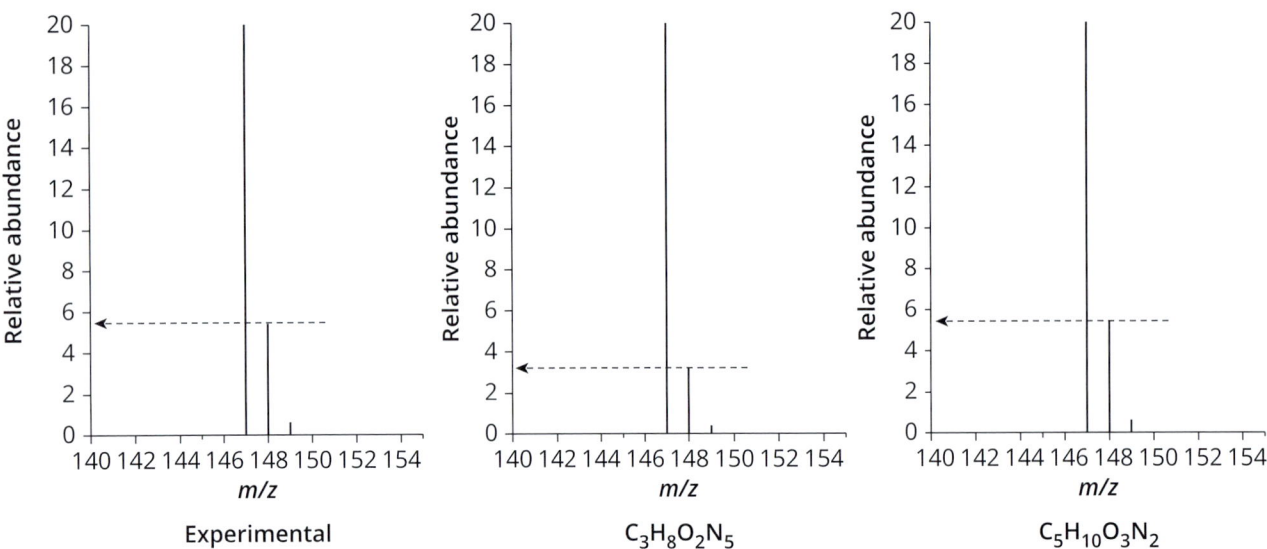

FIGURE 12.16 Comparison of the mass spectral isotope patterns from the experimental 'unknown' and those predicted for the two possible chemical formulae. $C_5H_{10}O_3N_2$ shows the closest match.

integer m/z value is odd, this predicts either an even number of nitrogen atoms in the experimental molecule or none. As both predicted formulae contain nitrogen, but only $C_5H_{10}O_3N_2$ contains an even number of nitrogen atoms, $C_3H_8O_2N_5$ can now be confidently discarded as an option.

c) After identifying the chemical formula, we can next predict the degree of unsaturation (DU) of the molecule (see Chapter 7, Checklist, point 5). For the formula $C_5H_{10}O_3N_2$, we calculate a value of 2. In theory this could indicate the presence of two double bonds, one ring structure and a double bond, or two ring structures. Given we predicted that the molecule only contains 5 carbon atoms, it cannot contain two ring structures. So, at least one double bond is likely either with an additional ring structure or more likely just two double bonds, given the relatively small size of the molecule based on its m/z value.

d) Let us next turn to the product ion spectrum (Figure 12.15) and perform chemical formula predictions for each of the fragments and estimate the chemical composition of the neutral losses. The results are summarized in Table 12.5.

The four main product ions produce unambiguous chemical composition information despite the higher ppm threshold used of 12 ppm (which can be common for tandem mass spectra). This is in part due to the fact that the chemical composition

TABLE 12.5 Chemical formula predictions and degree of unsaturation predicted for the compounds represented by the product ions of precursor 147.0758.

Product ions (precursor: 147.0758)	Predicted formulae (12 ppm)*	Rings & double bonds (product ion)	Neutral loss (predicted formula)	Rings & double bonds (Neutral loss)
130.0498	$C_5H_8O_3N$	2	NH_3 (17.0259)	0
101.0707	$C_4H_9ON_2$	1	CH_2O_2 (46.0051)	1
84.0433	C_4H_6ON	2	CH_5O_2N (63.03250)	0
56.0496	C_3H_6N	1	$C_2H_5O_3N$ (91.02620)	1

*Product ion spectra from CID fragmentation may be less accurate than precursor m/z values, and 12 ppm is often used here instead of 5 ppm accordingly.

is constrained by the precursor ion's composition and, as the molecular weight decreases, there are generally fewer chemical formula possibilities for a set mass accuracy threshold. Due to the possibility of charge migration and complex rearrangements, it is difficult to speculate on product ion structures despite this specificity. However, in combination with the degree of unsaturation predicted, this information helps us narrow down the structural possibilities.

e) We know that the loss of neutral small molecules is common in ESI fragmentation, so next we can look at the mass change from the precursor mass to each of the fragments and use accurate mass-based formula prediction to help identify these losses (see Table 12.5). Loss of NH_3 is the predominant fragment (base peak), confirmed by the accurate mass, which implies that the structure contains at least an NH_2 group. Loss of NH_3 often occurs where a primary amine or amide is the site of proton adduct formation. The fragmentation can occur via inductive cleavage as well as other mechanisms. The m/z value of 101.0707 represents a loss of 46.0051, which is equivalent to the two neutral small molecules $CO+H_2O$. Similarly, the m/z 84.0433 peak represents NH_3+CO+H_2O. The loss of $CO+H_2O$ is common for carboxylic acids, and so we have further evidence suggesting that the compound contains a carboxyl group and an NH_2 group.

f) At this stage, information about the type of analyte being investigated can help us determine the most appropriate way forward. For example, if the analyte is known to be a natural product or a previously identified synthetic product, then searching online databases can be useful (for more information about mass spectral databases see Appendix 1). ChemSpider (www.chemspider.com) reported 510 possible chemical structures for the formula $C_5H_{10}O_3N_2$—a dauntingly large number to investigate. However, a MassBank (https://massbank.eu/MassBank/) search using precursor *and* product ion m/z values narrowed down the number down to 110, with the highest ranked structure correctly predicting the analyte as *glutamine*. A Human Metabolome Database (www.hmdb.ca) search using the predicted formula ($C_5H_{10}O_3N_2$) led to four possible metabolites, one of which was glutamine.

g) The next step is to analyse a small number of selected standards of the highest-ranked structures from the spectral database searches, which in this case would include glutamine. The chromatographic retention time and CID fragmentation pattern (and potentially CCS values if ion-mobility MS was available) could then be used for confirmation of structural identity. A useful approach is to run an authentic standard separately, in addition to spiking it into the analyte sample. This will demonstrate that the analyte and standard have the same chromatographic retention time, which provides strong evidence for structural similarity, in addition to matching the fragmentation pattern in the tandem mass spectrum.

12.11 Chapter summary

In this chapter, we have discussed the interpretation of mass spectra generated by soft ionization using high-resolution mass spectrometers, focusing on protonated and deprotonated molecules. We have seen how soft ionization produces very little fragmentation, which enhances the signal-to-noise ratio for the analyte signal and provides a more sensitive and flexible analytical approach compared to EI. Analyte identification and structural assignment using soft ionization mass spectra are often most successful when independent analytical information and a priori knowledge are logically combined. Useful information can include chemical formula predication via accurate mass analysis, assessment of additional mass spectral information such as the presence of diagnostic isotopes, prediction of the number of carbon atoms, prediction of the degree of unsaturation, interpretation of structural components from product ion spectra, matching chromatographic retention times with standards, UV-vis absorbance, and CCS value matching with databases. Tandem mass spectrometry methods, such as collision-induced dissociation, can be used to produce fragment ions with more control than EI, increasing selectivity and sensitivity. The mechanisms of fragmentation can be more complex for protonated and deprotonated molecules, despite the tandem mass spectra showing fewer peaks and often looking simpler than for EI. Finally, data from the analysis of an authentic standard of the predicted compound is often essential for confirming assignment of the correct structure. Where possible, additional analysis, in particular by NMR and sometimes IR, can play a critical role in confirming MS-based assignments and is essential for elucidation of novel structures.

APPENDIX 1

Data tables

These data tables are provided as reference material that you may find useful when approaching structure elucidation yourself. As stressed a number of times in this book, we do not recommend becoming overly focused on referring to precise values in these tables (especially the NMR and IR data). Instead, we suggest that you consider these tables as guides to help you develop an understanding of the regions in which different chemical groups are likely to give responses. They contain sufficient detail to assist in your initial assessment of spectroscopic features.

Appendix 1.1 **NMR spectroscopy**

Table A.1 and Table A.2 provide a guide to the *typical* locations of the responses of chemical groups in ^{1}H and ^{13}C spectra respectively. However, remember that detailed structural features may cause peaks to move outside these common ranges, so these values are indicative rather than definitive.

More detailed NMR chemical shift information can nowadays best be obtained from chemical shift prediction tools rather than tabulated data. Some common commercial chemical structure drawing apps also contain rudimentary prediction capabilities, which

TABLE A.1 ^{1}H chemical shift correlations

The ranges reflect the *typical* shift regions observed for the groups indicated, but remember that specific structural features may cause resonances to fall outside these ranges.

Saturated Environment*	Shift range/ppm	Unsaturated Environment	Shift range/ppm
CH–O	3.2–3.9	=CH–O	5.8–6.4
CH–Cl	3.2–3.8	=CH–Cl	5.8–6.4
CH–N	2.3–2.9	=CH–N	5.5–6.1
CH–C	0.8–1.8	=CH–C	4.8–5.5
CH–C=C	1.8–2.5	=CH–C=C	5.5–5.9
CH–C=O	2.3–2.9	=CH–C=O	5.9–6.2
CH–Ar	2.5–3.0	=CH–Ar	6.1–7.0
C≡CH	1.8–3.1	Aromatic =CH	6.5–8.5
Acid COOH	10.0–12.0	Aldehyde CH=O	9.5–10.0

*For alkanes it is generally the case that a greater degree of substitution at a carbon will lead to an increase in the attached proton chemical shift. In a series such as R_1R_2**CH**–X (where X represents any functional group or heteroatom), subsequently replacing each R alkyl group with a proton will cause the **CH** proton shift to decrease by ~ 0.3 ppm with each replacement. Thus, within any defined CH$_n$–X chemical shift range, CH–X chemical shifts will tend to be higher than CH$_2$–X, which in turn are higher than CH$_3$X shifts, as illustrated by the simple example below where X= OH.

$$H_3C{\sim}OH \qquad H_3C{-}CH_2{-}OH \qquad \underset{H_3C{-}CH{-}OH}{\overset{CH_3}{|}}$$

3.4 ppm 3.7 ppm 4.0 ppm

TABLE A.2 ^{13}C chemical shift correlations

The ranges reflect the *typical* shift regions observed for the groups indicated, but remember that specific structural features may cause resonances to fall outside these ranges.

Saturated Environment*	Shift range/ppm	Unsaturated Environment	Shift range/ppm	Carbonyl	Shift range/ppm
C–O	55–75	=C–O	140–155	Ketone R$_2$C=O	205–220
C–Cl	45–55	=C–Cl	120–130	Aldehyde RHC=O	190–205
C–N	40–55	=C–N	110–140	Carboxylic acid –COOH	170–180
C–C	10–45	=C–C	110–140	Ester –COOR	165–180
C–C=C/O	30–55	Aromatic =C	120–160	Amide –CONH/R	165–180
		Alkyne C≡C	70–90		

*For alkanes it is generally the case that a greater degree of substitution at a carbon will lead to an increase in the carbon chemical shift, in a manner similar to that for protons noted in Table A.1. In a series such as R$_1$R$_2$R$_3$C–X (where X represents any functional group or heteroatom), subsequently replacing each R alkyl group with a proton will cause the **C** carbon shift to decrease with each replacement. Thus, within any CH$_n$–X chemical shift range, the expected shift sequence would follow **C–X > CH–X > CH$_2$–X > CH$_3$–X** as illustrated by the example below where X= OH and all the alkyl groups are methyl groups.

$$H_3C{\sim}OH \qquad H_3C{-}CH_2{-}OH \qquad \underset{H_3C{-}CH{-}OH}{\overset{CH_3}{|}} \qquad \underset{H_3C{-}C{-}OH}{\overset{H_3C\quad CH_3}{|}}$$

50 ppm 58 ppm 64 ppm 70 ppm

can provide useful guides. Some freely available online resources that enable you to predict shifts from structures are listed here and can be valuable educational tools:

- Spectral Database for Organic Compounds (SDBS, National Institute of Advanced Industrial Science and Technology, Japan) has data from MS, IR, ^{1}H and ^{13}C NMR spectra: https://sdbs.db.aist.go.jp.
- NMR Shift Database (nmrshiftdb2) is a web database of organic structures and their NMR spectra: https://nmrshiftdb.nmr.uni-koeln.de/.

Nevertheless, bear in mind that these databases provide *predictions* and will rarely provide exact matches to experimental data. Note especially that, in general, the prediction of carbon chemical shifts tends to be more accurate than that of proton shifts (protons tend to be exposed and hence their shifts are more influenced by solvation effects and intermolecular interactions, as well as by stereochemical features within a structure).

Appendix 1.2 **Infrared spectroscopy**

The more useful IR absorptions commonly observed in routine structure characterization are summarized in Table A.3. Extensive databases of IR spectra can be found in the *Spectral Database for Organic Compounds (SDBS)* noted in Appendix 1.1: https://sdbs.db.aist.go.jp.

TABLE A.3 Infrared correlation table

Functional Group	IR absorption region/cm⁻¹	Comment
Fingerprint region	600–1500	Region often crowded, but its features can be diagnostic for a structure
C–O	1050–1300	Strong for alcohols, ethers, and esters
C=O (full range)	1630–1850	Absorptions characteristic of carbonyl groups: Frequencies lowered by conjugation of C=O group, but increased by ring strain
C=O Acid anhydride	1740–1850	
C=O Acid chloride	1750–1815	
C=O Ester	1710–1750	
C=O Aldehyde	1680–1740	
C=O Ketone	1660–1720	
C=O Carboxylic acid	1680–1720	
C=O Amide	1630–1700	
Alkyne C≡C	2100–2250	Often weak, especially in more symmetrical environments
Nitrile C≡N	2200–2250	
CH stretch	2700–3100	Strong for sp^3, weaker for sp^2 CH
CH of aldehyde	2700–2900	Two weak bands, one usually close to 2720 cm⁻¹
CH of triple bond C≡CH	~3300	Strong and sharp
Carboxylic acid –OH	2700–3500	Characteristically broad
Alcohol –OH	3200–3600	Broadened by H-bonding
Amine –NH	3200–3500	Two bands for NH_2, one for NH

Appendix 1.3 **Mass spectrometry**

Here we provide some reference information that can help with the interpretation of mass spectra. Some of the more common fragment ions observed in EI-MS are summarized in Table A.4, and common neutral fragments lost from molecules in EI are listed in Table A.5.

TABLE A.4 The masses and structures of some common fragment ions (to 3 decimal places)

m/z	Structure	Interpretation/mechanism
29.039	$CH_3CH_2^+$	Ethyl group, σ, *i*
31.018	$H_2C=O^+H$	Primary alcohol, α
43.018	$H_3CC≡O^+$	Acetyl group, α
43.055	$[C_3H_7]^+$	Propyl/isopropyl group/alkane, σ, *i*
45.034	$H_3CHC=O^+H$	2-alcohol, α
48.985/50.982	$H_2C=Cl^+$	Primary chloride, α
55.055	$[C_4H_7]^+$	Butenyl group/alkene, σ, *i*

(continued)

TABLE A.4 (*continued*)

m/z	Structure	Interpretation/mechanism
57.034	$H_3CH_2CC≡O^+$	Propanoyl group, α
57.071	$[C_4H_9]^+$	Butyl group/alkane, σ, *i*
58.042		2-ketone, McLafferty
60.021		Carboxylic acid, McLafferty
69.071	$[C_5H_9]^+$	Pentenyl/alkene, σ, *i*
71.050	$H_3CH_2CH_2CC≡O^+$	Butanoyl group, α
71.086	$[C_5H_{11}]^+$	Pentyl/alkane, σ, *i*
72.058		3-ketone, McLafferty
74.037		Methyl ester, McLafferty
77.039		Phenyl group, α
78.034		Pyridyl group, α
83.086	$[C_6H_{11}]^+$	Hexenyl group/alkene, σ
85.065	$H_3CH_2CH_2CH_2CC≡O^+$	Pentanoyl group, α
85.102	$[C_6H_{13}]^+$	Hexyl group/alkane, σ, *i*
86.073		4-ketone, McLafferty
88.053		Ethyl ester, McLafferty
91.055		Benzyl group, α, σ, *i*
92.934/94.932	$H_2C=Br^+$	Primary bromide, α
105.034		Benzoyl group, α

TABLE A.5 The masses and structures of some common neutral fragment losses (to 3 decimal places)

Δm/z	Structure	Interpretation
1.008	H·	Loss of H, e.g. from aldehyde, α
2.016	H_2	Aromatic compounds
14.003	N	Aryl–NO
15.024	$H_3C·$	Loss of methyl group, α, σ
16.031	CH_4	
16.018	NH_2	Aromatic nitro/amines, N-oxides
17.027	NH_3	Amine: H-migration, then i
17.002	OH	Carboxylic acids, aromatics
18.011	H_2O	Alcohol: H-migration, then i
18.998	F	Fluoroalkanes
20.006	HF	Fluoride: H-migration, then i
26.016	HC≡CH	Eliminated from alkene or aryl
27.995	CO	Butenyl group/alkene, σ, i
28.031	$H_2C=CH_2$	Eliminated, e.g. from alicyclic ring
29.003	H·C=O	Aldehyde
29.039	$H_3CH_2C·$	Loss of ethyl group, α, σ
29.998	NO	Aromatic nitro compounds
31.018	$H_3CO·$	Loss of methoxy group, e.g. from methyl ester, α
34.969/36.966	Cl·	Chloride, i
34.005	H_2O_2	Polycarboxylic acids
41.039	$H_2C=CHH_2C·$	Loss of propenyl group, α, σ
43.055	$H_3CH_2CH_2C·$	Butanoyl group, α
63.961	SO_2	Aryl–SO_2OR
71.086	$[C_5H_{11}]^+$	Pentyl/alkane, σ, i
72.058		3-ketone, McLafferty
74.037		Methyl ester, McLafferty
77.039		Phenyl group, α
78.034		Pyridyl group, α
83.086	$[C_6H_{11}]^+$	Hexenyl group/alkene, σ
85.065	$H_3CH_2CH_2CH_2CC≡O^+$	Pentanoyl group, α
85.102	$[C_6H_{13}]^+$	Hexyl group/alkane, σ, i
86.073		4-ketone, McLafferty

(continued)

TABLE A.5 (continued)

Δm/z	Structure	Interpretation
88.053		Ethyl ester, McLafferty
91.055		Benzyl group, α, σ, i
92.934/94.932	$H_2C=Br^+$	Primary bromide, α
105.034		Benzoyl group, α

Appendix 1.4 Library searching

The use of database and mass spectral library searching for the identification of unknown compounds is becoming increasingly common and easier as the size of mass spectral libraries and instrument accuracy increase. Library searching can be used for both EI and soft ionization mass spectra although the databases are different due to the distinct type of mass spectra produced by each ionization type.

Appendix 1.4.1 EI spectral databases

The structural specificity of EI fragmentation has led to the development of mass spectral libraries, such as those provided by the National Institute of Standards and Technology (NIST), currently an agency of the U.S. Department of Commerce: https://chemdata.nist. gov/dokuwiki/doku.php?id=chemdata:start#libraries. Mass spectral database searching provides an attractive workflow for identifying previously characterized compounds, but the availability of searchable databases has not replaced the analyst. Whilst the NIST and other spectral libraries now contain hundreds of thousands of compound spectra, the ability to manually interpret EI fragmentation patterns remains an essential skill in the modern analytical chemistry laboratory.

Appendix 1.4.2 Databases for soft ionization mass spectra

There are a number of different ways to search online mass spectral databases. Accurate mass measurements can be used to search for corresponding chemical formulae, and known structures can be matched to these formulae. ChemSpider (www.chemspider.com), currently owned by the Royal Society of Chemistry, is an open-source chemical structure database containing over 120 million structures, properties, and associated chemical information. However, it is not designed for mass spectral searching specifically. MassBank (https://massbank. eu/MassBank/) is currently an open-source database of mass spectrometry reference spectra designed for direct searching by precursor m/z vales as well as product ion spectra. Peak lists can be uploaded directly. Combining precursor and product ion spectra in the same search provides significantly greater structural specificity. The Human Metabolome Database (www. hmdb.ca) is an electronic database currently containing detailed information about small molecule metabolites found in the human body and is an example of a database specific to a certain defined origin. It can be searched using MS (including MS/MS) and NMR spectra, and by chemical formula.

Solutions to self-study problems in Chapters 9 and 11

The tables below show the actual compounds we used to produce the spectra given in the self-study problems in Chapters 9 (Table A.6) and 11 (Table A.7)—that is, the 'correct structures'. In some cases, however, alternative regioisomers cannot be excluded reliably based on the spectra given, and these alternatives are also presented below. Differentiation of the regioisomers in Chapter 9 would often require use of more advanced NMR experiments, such as those introduced in Chapter 10. But in this context, any of these structures can be considered *consistent with the spectroscopic data* (and hence acceptable solutions to the problems). In other cases, the structure can exist in two enantiomeric forms (which are also shown for completeness), but the spectroscopic techniques used here are unable to distinguish these. Again, either form (or a combination of them, including the racemic mixture) would be consistent with the spectroscopic data.

TABLE A.6 Overview of solutions to self-study problems in Chapter 9

Example	Correct structure	Alternative isomers (if applicable)
1. 1-(4′-Bromophenyl) ethan-1-one		
2. Ethyl 2-methyl-3-oxobutanoate		Enantiomers
3. 1-(4-Methoxyphenyl) ethan-1-one		
4. 2-Acetoxybenzoic acid		
5. 4-Phenylbutanoic acid		
6. 2-(3′,4′-Dichlorophenyl) acetic acid		Regioisomers
7. Ethyl 4-(4-methoxyphenyl) butanoate		
8. 4-Oxo-4-phenylbutanoic acid		

9. (*E*)-3-(4-Methoxyphenyl) acrylic acid

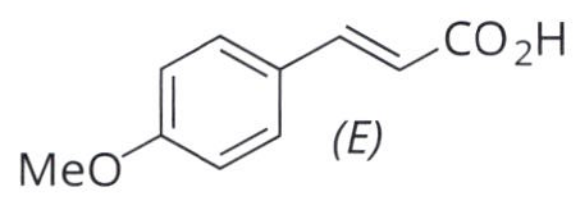

10. *Isobutyl* 4-phenylbuta-noate

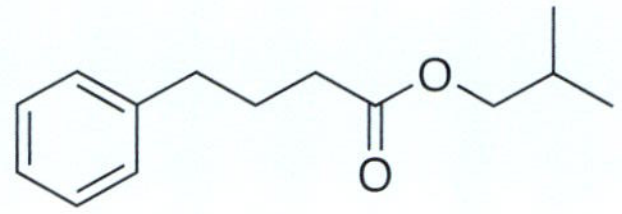

11. 4-Ethoxy-3-hydroxybenzaldehyde

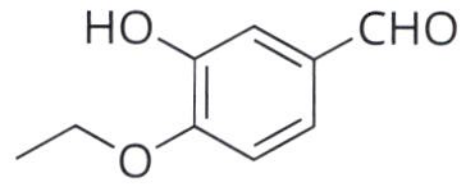

Regioisomers

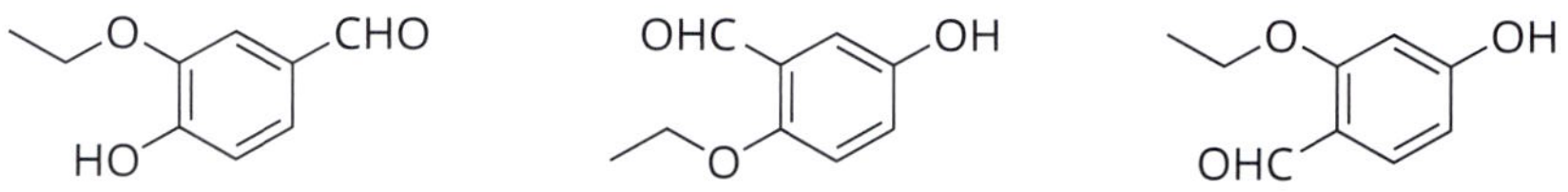

12. Ethyl 4-hydroxy-3-nitrobenzoate

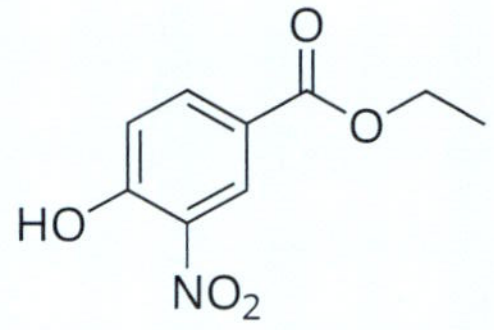

Regioisomers

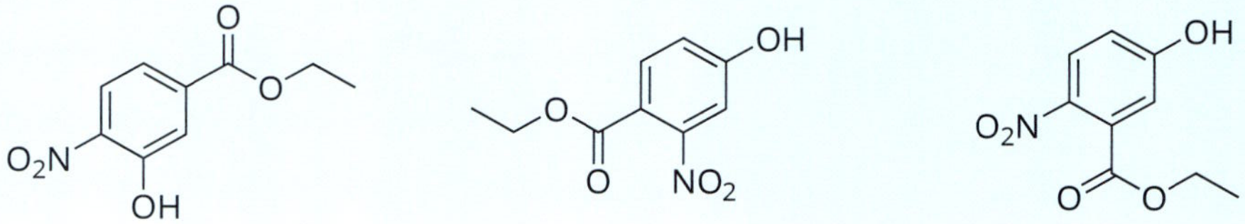

13. 2-(4-*Isobutylphenyl*) propanoic acid

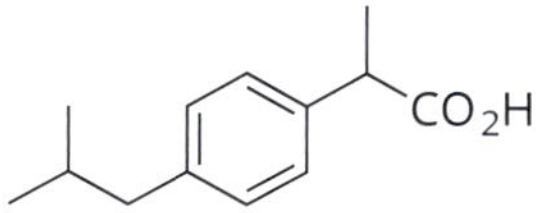

Enantiomers

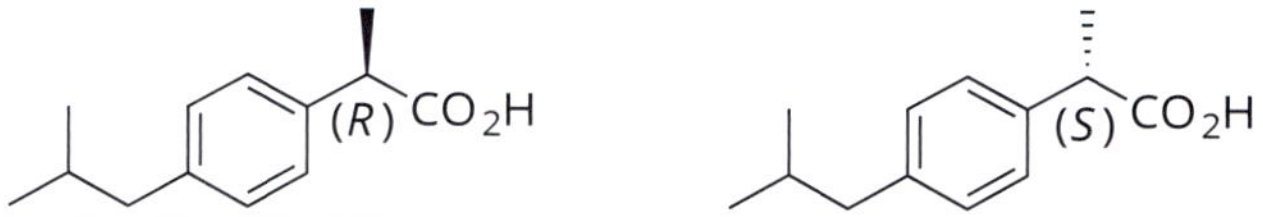

14. 2-Phenylbutanoic acid

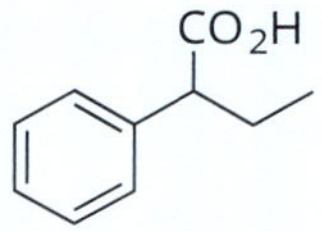

Enantiomers

The CH_2 protons adjacent to the stereogenic centre will be inequivalent

15. 1,4-Di-*tert*-butyl-2,5-dimethoxybenzene

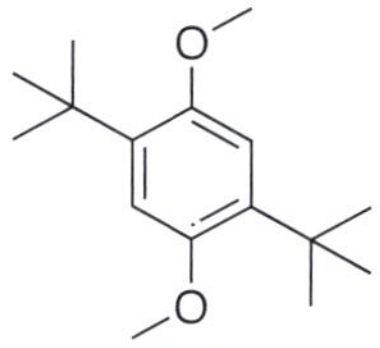

TABLE A.7 Overview of solutions to self-study problems in Chapter 11 on 2D NMR

Example	Correct structure	
1. **Shikimic acid**	^{1}H (ppm)	^{13}C (ppm)
2. **Benzylfuranyl methylketone**	^{1}H (ppm)	^{13}C (ppm)

Index

2 04